SACRED AND PROFANE DIMENSIONS OF LOVE IN INDIAN TRADITIONS AS EXEMPLIFIED IN THE *GĪTAGOVINDA* OF JAYADEVA

This volume is sponsored by the
Inter-Faculty Committee for South Asian Studies
University of Oxford

Oxford University South Asian Studies Series

Sacred and Profane Dimensions of Love in Indian Traditions as Exemplified in The *Gītagovinda* of Jayadeva

LEE SIEGEL

DELHI
OXFORD UNIVERSITY PRESS
BOMBAY CALCUTTA MADRAS

Oxford University Press, Walton Street, Oxford OX2 6DP

NEW YORK TORONTO
DELHI BOMBAY CALCUTTA MADRAS KARACHI
PETALING JAYA SINGAPORE HONG KONG TOKYO
NAIROBI DAR ES SALAAM
MELBOURNE AUCKLAND

and associates in
BERLIN IBADAN

First published 1978
First published in Oxford India Paperbacks 1990
Second impression 1992
SBN 0 19 562575 7

Printed by Pramodh P. Kapur at Rajbandhu Industrial Co.
C-61 Mayapuri, Phase-II, New Delhi 110064
and published by S.K. Mookerjee, Oxford University Press
YMCA Library Building, Jai Singh Road, New Delhi 110001

अद्वैतं सुखदुःखयोरनुगतं सर्वास्ववस्थासु यद्-
विश्रामो हृदयस्य यत्र जरसा यस्मिन्नहार्यो रसः ।
कालेनावरणात्ययात्परिणते यत्स्नेहसारे स्थितं
भद्रं तस्य सुमानुषस्य कथमप्येकं हि तत्प्राप्यते ।।

This state where there is no twoness in responses of
joy or sorrow,
Where the heart finds rest, where feeling does not
dry with age,
Where concealments fall away in time and the
essential love is ripened—
Sacred is this state of human fulfilment, which we
find once if ever.

The *Uttara-rāma-carita* of Bhavabhūti (I. 39)

CONTENTS

Acknowledgements ix

Introduction xi

1 THE MODALITIES OF LOVE 1

Sacred and profane dimensions of love . . . 1
. . . in Indian traditions . . . 13
. . . as exemplified in the *Gītagovinda* of Jayadeva 27

2 THE CONCEPT OF LOVE 43

The taste of love 43
The vocabulary of love 57
The mythology of love 71

3 THE LOVER AND BELOVED 89

Kṛṣṇa 89
Rādhā 110
The messenger 134

4 THE DYNAMICS OF LOVE 137

Separation 137
Union 159

5 THE MEANINGS OF LOVE 178

Allegory 178
Symbol 193

6 JAYADEVA 206

The poet 206
The saint 213

7 THE *GĪTAGOVINDA*: A TRANSLATION 233

Preface 233
The translation 240

Appendix: transliterated text 287

Bibliography of Works Cited 313

Index 323

ACKNOWLEDGEMENTS

This study was begun under the guidance of Professor R. C. Zaehner and it is completed in his memory: the guidance was always full of wisdom; the memory will always be full of love.

I am deeply grateful to Dr Richard Gombrich who, after Professor Zaehner's death, took over the supervision of my work. Generously, with energy, care and meticulous perceptivity, he has helped me with so many corrections and suggestions.

Friedhelm Hardy has given me access to his monumental research on the formation of the Kṛṣṇa-*bhakti* tradition. Paṇḍit Sadashiv Rath Sharma of Purī spent many hours telling me stories about the life of Jayadeva and discussing the *Gītagovinda* with me. Both these men have my warm gratitude.

Genuine appreciation must also be expressed for the financial assistance which I have received from the trustees of the Boden Fund, from the Committee for Faculty Research of Western Washington State College, and, above all, from Mr Robert Evans and my father.

There will always be a special kind of 'thank you' for Jaki, Dmitri and Sebastian.

And thank you, dearest Patricia, for proof-reading and so much more.

INTRODUCTION

Indeed, love becomes so universal a theme because of the remarkable variety of its worlds. Nothing else that unites human beings so emphatically declares at the same time the plurality of living; love may sometimes give us a marvellous degree of mutual consciousness but it also reinforces our most intractable solipsism.[1]

This is an exploration of various worlds of love, an inquiry into the conceptions of love in its sacred and profane dimensions as a universal theme in religious and secular literature. It is a study of the relationship between the sacred and the profane, the religious and the secular, the divine and the human. My strategy has been to proceed from the universal to the particular—to examine the way in which the generic experience finds expression in a specific context, the Indian traditions, and the way in which those traditions are exemplified in an individual text, the *Gītagovinda* of Jayadeva, a twelfth-century Sanskrit poem about the union, separation and reunion of Kṛṣṇa and Rādhā.[2] Simultaneously I have moved from the particular to the universal, attempting to make explicit that which is implicit in the *Gītagovinda* by considering it in its own cultural milieu and by comparing it with texts from other cultures.

The *Gītagovinda* is neither a religious allegory nor a purely secular erotic poem (although both these interpretations are common); the poet has juxtaposed conventional descriptions of carnal love-play with traditional expressions of devotion and in so doing has created an ambiguous relationship between the sacred and profane dimensions of love. The *Gītagovinda* is not a sacred *or* profane work, it is a sacred *and* profane work. The ambivalence reflects a coalescence of traditions: literary, erotic, folk, religious. The text provides a means of understanding some of the attitudes toward love in those various traditions. In addition to those influences *on* the text, I have considered the influences *of* the text, the place which the *Gītagovinda* has held in medieval Vaiṣṇavism, its sacredness to Caitanya and his

[1] John Bayley, *The Characters of Love*, 2nd edn (London, 1968), p. 5.

[2] I have found it necessary to translate the *Gītagovinda* because no *literal* translation of it into English has been made (although it has been translated into mediocre English 'poesy' quite useless for textual analysis).

followers and to the Vaiṣṇavas of the Sahajiyā school. And again, the text provides access to some of the attitudes toward love in those traditions.

The *Gītagovinda* continued the development of the Kṛṣṇa-*bhakti* religion which had found expression in South Indian texts, in the *Viṣṇu Purāṇa*, the *Bhāgavata Purāṇa* and other works. But it also created, or at least reflected, a new trend in that religious movement—the ascendancy of Rādhā. The characterizations of Rādhā and Kṛṣṇa manifest the interpenetration of the various traditions: Rādhā is at once the stereotypical heroine of court poetry, the ideal beauty of the erotic textbooks, the Goddess incarnate, the feminine power emanating from the God, and the female principle, and she also came to represent the devotee; Kṛṣṇa is at once the conventional lover of erotic poetry, the heroic warrior of the Epic and *Purāṇic* literatures, and the highest Godhead of the religious traditions. In this convergence, this coalescence, there is a reconciliation of the orthodox ideal of liberation from phenomenal existence through detachment and the heterodox ideal of devotion, loving attachment, as more important, more meaningful, than liberation. And these religious ideals, the Brāhmanical and the popular, were further reconciled with the secular, courtly ideal of taking delight in the empirical world, of fulfilment in joy and in love.

Chapter one is a consideration of the modalities of love as a universal theme (p. 1), as a specific issue in Indian literature (p. 13), and as the particular subject of the *Gītagovinda* (p. 27). Chapter two is a discussion of Indian concepts of love as they may be understood in light of *rasa* (a physiological experience, an aesthetic experience and a religious experience) (p. 43), in light of various Sanskrit words for 'love' (p. 57) and in light of the mythology of the love-god (p. 71). The third chapter is an examination of Jayadeva's depiction of Kṛṣṇa (p. 89), Rādhā (p. 110), and the love-messenger (p. 134). The fourth chapter deals with the separation (p. 137) and union (p. 159) of lovers, with the pattern of union–separation–reunion as an archetypal motif in the religious and secular literature of love. Chapter five is a study of the allegorical interpretations (p. 178) and symbolic implications (p. 193) of the *Gītagovinda*. As the work became accepted as a sacred text of the Vaiṣṇava devotional schools, hagiographic legends about Jayadeva proliferated in the vernacular languages. In Chapter six I have collected and analysed some of these implicitly interpretive stories (p. 213) and set them in contrast to a portrait of the professional poet which seems historically probable (p. 206). Chapter seven is my translation of the *Gītagovinda*.

In conclusion there can only be inconclusion. The text is equivocal; the context is evasive. But because the *Gītagovinda* is a poem, a literary work, and not a philosophical tractate, a theological work, its ambivalence is not its defect, but, on the contrary, it is its essential meaning and its power. The meaning of the specific text, like the nature of love itself, the universal theme with which it deals, is necessarily ambiguous and mysterious—love, in this, is I think like time, of which Saint Augustine said:

If no one asks me [what it is], I know: if I wish to explain it to one that asketh, I know not . . .[3]

[3] Augustine, *Confessions* XIV. 17, trans. E. B. Pusey, Everyman edn (London, 1907), p. 262.

1

THE MODALITIES OF LOVE

Oh, Love, that I might be love
To love thee with the love of Love!
Oh, Love, for Love's sake, make that I,
Being love know Love as fully as Love.[1]

SACRED AND PROFANE DIMENSIONS OF LOVE...

'... and they shall be one flesh.'[2]

Saint Augustine defined love (*amor*) as 'a certain life which couples or seeks to couple together some two things, namely him that loves, and that which is beloved'.[3] Love is a unitive, fusing force, an energy, then, through which the two seek to become one. It is a 'certain life', a particular mode of experiencing the self in relationship to someone or something else, experiencing the self drawn to that other, that beloved, merged with it, absorbed by it, or longing to be so. Augustine distinguished between spiritual and carnal love, between the sacred and the profane, but, from his own experience, he insisted that his definition held true for both, for *all* love.

> What about every kind of love? Does it not wish to become one with what it is loving? And if it reaches its object, does it not become one with it?[4]

The expression of love in these terms, this two–into–one motif,

[1] Hadewych, cited and trans. Kees Bolle, *The Freedom of Man in Myth* (Nashville, 1968), p. 124.

[2] Gen. 2: 24; cf. Matt. 19: 5; Mark 10: 7; 1 Cor. 6: 16; Eph. 5: 31.

[3] *On the Trinity* (*De Trinitate*), VIII, x, 14, trans. A. W. Haddan in *The Works of Aurelius Augustine*, vol. VII (Edinburgh, 1873), p. 221.

[4] *Divine Providence and the Problem of Evil* (*De Ordine*), II, xviii, 48, trans. Robert P. Russell in *The Writings of Saint Augustine*, vol. I (New York, 1948) (Fathers of the Church Series), p. 325.

occurs throughout the literature of secular love—Isolde speaks to Tristan:

We are one life and flesh.... You and I, Tristan and Isolde, shall for ever remain one and undivided!... I am yours... you are mine ... one Tristan and Isolde.[5]

And it occurs throughout the literature of religious love—the fourteenth-century mystic Gerlac Petersen speaks to God:

Thou givest me Thy whole Self to be mine whole and undivided.... Thou art in me and I in Thee, glued together as one and the selfsame thing which henceforth and forever cannot be divided.[6]

The distinction between the sacred and the profane does not seem to be in the expression of the love, in the feeling or yearning of the lover, but only in the nature of the beloved. Gerlac Petersen's love is 'sacred' because it is directed to God, consecrated in God, by God. It is 'hierophanous';[7] God, the sacred, becomes manifested within the self, the profane: 'I by Thy grace enjoy Thee in myself and myself in Thee.'[8] In these terms Isolde's love is 'profane' because Tristan is flesh and bone and blood; it is profane in the original sense of the word, *pro-fanum*, 'outside the temple': it is sensual, adulterous, carnal. But there are other terms in which her love might be considered sacred; Mircea Eliade has tried to offer a universal definition of 'sacred' as

an absolute reality... which transcends the world, but manifests itself in the world, thereby sanctifying it and making it real.[9]

By this definition Isolde's love is sacred, not because of the nature of the beloved, the object of love, but because of the nature of love itself as an 'absolute reality'. Love, in the medieval lyric and legend, was considered an ennobling, transmuting force,

[5] The *Tristan* of Gottfried von Strassburg, trans. A. T. Hatto (Harmondsworth, 1960), p. 282.

[6] Gerlaci Petri, *Ignitum cum Deo Soliquium*, xv, cited by Evelyn Underhill, *Mysticism*, Dutton Paperback edn (New York, 1961), pp. 427-8.

[7] Mircea Eliade, *The Sacred and the Profane*, trans. Willard R. Trask, Harper Torchbook edn (New York, 1959), p. 11: 'Man becomes aware of the sacred because it manifests itself, shows itself as something wholly different from the profane. To designate the *act of manifestation* of the sacred, we have proposed the term *hierophany*.'

[8] Petri, loc. cit. [9] Eliade, op. cit., p. 202.

more real than world, a 'higher estate', sanctifying souls. As God was Love, so Love was God. Love was sacramental and eternal. Love was 'the way, the truth and the life'.

The medieval religious poet used the language of profane, physical love, the language of the court, to describe the union of the soul and God; he did, nevertheless, remain heedful of the Pauline revelation that to go after flesh is to die: 'To be carnally minded is death; but to be spiritually minded is life.'[10] In ecclesiastical terms the sacred was in the eternal spirit and the profane was in the ephemeral flesh. But the medieval secular poet explored the possibilities of the sanctification of the physical man in love by using the language of sacred, spiritual love, the language of the Church, to describe the union of the soul with another soul, a total union: 'one life and flesh . . . one and undivided.' There is death, but death by love was described as an 'ecstasy', a 'sweet martyrdom'; it was Solomon's words that were recounted: 'Set me as a seal upon thy heart, as a seal upon thine arm: for love *is* strong as death.'[11] The secular poet did not necessarily reject the Church, the metaphysics of love, but he sought meaning and joy in human beauty, human love. 'It is no crime to love,' a twelfth-century poet sang; 'if it were, God would never have bound even divine things with love.'[12] Uc de Saint-Circ invoked his beloved Lady: 'I do not wish God to aid me nor give me joy and happiness except through you!'[13] And in the ballads of John Gower the sacred is transubstantiated into the flesh and blood of his beloved:

> . . . when I hear the chaplain read his litany and lesson, no other name but hers can I invoke, for my heart is so full of true love that all my devotion is fixed on her. God grant my prayer is not in vain.[14]

Kierkegaard described love as the place where 'the highest spiritual life is expressed in its antithesis, while sensuality claims

[10] Rom. 8: 6.

[11] Song of Songs 8: 6.

[12] *Carmina Burana*, cited and trans. James Wilhelm, *The Cruelest Month* (New Haven and London, 1965), p. 61.

[13] Cited by Denis de Rougemont, *Love in the Western World*, trans. Montgomery Belgion, Fawcett Premier edn (New York, 1966), p. 93.

[14] John Gower, 'Je cuide que ma dame . . .' in Brain Woledge (ed.), *The Penguin Book of French Verse*, vol. I (Harmondsworth, 1961), p. 231.

to represent the highest spiritual life'.[15] Sensual love seeks meaning and significance in the eternity of the sacred; spiritual love seeks meaning and impact in the immediacy of the profane. Throughout the history of the literature of love the one has encroached upon the other; the sacred and profane dimensions have become intermingled, one into the other and back again. As modes of loving, or experiencing love, the sacred and profane are not necessarily antonymous. There have been four fundamental patterns, four basic ways of considering the relationship between sacred love and profane love in that literary history of love: different and not analogous; different and yet analogous; homologous and yet differentiated; homologous and not differentiated:

1 *The sacred and profane dimensions of love as different and not analogous*

In Christian theology distinctions are made between *Eros*, Plato's word for 'love', and *Agape*, Paul's word for 'love'. *Agape* is clearly sacred: 'God is *Agape*';[16] *Eros* may be either profane, Plato's 'vulgar *Eros*', the love which leads and binds the soul to the sensible, material object, or it may be sacred, Plato's 'heavenly *Eros*', the love which leads and binds the soul to the supersensible, heavenly object. The Protestant theologian Anders Nygren has argued that profane love, sensual love, has *nothing* to do with the sacred and does not merit any place in a discussion of religious love: 'Between Vulgar Eros and Christian Agape there is no relation at all.'[17] He suggests, furthermore, that even the 'heavenly *Eros*', even its most sublimated and spiritualized form, is profane in that it is always necessarily 'egocentric' in contrast to *Agape* which is always 'theocentric'. *Eros* is acquisitive desire, an appetite dependent upon external motivation, a response to a quality, in contrast to *Agape* which is sacrificial giving, gratuitous, unmotivated, a creation of a quality. Nygren insists that they are rivals, fundamentally op-

[15] 'In Vino Veritas', cited by Denis de Rougemont, *Love Declared*, trans. Richard Howard, Beacon edn (Boston, 1964), p. 13.

[16] John 4: 8 and 16.

[17] Anders Nygren, *Agape and Eros*, trans. Philip S. Watson, Harper Torchbook edn (New York, 1969), p. 51 (cf. p. 660).

posed.[18] In this Protestant context the language of human love, of sexual union, cannot represent the relationship between man and God. 'Christian love', Nygren wrote, 'is wholly other than human love . . .'.[19] The chasm between the profane and sacred is bridged neither by analogy nor by sublimation, is crossed by no power or energy or love in man, but only by the power and love which is God. Nygren, as a spokesman for Luther, attempts to desexualize and dehumanize religious love, love as understood in Christian theology prior to the Reformation. In that theology, that Catholic tradition, the love of a man for a woman, a woman for a man, the longing for union, becoming 'one flesh', can, either by analogy or by sublimation, be understood as holy, as sacred.

2 *The sacred and profane dimensions of love as different and yet analogous*

This is the 'great mystery' described by Paul:

. . . as the Church is subject unto Christ, so *let* the wives *be* to their own husbands in everything. Husbands, love your wives, even as Christ also loved the church, and gave himself for it. . . . For this cause shall a man leave his father and mother, and shall be joined unto his wife, and they two shall be one flesh. This is a great mystery: but I speak concerning Christ and the church. . . .[20]

Human love is a simile for divine love; the chasm remains between the sacred and the profane, a reflection of the gap between spirit and flesh, eternity and temporality, heaven and earth, but the human is *like* the divine; the dynamics of love are the same in both realms. On the authority of Paul orthodox theology could be expressed in the language of human love, the *Song of Songs* could be interpreted as a sacred epithalamion, a description of the nuptial rites between Christ and the Church (as the collective body of souls loving Christ), an allegory of love. And in the third century Origen interpreted the *Song of Songs* not only as a marriage song for Christ and the Church, but also as a celebration of the union of the individual soul and the Word:

. . . and let us understand that by the 'mouth' of the Bridegroom is meant the power by which He enlightens the mind and . . . makes

[18] Ibid., p. 210, *passim*. [19] Ibid., p. 726. [20] Eph. 5: 24–32.

plain whatever is unknown and dark to her. And this is the truer, closer, holier kiss, which is said to be granted by the Bridegroom-Word of God to the Bride—that is to say to the pure and perfect soul.[21]

Human love, physical passion, became the perfect analogy for divine love, the spiritual passion.

This is absolutely appropriate for just as the human body knows no sensation comparable in sheer joyful intensity to that which the sexual act procures for a man and a woman in love, so must the mystical experience of the soul in the embrace of God be utterly beyond all other spiritual joys. The sexual image is, moreover, particularly apt since the man both envelops and penetrates the woman, is both within and without her, just as God Who dwells at the deepest point of the soul also envelops it and covers it with His infinite love.[22]

The mystical exegetes needed to be cautionary, to make it clear that *likeness* is not *sameness*, that although sacred and profane love are analogous they are different. Richard of Saint Victor spoke of the degrees of love within each dimension:

. . . they are not the same in spiritual and carnal desires. In spiritual desires, the greater the love the better it is; in carnal desires, the greater the love the worse it is.[23]

Gregory of Nyssa, insisting on the separation between sacred and profane, divine and human, explained that man has two distinct sense-systems—the internal, spiritual senses and the external, physical senses. The *Song of Songs* is addressed to the former.[24] Gregory made no distinction between *Eros* and *Agape*; the two words were used interchangeably. He drew not only on the language of human love to describe the passions of the soul but even upon the conceits of 'pagan' love—Christ pierces the soul with the arrow of Eros.[25]

As secular language was used to describe religious love, so too

[21] Origen, *The Song of Songs, Commentary and Homilies*, trans. R. P. Lawson (London, 1957), p. 62.

[22] R. C. Zaehner, *Mysticism, Sacred and Profane* (New York, 1961), pp. 151–2.

[23] *De IV Gradibus Violentae Caritatis*, 18, cited by N. J. Perella, *The Kiss Sacred and Profane* (Berkeley and Los Angeles, 1969), p. 35.

[24] See Jean Daniélou (ed.), *From Glory to Glory, Texts from Gregory of Nyssa's Mystical Writings* (London, 1962).

[25] See Nygren, op. cit., pp. 433–4.

was religious language used to describe secular love. A twelfth-century poet sang:

> Loving torments [*crucior*] me, I die
> From the wound in which I glory![26]

Glory, sweet martyrdom, the crucifixion of love—Christian sentiment infused the secular love lyric of the medieval period; and in turn the medieval mystics seem to have borrowed from the rhetoric of courtly love. Back and forth between the sacred and profane, there seems to have been mutual influence. As the religious poets took the analogy of love from the biblical commentators and wrote songs addressed to God, love songs expressing the longing of their feminine souls for union with God, a literature emerged based not so much upon Pauline analogy as upon Platonic sublimation. Sacred love was realized through a concentration of the full power of their love upon God; the sexual instinct, the need to overcome isolation, the craving for union, all of the phases of human passion were directed to God. The religious poet could write a carnal, sensual, passionate love song because he was actually, utterly and *completely* in love with God. And in that the profane became sacred:

> Oh flame of love so living,
> How tenderly you force
> To my soul's inmost core your fiery probe!
> Since now you've no misgiving,
> End it, pursue your course
> And for our sweet encounter tear the robe![27]

3 *The sacred and profane dimensions of love as homologous and yet differentiated*

Love in this case is qualified as sacred or profane depending on its object; the 'vulgar' and 'heavenly' *Eros* are distinguished solely on the basis of the level upon which they are experienced; the experience of *Eros* itself is uniform—it is *desire*. *Eros* was considered by Plato to be a force which leads the soul in the direc-

[26] *Carmina Burana*, cited and trans. by Perella, op. cit., p. 106.
[27] St. John of the Cross, *Poems*, ed. with translations by Roy Campbell, Penguin edn (Baltimore, 1968), p. 45.

tion of the Ideal world, which leads from the profane to the sacred through a process of sublimation:

... the true order of going, or being led by another, to the things of love [*eros*], is to begin from the beauties of earth and mount upwards for the sake of that other beauty, using these as rungs of a ladder, and from one going on to two, and from two to all fair forms, and from fair forms to fair practices, and from fair practices to fair notions, until from fair notions he arrives at the notion of absolute beauty, and at last knows what the essence of beauty is. This ... is that life above all others which man should live, in the contemplation of beauty absolute.[28]

Aristotle expanded the Platonic theory to describe *Eros* as a force in all things, that longing of all things to reach out and become like higher things. The theory was modified further by the Neoplatonists so that *Eros* became the universal force, the primary energy of existence, the power of cohesion in the universe, the force 'that moves the sun and the other stars'.[29]

For Saint Augustine, as we have seen, love, which is a longing, a desire for union, is the same whether it is directed downward toward the temporal world or upward toward the eternal God. But he distinguished between the two levels of love as the profane *cupiditas* and the sacred *caritas*:

Love, but see to it *what* you love. Love to God and love to neighbor is called *caritas*; love of the world and love of temporal things is called *cupiditas*.[30]

He had been initiated into love through *cupiditas*, learned and felt the power of love at first through his senses:

And what was it that I delighted in, but to love, and be beloved? but I kept not the measure of love ... but out of the muddy concupiscence of the flesh, and the bubblings of youth, mists fumed up which beclouded and overcast my heart. ... I boiled over in my fornications. ...[31]

[28] The *Symposium*—Socrates is relating the lessons about love taught to him by Diotima; in Plato's *Euthyphro, Crito, Apology and Symposium*, the Jowett translation, Gateway edn (Chicago, 1953), p. 126.

[29] The last line of Dante's *Divina Commedia*.

[30] Cited by Nygren, op. cit., p. 495.

[31] Augustine, *Confessions*, II. 2, trans. E. B. Pusey, Everyman edn (London, 1907), pp. 21–2.

But then, through a process of sublimation, all the desire in his flesh was directed upward; still the ardorous passion, the fervour of his love abided:

Too late loved I thee, O Thou Beauty of ancient days, yet ever new! too late I love Thee! And behold, Thou wert within, and I abroad, and there I searched for Thee; deformed I, plunging amid those fair forms, which Thou hadst made. Thou wert with me, but I was not with Thee. Things held me far from Thee, which, unless they were in Thee, were not at all. Thou calledst, and shoutedst, and burstest, my deafness. Thou flashedst, shonest, and scatteredst my blindness. Thou breathedst odours, and I *drew in breath* and *pant for Thee*. I tasted, and *hunger and thirst*. Thou touchedst me, and I burned for Thy peace.[32]

As Plato, the Neoplatonists, Dante, Augustine and others saw sexual love as a degraded form of spiritual love, Freud saw spiritual love as a 'perverted' form of sexual love, i.e., as 'aim-inhibited sexuality', the sexual impulse, libidinous drive, from which the need for actual sexual gratification has been removed:

Libido is an expression taken from the theory of the emotions. We call by that name the energy, regarded as a quantitative magnitude . . . , of those instincts which have to do with all that may be comprised under the word 'love'. The nucleus of what we mean by love naturally consists (and this is what is commonly called love, and what the poets sing of) in sexual love with sexual union as its aim. But we do not separate from this . . . on the one hand, self-love, and on the other, love for parents and children, friendship and love for humanity in general, and also devotion to concrete objects and to abstract ideas [e.g., sacred love, love for God]. Our justification lies in the fact that psychoanalytic research has taught us that all these tendencies are an expression of the same instinctual impulses; in relations between the sexes these impulses force their way towards sexual union, but in other circumstances they are diverted from this aim or are prevented from reaching it, though always preserving enough of their original nature to keep their identity recognizable (as in such features as the longing for proximity and self-sacrifice).[33]

Freud also used the word *Eros* to discuss sexuality but it was to suggest something more far-reaching than *libido* which is a

[32] Ibid., x. 38, p. 227.

[33] *Group Psychology and the Analysis of the Ego*, vol. XVIII in the Standard Edition of *The Complete Psychological Works of Sigmund Freud*, trans. James Strachey (London and New York, 1921), pp. 90–1.

tension in the organism which demands release and leads to muscular actions which will reduce that tension; *libido* is analogous to hunger. *Eros* is more encompassing; it is a cohesive power, it 'seeks to force together and hold together the portions of living substance',[34] it is the life instinct, the bond of union in existence, at work within the organism and between organisms.

In *Beyond the Pleasure Principle* Freud cautiously and tentatively suggested that the tendency of the sexual instinct is 'to restore an earlier state of things',[35] to return to the primordial unity, a state of equilibrium in which existence is not sexually differentiated.

> Shall we follow the hint given us by the poet-philosopher, and venture upon the hypothesis that the living substance at the time of its coming to life was torn apart into small particles, which have ever since endeavoured to reunite through the sexual instincts?[36]

The 'poet-philosopher' is Plato, the 'hint' is the speech Plato puts into the mouth of Aristophanes in the *Symposium*. Aristophanes explains love in cosmogonic terms; he says that originally there were androgynous beings which were divided into two, into male and female. The two endeavour to reunite:

> . . . human nature was originally one and we were a whole, and the desire and pursuit of the whole is called love.[37]

When love is seen thus as the urge for primal unity manifested in sexuality, a biological longing for the uniformity which preceded the present and profane state of man, when human love takes on cosmogonic value, it becomes sacralized as such. Human sexuality becomes the microcosmic expression of the cosmic process. When the union of the man and the woman is homologized and identified with the union of all that may be qualified as male and female, sexual union becomes an erotic theurgy, a carnal hierophany, an act of cosmic significance. This identification is expressed in ritual sexuality, nuptial rituals, rituals in which the creation of the universe is re-enacted on the microcosmic plane in sexual union:

> In premodern societies, sexuality, like all the other functions of life,

[34] Sigmund Freud, *Beyond the Pleasure Principle*, trans. James Strachey, Bantam edn (New York, 1959), p. 106.

[35] Ibid., p. 100.
[36] Ibid., p. 102.
[37] Plato, op. cit., p. 103.

is fraught with sacredness. It is a way of participating in the fundamental mysteries of life and fertility.[38]

And it is expressed in the ritual prostitution which makes the sexual act a 'sacrifice' to the god—as a sacrifice 'the sexual act, particularly, can take on a sacral significance'.[39] When sexuality becomes a sacrifice, a giving completely of the body and of the soul, a yielding to ravishment, the distinction between sacred and profane love disappears:

The sacrifice is the point not only at which the profane and the sacred touch, but at which they permeate one another indissolubly.[40]

4 *The sacred and profane dimensions of love as homologous and not differentiated*

In the Kabbalistic tradition of Judaism the cosmic mystery was a sexual mystery and the sexual mystery was a cosmic mystery; sexuality was attributed to God and God's holiness was attributed to sexuality. Sexual union (between a man and his wife) was held to be the earthly actualization of the relationship between God and the *Shekhinah* (literally, 'in-dwelling', God's Divine Presence, the female part and counter-part of God). Sexual union was described as 'the glad performance of a religious precept';[41] it was considered a man's sacred duty to give his wife sexual pleasure for two reasons: according to the *Zohar*:

One is that the pleasure is a religious pleasure, and one which gives joy to the Shekhinah also, and what is more, by its means he spreads peace in the world. . . .[42]

Sexual union not only symbolizes divine union, it prompts and causes it; the profane union of the man and the woman is sacred, full of redemption, inspiring the perfect union of the male and female principles. 'The secret of the matter', it is

[38] Mircea Eliade, *Rites and Symbols of Initiation*, trans. Willard R. Trask, Harper Torchbook edn (New York, 1965), p. 25.

[39] Ernst Cassirer, *Mythical Thought*, vol. II of *The Philosophy of Symbolic Forms*, trans. Ralph Manheim (New Haven and London, 1955), p. 221.

[40] Ibid.

[41] *Zohar*, trans. Harry Sperling, Maurice Simon and Paul Levertov (London, 1949), vol. I, p. 159.

[42] Ibid., I, pp. 158–9.

written in the *Zohar*, 'is that wherever male and female are not together, blessings do not rest.'[43]

Distinction between sacred and profane love is largely distinction between spirit and flesh, eternity and temporality. Sexual love is profane in that it longs for temporal flesh, but it becomes sacred when the body is conceived as the materialization of spirit, the sensible expression of spirit, when the temporal encloses the eternal, when the finite encircles the infinite, when there is an apotheosis of the senses, a fleshly assumption: 'We need pray', Thoreau wrote, 'for no higher heaven than the pure senses can furnish, a purely sensual life.'[44] Matter matters. The senses make sense. Humanity is divinity. With this voluptuous vision of the holiness of the erotic life Walt Whitman sang, 'I am the poet of the Body and I am the poet of the Soul'.[45] The two were one, made one by love for love:

> Behold, the body includes and is the meaning, the main concern, and includes and is the soul;
> Whoever you are, how superb and how divine is your body, or any part of it.[46]

This sacralization of flesh and blood, this establishment of the divinity of the body, was celebrated no less by Blake:

> 1. Man has no Body distinct from his Soul; for that called Body is a portion of Soul discerned by the five Senses, the chief inlets of Soul in this age.
> 2. Energy is the only life, and is from the Body; . . .
> 3. Energy is Eternal Delight.[47]

For Blake, as for Whitman, the ultimate and essential 'song of liberty' echoes: '. . . everything that lives is Holy'.[48] *All is sacred.* That revelation is at the heart of Indian speculation: '*sarvam khalu idaṃ brahma*',[49]—all this, this whole universe, is truly

[43] Ibid., IV, pp. 345–55.

[44] *A Week on the Concord and Merrimack Rivers*, cited by Norman O. Brown, *Life Against Death*, Sphere Books edn (London, 1968), p. 270.

[45] 'Song of Myself', XXI, in *A Choice of Whitman's Verse*, selected by Donald Hall (London, 1968), p. 42.

[46] Idem., 'Starting from Paumanok', XIII, p. 110.

[47] 'Marriage of Heaven and Hell' in *Selected Poems of William Blake*, ed. F. W. Bateson (London, 1957), p. 64.

[48] Ibid., p. 70. [49] *Chāndogya Upaniṣad* III. 14. 3.

brahman. And *brahman*, I believe, is the closest Sanskrit equivalent to 'the sacred'.

. . . IN INDIAN TRADITIONS . . .

How can I describe
what is love
and when it is born
and where it is seen
and who found it
and how?[50]

Brahman, because it is 'all this', infinite and eternal, impersonal, cannot be loved. The central teaching of the *Upaniṣads* is that the *brahman* is the *ātman*, the Self, and the goal of the teaching is liberation (*mokṣa*) through the realization of that identity. *Brahman* cannot be loved because the very idea of *brahman* precludes any idea of love, of distinction between subject and object, lover and beloved. Love as a force which 'couples or seeks to couple some two things' into one cannot exist because there is no two-ness, only unity. Love is quite irrelevant.

If *brahman*, as the eternal, ultimate reality and infinite, absolute unity behind all appearance, is the sacred, the profane is the appearance, temporal and finite. The goal, in the teachings of the *Upaniṣads*, is absorption in the sacred, liberation from the profane (although the profane may be an emanation of the sacred if *all* is *brahman*), liberation from nature, the objective world, that which is called *prakṛti* in the Sāṃkhya systems and *māyā* in the Vedānta system, the two major systems of liberation to arise out of the speculations of the *Upaniṣads*. Love has no place in either system. Love is attachment to and in *saṃsāra*, empirical existence, the appearance. Love is itself profane. It plays no part in the Sāṃkhya-*yogin*'s effort to isolate his individual, eternal self from all that is temporal or in the Vedāntin's effort to fully realize that his Self is identical with the godhead and that all else is illusory.

The man of wisdom, according to the *Bhagavad-Gītā*, 'has no love [*sneha*] for any thing':[51] love is a snare. Unattached devo-

[50] Caṇḍīdāsa, trans. D. Bhattacharya, *Love Songs of Chandidās* (London, 1967), p. 81.

[51] *Bhagavad-Gītā* II. 57.

tion to God leads to liberation, but not passionate love, nothing like human love, the love which a man may have for a woman or a woman for a man. Renunciation (*tyāga*) and passionlessness (*vairāgya*) are the ideals in the *Bhagavad-Gītā*: 'sacred love' would be *bhakti* as total devotion to the Bhagavat, the Lord, intentness on God with the senses curbed, love as gratuitous worship made sacred by virtue of its object, Kṛṣṇa; 'profane love would be any desire, purposeful, passionate, sensual longing And these two are different and not analogous.[52] With the development of theistic Kṛṣṇa-*bhakti* cults however, th idea of analogy as well as homology between profane and sacred, human and divine, carnal and spiritual, love developed God and the soul were distinctly two always and essentiall longing to be merged into one *like* a man and a woman or eve *as* a man and a woman.

The use of the human, sexual metaphor to describe th relationship between the 'person' and the Self in the *Bṛhadāra ṇyaka Upaniṣad* foreshadows, perhaps, the fuller 'analogy of love'

Just as a man, closely embraced by his loving wife, knows nothin without, nothing within, so does this 'person', closely embraced by th Self that consists of wisdom, know nothing without, nothing within.[53]

The experience of absorption in the Self is compared to th experience of the suspension of ego-activity in the culminatic of coition. But the use of the analogy seems purely rhetorica love is not the issue; the 'person' is free from attachment:

That is his [true] form in which [all] his desires are fulfilled, in whi Self [alone] is his desire, in which he has no desire, no sorrow.[54]

There is no equation of the Self and the person with the belov and the lover. The Self, one and the same as *brahman*, is

[52] The analogy is suggested but remains undeveloped: Arjuna begs Kṛṣṇa bear with him as a father with a son, a friend with a friend, *a lover with a belo* (*Bhag.-G.* xi. 44).

[53] *Bṛhadāraṇyaka Upaniṣad* iv. iii. 21 (Zaehner trans.):

. . . tad yathā priyayā striyā sampariṣvakto na bāhyaṃ kiṃ cana veda nāntaram; evam evāyam puruṣaḥ prājñenātmanā sampariṣvakto na bāhyaṃ kiṃ cana veda nāntaram . . .

[54] Ibid.

. . . tad vā asyaitad āpta-kāmam ātma-kāmam akāmaṃ rūpaṃ śokāntaram.

reality superseding the phenomenal dimensions of love, superseding any emotion. If the lover thinks the beloved herself is dear, he is mistaken:

The husband is not dear for the sake of the husband, the husband is dear for the sake of the Self; the wife is not dear for the sake of the wife, the wife is dear for the sake of the Self.[55]

The same *Upaniṣad* formulates a correspondence between microcosm and macrocosm, between human sexuality and cosmogonic process, a correspondence which suggests homology and the sacralization of human sexuality. There is a creation myth very similar to the one which Aristophanes recounted to define 'love' in the *Symposium*:

In the beginning this [universe] was the Self alone,—in the likeness of a man. Looking around he saw nothing other than [him]self. . . . He found no pleasure at all. So, [even now] a man who is all alone finds no pleasure. He longed for a second. Now he was the size of a man and a woman in close embrace. He split this Self in two: and from this arose husband and wife. Hence we say, 'We two are like half a potsherd', as indeed Yājñavalkya used to say. That is why space is filled up with woman. He copulated with her, and thence were human beings born.[56]

Sexual union is based upon a cosmic paradigm and can be a reenactment of sacred history. The correspondence between individual and universal activity, a primary axiom of sympathetic magic, could be used to attain certain ends, to gain access to powers at work in the cosmos. To ensure the birth of a male child, the man was instructed, in this *Upaniṣad*, to say to the woman as he embraced her:

[55] Ibid., IV. v. 6:
. . . na vā are patyuḥ kāmāya patiḥ priyo bhavati ātmanas
tu kāmāya patiḥ priyo bhavati; na vā are jāyāyai kāmāya
jāyā priyā bhavati ātmanas tu kāmāya jāyā priyā bhavati . . .

[56] Ibid., I. iv. 1–3 (following Zaehner's trans.):
ātmaivedam agra āsīt puruṣavidhaḥ; so'nuvīkṣya nānyad
ātmano'paśyat . . . . sa vai naiva reme; tasmād ekākī na
ramate; sa dvitīyam aicchat; sa haitāvān āsa yathā
strī-pumāṃsau sampariṣvaktau; sa imam evātmānaṃ
dvedhāpātayat; tataḥ patiś ca patnī cābhavatām; tasmāt
idam ardha-bṛgalam iva svaḥ iti ha smāha yājñavalkyaḥ;
tasmād ayam ākāśaḥ striyā pūryata eva; tāṃ samabhavat;
tato manuṣyā ajāyanta.

I am the life-breath, you are the speech; you are the speech, I am the life-breath; I am the *Sāma* [*Veda*], you are the *Ṛg* [*Veda*]; I am the heaven, you are the earth.[57]

He was to conceive, furthermore, of her genitals as the sacrificial altar—her pubic hair as the sacrificial grass, her vaginal labia as the sacrificial fire.[58] The close connection between sexuality and sacrifice was reiterated:

Woman is a fire, Gautama: the phallus is her fuel; the hairs are her smoke; the vulva is her flame; when a man penetrates her, that is her coal; the ecstasy is her sparks.[59]

Sexuality was sacred as a sacrifice, as a ritual. In various Tantric schools the sexual act was also ritualized and also conceived of as the microcosmic enactment of a macrocosmic process.

In the Tantric literature the universe is differentiated into male and female aspects: Śiva and Śakti on the cosmic level (in the Hindu *Tantras*; Upāya and Prajñā in the Buddhist *Tantras*) and man and woman on the human level. The differentiation runs through all phenomenological levels—social, physiological, geographical, chemical, etc. In the ritualized sexual union of the man and the woman all the polar opposites are to be united —there is to be a re-conquest of the primordial unity, the ultimate unified reality behind appearance. Sexual union is used as a means to transcend the phenomenal world.

The male and female represent in the visible world the division which is present in the nature of the absolute as Śiva and Śakti, and the perfect union of the Śiva and the Śakti is the highest reality [which is non-dual]. Within the physical body of man and woman reside the ontological principles of Śiva and Śakti; therefore to realise the absolute truth, or in other words, to obtain the highest spiritual experience man and woman must first of all realise themselves as manifestations of Śiva and Śakti and unite together physically, mentally and

[57] Ibid., VI. iv. 20:

amo'ham asmi sā tvam; sā tvam asi amo'ham; sāmāham asmi ṛk tvam; dyaur aham pṛthivī tvam. . . .

[58] Ibid., VI. iv. 3.

[59] Ibid., VI. ii. 13 (Zaehner trans.):

yoṣa vā agniḥ gautama; tasyā upastha eva samit; lomāni dhūmaḥ; yonir arciḥ; yad antaḥ karoti te'ṅgārāḥ; abhinandā visphuliṅgāḥ. . . .

spiritually, and the supreme bliss that proceeds from such union is the highest religious gain.[60]

But sacralized sexuality is not necessarily 'sacred love'. Love plays no part in the *Tantras*:

Far from according any value to any human relationship, the Tantric *sādhaka* [initiate] emphatically rejects all such forms of relationship, including human love in its psychological and social aspects. The female partner, therefore, ought to be considered under a purely sexual angle, as a manifestation of divine Energy or Female Principle, rather than as a woman; love in the full human sense, being not permissible to a Yogi. . . . The love-symbolism of the Tantric schools tends to isolate the sexual act from its human context and ignore the personal aspect of love-relationship.[61]

But within Tantra an idea persisted which (in some Sahajiyā schools) could be a basis for establishing a homology between sacred love and profane love, for abolishing all distinction between the human and the divine: 'Such is *nirvāṇa*, such is *saṃsāra*', it is written in the *Hevajra Tantra*. 'There is no *nirvāṇa* other than *saṃsāra* we say.'[62] There is no sacred other than the profane.

In Śaivite mythology, as in the *Tantras*, the sexual energy is in itself sacred as the fundamental cosmic force. Asceticism and sexuality are inextricably related:

. . . *tapas* (asceticism) and *kāma* (desire) are not diametrically opposed. . . . They are in fact two forms of heat, *tapas* being the potentially destructive or creative fire that the ascetic generates within himself, *kāma* the heat of desire. Thus they are closely related in human terms. . . .[63]

The two are homologous and yet differentiated. Śaivite *yogic* discipline is a process of sublimation of *kāma*, of *eros*; but it is directed inward rather than upward as it is in Platonic doctrine.

[60] S. B. Das Gupta, *Obscure Religious Cults*, 2nd edn (Calcutta, 1963), p. xxxvi.

[61] Charlotte Vaudeville, 'Evolution of Love-Symbolism in Bhagavatism', *Journal of the American Oriental Society*, LXXXII (1962), p. 32.

[62] *Hevajra Tantra* II. iv. 32 (Snellgrove trans.):

evam eva tu saṃsāraṃ nirvāṇam evam eva tu;
saṃsārād ṛte nānyan nirvāṇam iti kathyate.

[63] Wendy O'Flaherty, 'Asceticism and Sexuality in the Mythology of Śiva', Pt. I, *History of Religions*, VIII, no. 4, May 1969 (pp. 300–37), p. 301.

Sublimation, curbing the senses, overcoming carnal desire, is a fundamental ideal throughout Hinduism. Vātsyāyana says that he composed the *Kāmasūtra* only after observing strict celibacy.[64] Passionlessness remained the goal and model even in the sexual science; the art and science of love was to be applied with restraint, never with excessive passion (*atirāga*), always with the senses under control (*jitendriyaḥ*).[65] Vātsyāyana is concerned with *kāma* (love and pleasure) in the context of the other aims of life—*dharma* (duty, righteousness, virtue) and *artha* (material gain)—and he explains that love must be pursued only in such a way as not to interfere with these other aims.[66] There must be, furthermore, special consideration for one's own most appropriate end, one's own duty (*svadharma*): he says that for a king *artha* is best; for a courtesan *kāma* should be of prime importance.[67] There is no conception of sacred love here, but if the courtesan pursues *kāma* well and fully, fulfilling her *svadharma*, she will eventually (perhaps in several lifetimes) gain release from the world, the profane.

In the *Mahābhārata* and the *Rāmāyaṇa* love in the man is generally considered a foible; woman is a chattel:

> The woman is, indeed, an object of the senses (*indriyārtha*), an instrument of pleasure . . . she is merely one among the needs of life, such as a seat, a bed, a vehicle, a house, corn, etc.[68]

Sexual love is a pastime for the hero, an amusement, profane but only dangerously so if the man is too attached to it. Conjugal love was idealized for the woman: in the wife, love, as devotion to the husband, is considered the highest realization of *dharma* and as such is sacred:

> While keeping the element of fondness and tenderness which makes it really human, their [i.e. the Epic wives'] conjugal love contains a quality of absolute, nearly super-human purity, of an intensity and perfect self-surrender which makes it a truly religious ascesis. One may say that the feminine ideal of conjugal love, as exemplified in the Epic, is essentially religious in character and worthy of a higher object than the very imperfect human being towards whom it is directed. . . . But it is the wife, and she alone who is really transformed

[64] *Kāmasūtra* VII. ii. 57. [65] Ibid., VII. iii. 58–59.
[66] Ibid., I. ii. 51. [67] Ibid., I. ii. 15–17.
[68] J. J. Meyer, *Sexual Life in Ancient India* (London, 1930), p. 531.

and elevated by it, and who makes it, so to speak, her own *sādhanā*. The pure Hindu wife, the *Satī*, is already a type of *bhakta*.[69]

In classical Sanskrit court poetry love, deeply sensual and passionate love, became celebrated with a sense of refinement and delight which justified it at least as a worthy pursuit of gentlemen and the most significant pursuit of a woman. The goal was no longer a liberation from *saṃsāra* (at least not immediately) but rather a revelation of the luxurious joys, elegant felicities, marvellous beauties, of *saṃsāra* no matter how fleeting.

> . . . since the fashion in Indian philosophy in the classical period was for monism, it will be apparent that to authors who were philosophically inclined it must have seemed that there was something unreal about beauty. . . . One can see it in the popular concept of *māyā*. Beauty was not, however, despised on this account; for the same suspicion of unreality attached to the whole of worldly existence. And there is ample evidence to show that those who pursued a worldly life, a class that included nearly all the nobility and the poets, sought out what affected them as beautiful to the best of their ability despite its transience.[70]

The poet Bhartṛhari described the two paths that one might take in *saṃsāra*: the one is sacred and the other profane, but they are hardly at odds:

> In this world [*saṃsāra*] which is without-substance and ever changing there are two paths: sometimes let it be the occasion for the sages' religious-devotion which is lovely because it overflows with the nectarous waters of the knowledge of truth; or if not [let it be the occasion] for the lusty undertaking of touching with one's palm that hidden [part] in the firm laps of lovely-limbed women, loving-women with great expanses of breasts and thighs.[71]

The sage is pretentious, however, and deceives himself and others when he reviles young girls, for on account of his religious-austerity, his *tapas*, he gets the heavenly reward of

[69] Vaudeville, *op. cit.*, p. 33.

[70] Daniel Ingalls, 'Words for Beauty in Classical Sanskrit Poetry', *Indological Studies in Honor of W. Norman Brown* (New Haven, 1962), p. 107.

[71] Bhartṛhari 88:

samsāre'sminn asāre pariṇati-tarale dve gatī paṇḍitānāṃ
tattva-jñānāmṛtāmbhaḥ-pluta-lalita-dhiyāṃ yātu kālaḥ kadācit;
no cen mugdhāṅganānāṃ stana-jaghana-ghanābhoga-sambhoginīnāṃ
sthūlopastha-sthalīṣu sthagita-kara-tala-sparśa-lolodyatānāṃ.

heavenly nymphs.[72] In another verse Bhartṛhari described the beloved's hair as 'tied up' (*saṃyamina*), a word which in the religious context means 'self-control', and he praised her 'ears' (*śruti*) and her 'teeth' (*dvīja*), words which might refer to the *Vedas* and 'the twice-born, the initiate', respectively; her breast is said to be a resting place for 'pearls' (*mukta*) or 'those who are liberated'.[73] In this kind of punning the profane mocks the sacred and this is typical of the court poetry. Society seems to have been systematically segmented, with the various segments serving each other, so that the man of the court could devote himself to wonderful profanities while the priests took care of austere sacraments. The tutelary deity would certainly be honoured but it was the passionate woman who was adored.

Bearing dishevelled, shaking rows of curls, a quivering ear-ring, having its forehead-mark slightly smeared by tiny networks of sweat, having its eyes wearied in the end of love-making, the face of the slender-woman, during the love-making-in-which-she's-on-top, may it protect you for a long time! *What need is there of deities, Hari, Hara, Brahmā, and the others?*[74]

The court poets took particular delight in describing the love-play of the gods, presenting the gods humanized in love in a way which might suggest a correlative vision of man divinized in love. The mythology of Kṛṣṇa, in particular, lent itself to this; the motif had been established at least by the third century A.D.:

When Yaśodā [Kṛṣṇa's foster-mother] said, 'Dāmodara [Kṛṣṇa] is still a child', the women of Vraja laughed secretly directing their eyes to the mouth of Kṛṣṇa.[75]

[72] Bhartṛhari 120:
sva-para-pratārako'sau nindati yo'līka-paṇḍito yuvatīḥ;
yasmāt tapaso'pi phalaṃ svargaḥ svarge'pi cāpsarasaḥ.

[73] Ibid., 139.

[74] *Amaruśatakam* 3:
ālolām alakāvalīṃ vilulitāṃ bibhrac calat-kuṇḍalaṃ
kiṃcin-mṛṣta-viśeṣakaṃ tanutaraiḥ svedāmbhasāṃ jālakaiḥ;
tanvyā yat suratānta-tānta-nayanaṃ vaktraṃ rati-vyatyaye
tat tvāṃ pātu cirāya kiṃ hari-hara-brahmādibhir daivataiḥ.

[75] Hāla *Sattasaī* 112 (trans. from Prakrit by Friedhelm Hardy):
ajja vi vālo Dāmo-
aro tti ia jaṃpie Jasoāe;
Kaṇhamuhapesiacchaṃ
ṇihuaṃ hasiaṃ Vaavahūhiṃ.

The love-sports of the youthful Kṛṣṇa, a feature of the court poetry and the *Purāṇas*, seem to have been a popular motif in the folk tradition:

> The pilgrims in the street have warded off the painful cold with their broad quilts sewn of a hundred rags; and now with voices clear and sweet they break the morning slumber of the city folk with songs of the secret love of Mādhava [Kṛṣṇa] and Rādhā.[76]

In the history of Kṛṣṇaism profane love, sensual, human longing, became clearly related to sacred love. Within the Kṛṣṇa-cults a love-mysticism, a love-symbolism, developed. The human and divine became inextricably interwoven in love, in *bhakti*, no longer *bhakti* simply as 'devotion', but *bhakti* as fervent, passionate love.

> Such a bhakti, as described in the *Bhagavata* [*Purāṇa*] or the *Shandilya-sutra*, is not worship out of a sense of duty or mere meditation on God or mere singing of His name, but it is deep affection (*anurakti*). It is therefore neither knowledge nor any kind of activity, but is a feeling.[77]

The mythology of Viṣṇu incarnated as the irresistible and playful cowherd-lover was a sacralization of human love. God became understood as transcendent and yet immanent, wholly distinct from the self and dearly personal, to be desired with all one's heart. Passionate, human love, if it was directed toward Kṛṣṇa, was sacred.

> The ideal of desirelessness and absolute self-control is replaced by that of participation in a drama of divine joy, and the desires are given full play in the direction of God. Desires are not to be distinguished; only their directions are to be changed.[78]

Attachment became a means rather than an obstruction to liberation. Sacred and profane love were homologous and yet differentiated in respect of the object of the love. Kṛṣṇa emerged in the *Purāṇas* as a carnal, sensual god, dancing amidst the

[76] Ḍimboka in Vidyākara's *Subhāṣitaratnakoṣa* 980 (Ingalls trans.):
rathyā-kārpaṭikaiḥ paṭac-cara-śata-syūtoru-kanthā-bala-
pratyādiṣṭa-himāgamārti-viśada-prasnigdha-kaṇṭhodaraiḥ;
gīyante nagareṣu nāgara-jana-praty-ūṣa-nidrā-nudo
rādhā-mādhavayoḥ paras-para-rahaḥ-prastāvanā-gītayoḥ.

[77] S. N. Das Gupta, *Hindu Mysticism*, 2nd edn (New York, 1959), p. 145.

[78] Ibid., p. 143.

cowherd-women, enchanting them with his song, his glances, his caresses. When the women heard his flute, late at night, they left their husbands and fathers, to come and dance with Kṛṣṇa:

One joined gently, gently into his song; and another girl remembered no one but him, all her attention given [to it]. One girl said 'Kṛṣṇa! Kṛṣṇa!' and then felt embarrassed; another one, blind from love, went to his side without any shame.[79]

The supreme Bhagavat took on the form of the lover so that the devotee could gain access to the sacred, the infinite and eternal, through the expression, rather than the suppression of earthly desires:

Another girl remained inside her house, when she saw her father outside, and with her eyes closed, meditated on Govinda, getting entirely absorbed into him. The heap of her good deeds vanished through the intense pleasure of thinking of him, all her sins removed through the great pain of not meeting him. Another cowherd girl obtained liberation [*mukti*] in her death while she thought [of him] as the cause of the world, as being essentially *para-brahman*.[80]

And the cowherd-women were to be exemplary for all 'embodied creatures', for all devotees:

He indwells (and moves) among the cowherds' daughters and their husbands as he does among all embodied creatures, looking after them the while and partaking of a body here on earth in the exuberance of his being (*krīḍanena*). It is out of sheer grace towards his creatures that he takes on a human body.[81]

[79] *Viṣṇu Purāṇa* v. xiv. 18–19 (Hardy trans.):
śanaiḥ śanair jagau gopī kācit tasya padānugā;
dattāvadhānā kācic ca tam eva manasāsmarat.
kācit kṛṣṇeti kṛṣṇeti coktvā lajjām upāyayau;
yayau ca kācit premāndhā tat-pārśvam avilajjitā.

[80] Ibid., v. xiv. 20–22 (Hardy trans.):
kācid āvasthasyāntaḥ sthitvā dṛṣṭvā bahir gurum;
tan-maya-tvena govindaṃ dadhyau mīlita-locanā.
tac-cintā-vipulāhlāda-kṣīṇa-puṇya-cayā tathā;
tad-aprāpti-mahā-duḥkha-vilīnāśeṣa-pātakā.
cintayantī jagat-sūtiṃ para-brahma-svarūpiṇam;
nirucchvāsatayā muktiṃ gatānyā gopa-kanyakā.

[81] *Bhāgavata-Purāṇa* x. xxxiii. 36–37 (Zaehner trans. in *Concordant Discord* [Oxford, 1970], p. 159):
gopīnāṃ tat-patīnāṃ ca sarveṣām eva dehināṃ;
yo'ntaś carati so'dhyakṣaḥ krīḍanen.eha dehabhāk.

He was to be loved by the devotee as a woman loves, longs for, passionately burns for, a man:

Only he who follows the way of *rāga* [passion] and worships Kṛṣṇa in profound love can gain the *mādhurya* [sweetness] of Kṛṣṇa. . . .[82]

The sentiment was cultivated among the Āḻvārs in the south; Nāmmāḻvār sang:

Tossing about restlessly, with a mind that has melted, singing again and again and shedding tears, calling upon You as Narasimha and seeking You everywhere, this beautiful maid [the heart of the devotee] is languishing.[83]

And it proliferated throughout India in the vernacular languages; the poetess Mira-bai, after the death of her husband, directed all her passionate love to Kṛṣṇa:

> I am fascinated by the beauty of Mohan:
> In the bazaar and by the way he teases me.
> I have not learned the sweet desire of my beloved.
> His body is beautiful and his eyes are like lotus flowers.
> His glance is very pleasing, and his smile is very sweet.
> Near the bank of the river Jumna he is grazing the cows,
> And sings a sweet song to the flute.
> I surrender myself, body and soul and wealth to the Mountain-
>   holder
> Mira clasps his lotus feet.[84]

In the Vaiṣṇava traditions the rapture and ardour, the convulsive ecstasy and miserable delight of love became an ideal. Liberation came through loving, desiring, Kṛṣṇa with no interest in liberation. Liberation was not the goal of love, but its by-product.

There were two major trends in medieval Vaiṣṇavism, the Vaiṣṇavism of Caitanya and that of the Sahajiyās.

---

anugrahāya bhūtānaṃ mānuṣaṃ deham āsthitaḥ
bhajate tādṛśīḥ krīḍā yāḥ śrutvā tat-paro bhavet.

[82] *Caitanya-Caritāmṛta* of Kṛṣṇadāsa-Kavirāja, *Madhyalīlā* xxi. 100, cited by E. Dimock, *The Place of the Hidden Moon* (Chicago, 1966), p. 194.

[83] *Tiruvāymoli* ii. iv. 1, cited in W. T. de Bary (ed.), *Sources of the Indian Tradition*, vol. i (New York, 1958), p. 351.

[84] Mira-bai trans. from Hindi by Keay in J. B. Alphonso-Karkala (ed.), *Anthology of Indian Literature*, Penguin edn (Harmondsworth, 1971), p. 540.

The Bengal School of Caitanya, no doubt, condemns direct erotic practice, but it encourages vicarious erotic contemplation. It emphasizes the inward realization of the divine sports in all their erotic implications as the ultimate felicitous state. . . . The dogma is implicitly accepted that Kṛṣṇa is the only male in the universe, and that the highest ideal of the devotee, like that of Rādhā, is the desire of a woman eternally seeking to satisfy her lover, who frankly, but divinely, thirsts after womanly charms of adolescence and youth.[85]

The vision begins with the monism of the *Upaniṣads*: All is One; but this One, *brahman*, divided itself into two. Kṛṣṇa, the Bhagavat, and Rādhā, his *hlādinī-śakti* ('infinite capacity for imparting ravishing sweetness'):[86]

. . . Radha-Krishna are one and the same eternally. But, for divine sport, they separated themselves into two bodies.[87]

The conception is of an eroticized universe; the love-play of Rādhā and Kṛṣṇa is a cosmic principle, all-pervasive, the underlying truth of all being. The devotee awakened to this enjoys the love of Rādhā and Kṛṣṇa. He identifies himself with Rādhā, feminizes his soul that it may long for Kṛṣṇa and be ravished by him:

The heart of Lord Chaitanya is a picture of Sri Radhika's sentiment and the feelings of pleasure and pain arise constantly therein in accordance with Her predicament.[88]

And he behaved like a mad man and raved as such. Just as Sri Radha raved in *Vraja* at the sight of Uddhava, so did he rave day and night.[89]

And in the last years the Lord's [Caitanya's] soul was filled with love for Lord Krishna. And the Lord went mad with love. And he was unconscious whether it was night or day. And he wept and laughed and danced and sang. And at the very next moment he seemed in deep grief.[90]

85 S. K. De, *Early History of the Vaiṣṇava Faith and Movement in Bengal*, 2nd edn (Calcutta, 1942), p. 419.

86 Zaehner, *Concordant Discord*, p. 160.

87 Rūpa Gosvāmin quoted in the *Caitanya-Caritāmṛta*, *Ādi-Līlā* IV; trans. Nagendra Kumar Ray, 2nd edn (Purī, 1959), vol. I, p. 43.

88 *Caitanya-Caritāmṛta* (Ray trans.), op. cit., *Ādi.* IV (vol. I, p. 47).

89 Ibid., *Ādi.* XIII (vol. I, p. 158).

90 Ibid., *Madhya.* I (vol. II, p. 5).

But the deep grief felt by the cowherd-women, by Rādhā, by Caitanya, by the devotee, is also the deepest joy:

Painful indeed is the love for the Lord. . . . But it is sweet also in the extreme . . . it defies the very sweetness of nectar in ecstasy of union.[91]

Actual union is avoided because if the beloved and the lover become one there is no longer any active love and active love, *bhakti*, was considered the highest realization, higher than, and yet the same as, liberation.

The Vaiṣṇava Sahajiyās coalesced the orthodox devotionalism of Caitanya with the esoteric teachings of the *Tantras*: Rādhā and Kṛṣṇa became the Sahajiyā equivalents of Śakti and Śiva in the Hindu *Tantras* and Prajñā and Upāya in the Buddhist *Tantras*. All men were considered the physical manifestations of Kṛṣṇa, with the goal of realizing that Kṛṣṇa-hood, and all women were considered physical manifestations of Rādhā, with the goal of realizing that Rādhā-hood. The goal was to be achieved in love:

The essence of beauty springs from the eternal play of man as Krishna and woman as Radha. Devoted lovers in the act of loving seek to reach the goal. Who is devoted to whom and how is of no interest. Dedicate your soul to the service of loving. . . . Says Chandīdās: Listen, O, brother man, Man is the greatest Truth of all, Nothing beyond.[92]

Beloved, it hurts me to let secrets out. With you and me and our love between us, we are transformed into Radha and Krishna, All into one. Passion rises from the natural joy born in the heart. We are all one with the eternal love that dwells in Braja. . . .[93]

The Sahajiyās did not make the distinction between sacred and profane love; the divine was for them the fullest blossoming and finest distillation of the profane. The goal was realization of the sacredness of humanness in love, and the sexual longing of the cowherd-women, of Rādhā, took on new meaning:

It can be read as suggesting not that worshipful longing is the end of man but that if union with the divine could be attained it would reward one with the joy of human sexual union raised to the nth degree. And if, unlike Christians and orthodox Vaiṣṇavas, one were

[91] Ibid., *Antya*. I (vol. IV, p. 19).
[92] Caṇḍīdāsa (trans. Bhattacharya), op. cit., p. 105.
[93] Ibid., p. 99.

to hold that there is no qualitative difference between human and divine, such union would seem possible. And under such conditions, there would no longer be a distinction between sacred and profane, between carnal and spiritual love. Longing for union then becomes not the end of man but the means to the end; sexual union becomes not denial of the ideal relationship between human and divine, but affirmation of it.[94]

Love, 'using one's body as a medium of prayer and loving spontaneously',[95] was for the Sahajiyās the highest truth, encompassing the sacred and the profane:

I would make my residence in the City of Love [*prema*]. I shall build there a hut with love. I shall make love my neighbor and part company with all else. My door shall be of love, and love too shall be my roof. I shall pass time in the sweet repose of love and I shall sleep on a bed of love and have love for my pillow. I shall lie idly clasping the pillow of love and shall be a playmate of love. I shall bathe in the lake of love and shall wear the collyrium of love. Love will be my religion [*dharma*], love will be my service, and I shall dedicate myself to love. I shall make a nose-ring of love, which will wave to and fro by the corner of the eye. Says Chandīdāsa, 'I too will wear the collyrium of love'.[96]

But the path of love is dangerous for love is a snare, a bondage to things finite and temporal. How can liberation from the world be attained in attachment to the world? The answer to the paradox was the 'secret' of *sahaja*:

You must make the frog dance before the serpent and then only are you true lovers. The skilful man, who can wreathe the peak of mount Sumeru with thread and ensnare the elephant into the web of the spider, becomes eligible for such a secret love.[97]

The Vaiṣṇava Sahajiyās regarded the twelfth-century poet Jayadeva as their *Ādi-guru*, their first teacher, and his poem, the *Gītagovinda*, as a revelation of their doctrine of 'secret'

[94] Edward Dimock, 'Doctrine and Practice among the Vaiṣṇavas of Bengal', in Milton Singer (ed.), *Krishna; Myths, Rites and Attitudes*, Phoenix edn (Chicago, 1968), p. 62.

[95] Caṇḍīdāsa (trans. Bhattacharya), op. cit., p. 82.

[96] Caṇḍīdāsa in *Vaishnaba Lyrics*, ed. and trans. Matilal Das, cited by Kanwar Lal, *The Religion of Love* (Delhi, 1971), p. 87.

[97] Caṇḍīdāsa, cited by S. B. Das Gupta, *Obscure Religious Cults*, p. 424.

love.[98] And Caitanya 'tasted day and night the songs of Jayadeva'[99] and received from them inspiration for his love. Various traditions, religious, literary, folk, erotic, dramatic, converge and coalesce in the *Gītagovinda*. It is a work ambiguously and ambivalently about the sacred and profane dimensions of love.

. . . AS EXEMPLIFIED IN THE *GĪTAGOVINDA* OF JAYADEVA

> I can only write a short note to-day [January 22, 1802] to accompany the enclosed [copy of the *Gītagovinda* translated by Dalberg], which will be sure to give you pleasure. . . . What struck me as remarkable are the extremely varied motives by which an extremely simple subject is made endless.[100]

The *Gītagovindakāvyam* or 'poem[101] about Govinda (Kṛṣṇa as a cowherd) in songs' of Jayadeva, describing the separation and union of Rādhā and Kṛṣṇa, marks the culmination of the classical Sanskrit poetic tradition and the beginning of the real flowering of the Indian vernacular poetic and the medieval Vaiṣṇava devotional traditions.

GG intro

The work opens with several benedictory verses: a verse recollecting a previous union of Rādhā and Kṛṣṇa (I. 1),[102] a verse in which the poet and the theme are introduced (I. 2), a description of the poet in comparison with his contemporaries and predecessors (I. 4) and a verse in which the essential ambi-

[98] See De, op. cit., p. 436; Dimock, *Place of the Hidden Moon*, pp. 56–7.

[99] *Caitanya-Caritāmṛta* (trans. Ray), *Ādi*. XIII (vol. I, p. 158).

[100] Letter from Goethe in the *Correspondence Between Schiller and Goethe*, vol. II, trans. L. Dora Schmitz (London, 1909), p. 395.

[101] 'Poem' is perhaps a misleading translation of '*kāvya*'. Richard Gombrich has explained the term: *Kāvya*, like poetry, is usually in verse but not coextensive with verse; not all verse is *kāvya*, nor is all *kāvya* in verse. . . . perhaps the commonest definition is in terms of a quality called *rasa*. This word is a metaphor from tasting, and means 'flavor'. . . . *Kāvya*, then, or poetry, is any coherent speech or writing which is informed by flavor. A flavor is an emotion or sentiment, not experienced directly as in real life, but aesthetically, so that it affords a calm enjoyment, a dispassionate pleasure in the passions . . .' (A preface to his translation of the *Caurapañcāsikā* in *Mahfil*, vol VII [East Lansing, Michigan, 1971], p. 175.)

[102] The numbers given in parentheses throughout this study refer to my translation of the *Gītagovinda* (Chapter VII, below) and the transliteration of the text given in the Appendix.

valence between the sacred and profane dimensions of love is presented:

If your mind is passionate in remembrance of Hari, if it is curious about the amatory arts, then listen to Jayadeva's eloquence. . . . [I. 3]

Remembrance (*smaraṇa*) of Kṛṣṇa is devotional activity, sacred; the amatory arts (*vilāsa-kalā*) are sensual, sexual, profane. Jayadeva immediately makes it clear that the *Gītagovinda* is about *both*.

Jayadeva and most Vaiṣṇava poets, especially the Bengali Vaiṣṇavas, have grasped the physical as the concrete form of the spiritual. Thus eroticism is at once sensual and spiritual.[103]

Kṛṣṇa as the supreme and transcendent Lord is praised in each of his incarnations or 'descents' (*avatāras*) (I. 5–16) and he is celebrated through mythological reference as hero and lover and Lord (I. 17–25). The spring setting, the erotic season in which Kṛṣṇa plays in love and separated lovers grieve, is established (I. 27–37), and Kṛṣṇa's love-play with the cowherdesses is described to Rādhā by her companion (I. 38–47). Rādhā, separated from Kṛṣṇa on account of her jealousy (II. 1), recounts the time when she was first united with Kṛṣṇa (II. 2–20). Kṛṣṇa, becoming disenchanted with the other women (II. 21), begins to long for Rādhā, to suffer over his separation from her (III). The friend describes Rādhā's misery to Kṛṣṇa (IV) and then, in an attempt to fulfil his subsequent request that she bring Rādhā to him (V. 1), the friend urges Rādhā to join her lover by telling her of Kṛṣṇa's sorrows and by reminding her of the delights of love in union (V). But Rādhā is too weak from longing to move and so the friend returns to Kṛṣṇa to try, again by describing Rādhā's suffering, to convince him to go to his beloved (VI). Kṛṣṇa delays, does not keep the tryst, and Rādhā laments (VII. 1–11). When she sees her friend return without Kṛṣṇa she imagines that he is making love with another woman (VII. 12 ff.). At dawn Kṛṣṇa arrives and bows before her, but she scolds him in a fit of pique and jealousy (VIII). Kṛṣṇa withdraws and the friend chides Rādhā for her pride and vain misery and once again urges Rādhā to follow Kṛṣṇa (IX). Kṛṣṇa returns to

[103] Ranajit Sarkar, 'Gītagovinda: Towards a Total Understanding', *Publikaties van het Instituut voor Indische talen en culturen No. 2* (Rijksuniversiteit te Groningen, 1974), p. 8.

appease Rādhā with loving words and tender flatteries and he then goes to the trysting place (x). Yet again the friend exhorts Rādhā to pursue him (xi. 1–22). Rādhā does so and is enraptured by his beauty (xi. 23 ff.). Kṛṣṇa beckons her to the bed, to his embraces (xii. 1–9) and they make love (xii. 10–12). In the morning Rādhā, dishevelled from the playful battle of love (xii. 13–15), asks Kṛṣṇa to dress her, to re-do her ornaments and make-up, and he complies (xii. 16–25). The final verses echo the theme and declare the sweetness and excellence of the poet and the poem. Again the *Gītagovinda* is described as devotional *and* erotic *and* literary:

Skill in the arts of the Gāndharvas [i.e., the musical arts], meditation consecrated-to-Viṣṇu, playful-creation in poems which are literary-works on the truth of discrimination in erotics—may wise people joyfully purely-understand all that according to the *Srī Gītagovinda* of the poet and scholar Jayadeva whose soul is solely directed to Kṛṣṇa. [xii. 28]

The 'extremely simple subject is made endless', the sorrows of separation and the joys of union are sung again and again. The conventions and conceits are primarily those of courtly love poetry, secular, erotic *kāvya*, but the fact that Kṛṣṇa is the lover and that the poet declares again and again his devotion to Kṛṣṇa suggests that Jayadeva's attitude and achievement were similar to those of his contemporaries in Europe:

The [twelfth-century Latin] poets achieved the kind of mean [between the sacred and profane, the metaphysical and physical] by refusing to give either world absolute priority, and suggesting that both have certain values. Metaphysics is not disregarded, neither however is human beauty.[104]

The poem has been interpreted as purely secular, but more often as purely religious, mystical, a sacred allegory of divine love. But the 'problem' of interpreting the work arises with the limiting assumption that it must be *either* sacred *or* profane, an assumption which obscures the essential achievement of the poem—its ambiguity. The *Gītagovinda* is sung in homage of Kṛṣṇa, the Lord of the World (*jagad-īśa*) (i. 5–16), as an act of remembrance of the transcendent godhead incarnated for the

[104] Wilhelm, op. cit., p. 142.

sake of love as Kṛṣṇa who has an 'erotic disposition' (*nidhuvana-śīla*) (II. 18). The ambiguity of the whole is achieved largely through the ambiguities inherent in the vocabulary of love, a vocabulary equally suggestive of sensual passion and religious devotion. There is a linguistic reconciliation of devotional activity with worldly, carnal activity. For example, Jayadeva uses the word 'remembrance' (*smaraṇa*) in a profane way, as an erotic activity, in the fifth song (II. 2–8)—Rādhā *remembers* her love-making with Kṛṣṇa. But then in the last verse of the song (II. 9) the word acquires devotional significance—Jayadeva and the virtuous *remember* Kṛṣṇa's lotus-feet.

The ambiguity in the language reflects the psychological correlation and interpenetration of the sacred and profane dimensions of love. This kind of ambiguity runs similarly through John Donne's poetry:

> . . . in Donne's 'Songs and Sonnets,' his poems of profane love, the metaphoric gloss is constantly drawn from the Catholic world of ecstasy, canonization, martyrdom, relics; while in some of his 'Holy Sonnets' he addresses God in violent erotic figures. . . . The interchange between the spheres of sex and religion recognizes that sex is a religion and religion is a love.[105]

In the same way that Donne uses Catholic terminology, Jayadeva uses the vocabulary of the orthodox religious systems; the go-between speaks to Rādhā:

> When he was together with you before the perfections [*siddhi*] of love were attained; truly there in love's great-pilgrimage-place [*mahātīrtha*] in the grove again Mādhava, meditating [$\sqrt{}$*dhyai-*] on you, constantly chanting [$\sqrt{}$*jap-*] also a string of sacred-sounds [*mantra*] as an invocation to you alone, desires again the nectar of ardent embraces of the pitchers of your breasts. [v. 7]

The lover, thinking of his beloved in a grove and muttering her name, becomes like the *yogī* in a sacred pilgrimage place praying, meditating, chanting and attaining beatitude.[106] The ambi-

[105] René Wellek and Austin Warren, *Theory of Literature* (New York, 1949), pp. 213–14.

[106] The *Rasika-rañjana* of Rāmacandra and the *Śṛṅgāra-vairāgya-taraṅgiṇī* of Somaprabhācārya are poems simultaneously about love and renunciation, achieving total ambiguity through the use of puns. This particular ambiguity, the play between the erotic and ascetic, was a conventional motif in Sanskrit court poetry.

guity in the language of the *Gītagovinda*, like the ambiguity in Donne's 'Songs and Sonnets', is the poetic power, the force that raises it out of the domain of a static theology (which has an obligation to be consistent) and invests it with a psychologically validating contradictoriness—the poet, unlike the theologian, has every right to be self-contradictory and inconsistent. The leap may be made from the sacred to the profane and back again impulsively. Such vacillation is typical of Sufi poetry:

> Unless we have some clue to the writer's intention, it may not be possible to know whether his beloved is human or divine—indeed the question whether he himself knows is one which students of Oriental mysticism cannot regard as impertinent.[107]

The ambiguous interpenetration of the sacred and profane spheres in the *Gītagovinda* reflects a blending of traditions.

1 *The literary tradition*

Sanskrit court poetry is generally impersonal, learned, artificial, and above all conventional, aiming at clever and playful use of the conventions. The dynamism of the aesthetic experience was meticulously studied and systematically classified by rhetoricians following the poets, and the poets in turn scrupulously followed the theorists. Such close affinity and causal connection between rhetoric and poetry, theory and practice, yielded a highly stylized and refined literature. *Kāvya* was for *kāvya*'s sake—aesthetic truth took precedence over psychological, historical, religious, social or factual truth; and this truth was delineated and codified in the rhetorical texts.

The poet had a particular position in society and he addressed a very particular audience of educated and sophisticated connoisseurs. *Kāvya* was a social institution. 'Esthetic institutions are not based upon social institutions: they are social institutions of one type and intimately connected with those others.'[108] *Kāvya* was produced by, served, and reflects the courtly tradition, the feudal system, the life of the nobility. *Kāvya* was meant to delight and instruct. The attitude toward pleasure, and

[107] R. A. Nicholson, *Studies in Islamic Mysticism* (Cambridge, 1921), pp. 163–4.

[108] Adolph Siegfried Tomars, *Introduction to the Sociology of Art*, cited by Wellek and Warren, op. cit., p. 89.

particularly the pleasures of love, the estimation of the value of the senses, was distinctly positive in the literary tradition:

> The celebration of festivals with pomp and grandeur, the amusements of the court and the people, the sports in the water, the game of swing, the plucking of flowers, song, dance, dramatic performances and other diversions, elaborate descriptions of which form the stock-in-trade of most Kāvya-poets, bear witness not only to this new sense of life but also to the general demand for refinement, beauty and luxury. The people could enjoy heartily the good things of this world, while heartily believing in the next. If pleasure with refinement was sought for in life, pleasure with elegance was demanded in art.[109]

As an elite social institution, this patronized courtly literature, although very often dealing with gods, remained essentially concerned with the profane, with worldly activity.

> It is clear that *kāvya* in such a setting as this will be fundamentally secular in outlook. . . . *Kāvya* in the service of religion, though occasionally found, is a secondary and sporadic phenomenon.[110]

'Secular in outlook' does not mean secular in theme, in subject —the profane manifestations of the sacred, particularly the sexual life of the gods, was a conventional theme. And sexual life generally, the sexual life as an expression of love, was a conventional theme, a dominant subject of *kāvya*. But while *kāvya* is extensively love-poetry, it is not so in the traditional European sense:

> The whole attitude and training of the Indian poet was such that the idea of using poetry to express the poet's own private feelings hardly ever occurred to him, and probably never in the case of love. . . . This situation may to some extent be due to the fact that he was composing in an elaborate, and in many ways artificial, language of learning which may not be the ideal type of medium to express emotions which come 'straight from the heart'. But the more important point, really, is that the romantic conception of private, personal poetry . . . was not among the cultural attitudes of the times. In general, it can be said that Sanskrit love-verses are verses about love, not the verses of a lover.[111]

[109] S. K. De, *Treatment of Love in Sanskrit Literature*, 2nd edn (Calcutta, 1959), p. 18.

[110] A. K. Warder, *Indian Kāvya Literature*, vol. I (Delhi, 1972), pp. 14–15.

[111] John Brough, 'Introduction' to his *Poems from the Sanskrit*, Penguin edn (Harmondsworth, 1968), p. 40.

There is a tension in Sanskrit love-poetry between what is expressed and how it is expressed (the separated lovers' pain described with delight and detachment), a tension between content and form, spontaneity and conventionality, feeling and thought. In the *Gītagovinda*, for example, Rādhā laments, sings her great sorrow to her friend (II. 20) and the friend in turn describes Rādhā's misery to Kṛṣṇa (IV. 10, 20) and Kṛṣṇa ecstatically praises Rādhā's beauty (X. 15); in each of these verses there is a pun—Jayadeva names the metre in which the verse is written and thereby creates two levels, the level *in* the verse and the level *of* the verse. The passion, pain, ecstasy of the words as uttered by Rādhā or her friend are juxtaposed to the control, humour and cleverness of the same words as uttered by Jayadeva. The poet's concern with Rādhā and Kṛṣṇa is juxtaposed to the concern with poetic process as a technical skill. He delights furthermore in the use of erudite grammatical forms, words formed according to special grammatical rules, words which would test the readers' knowledge of the grammatical texts.[112]

Although intellectual, learned and suggestive, the approach in Sanskrit love-poetry generally, and in the *Gītagovinda* specifically, is ornamental and sensual rather than discursive or philosophical. The poem *is* rather than *means*.[113]

The painter Degas once told the poet Mallarmé that he had a 'good idea for a poem'; the poet scoffed, 'poetry is not written with *ideas*, it's written with *words*'.[114] The *Gītagovinda* is constructed with words in this sense, words as sound, and sound as meaning. There is an orchestration of syllables, a grand play with the auditory aspects of the vocabulary: *madhukaranikarakarambitakokilakūjitakuñjakuṭīre* (I. 28).[115]

The *Gītagovinda* is very clearly in the main current of the literary tradition; the union, separation and reunion of lover

[112] For example, *yauvata*, 'a number of young girls' (X. 15) is formed according to Paṇini IV. ii. 38.

[113] W. K. Wimsatt, *The Verbal Icon*, British edn (London, 1970), p. 80: ' "A poem should not mean but be." It is an epigram worth quoting in every essay on poetry.' The epigram is from Archibald MacLeish's 'Ars Poetica'.

[114] Philip Wheelwright, *Metaphor and Reality* (Bloomington, 1967), p. 49.

[115] In form and content so like Tennyson's:

The moan of doves in immemorial elms,
The murmuring of innumerable bees.

and beloved is the standard theme; Jayadeva uses the conventional, typical, prescribed vocabulary, ornaments, tropes, metaphors, conceits.

Jayadeva . . . the last great Sanskrit poet of the highest artistic accomplishment, in whom Sanskrit love-poetry, both in its technical and emotional aspects, reaches its climax . . . prides himself upon the grace, beauty and music of his diction as well as upon the delicacy of his sentiments. . . . Jayadeva's achievement lies more in the direction of form than in the substance of his poem. It presents hardly any new ideas; it scarcely describes any situation or emotion which earlier love-poets have not familiarized; it only makes skilful poetic use of all the conventions and traditions of Sanskrit love-poetry.[116]

The subject, the love-play of Rādhā and Kṛṣṇa, had been conventionalized at least a century before the *Gītagovinda*.[117]

'Go on ahead, milkmaids, taking home the pots already full.
Rādhā will follow later when the older cows are milked.'
May Krishna, who by subterfuge thus made the cattle station
 deserted but for Rādhā and for him,
the god, the foster-son of Nanda,
steal away your ills.[118]

By a subterfuge the love-play of Rādhā and Kṛṣṇa prevails. This motif opens the *Gītagovinda*:

'The sky is densely clouded, the forest grounds are dark with *tamāla* trees; at night he [Kṛṣṇa] is afraid. Rādhā, you alone must take him home.' This is Nanda's command. [But] Rādhā and Mādhava stray to a tree in the grove by the path and on the bank of the Yamunā their secret love-games prevail. [I. 1][119]

[116] De, *Treatment of Love*, pp. 54–5.

[117] See poems 131, 136, 139, 147, 808, 980, in Vidyākara's *Subhāṣitaratnakoṣa* which was compiled shortly before A.D. 1100.

[118] Ibid., 139 (Ingalls trans.):

agre gacchata dhenu-dugdha-kalaśānādāya gopyo gṛhaṃ
dugdhe vaskayaṇīkule punar iyaṃ rādhā śanair yāsyati;
ity anya-vyapadeśa-gupta-hṛdayaḥ kurvan viviktaṃ vrajaṃ
devaḥ kāraṇa-nanda-sūnur aśivaṃ kṛṣṇaḥ sa muṣṇātu vaḥ.

[119] De, *Vaiṣṇava Faith and Movement*, p. 8, suggests that this verse refers to an episode described in the fifteenth chapter of the 'Kṛṣṇa-janma-khaṇḍa' of the *Brahmavaivarta Purāṇa* (*c.* 750–1550 A.D.) but dates of portions of that *Purāṇa* are so unreliable that it is equally likely that the episode in the *Purāṇa* was inspired by the *Gītagovinda*.

Conventionally the clouds suggest that it is the monsoon, 'a season for love-making unequalled by any other except early spring'.[120] The building up of dark images—the clouds, the forest, the *tamāla* with its dark trunk, the night, the dark waters of the Yamunā, dark-bodied Kṛṣṇa—creates an interplay between the fearful and the erotic moods or sentiments. Kṛṣṇa pretends that the darkness is causing him fear while in fact it is causing desire, love for Rādhā. Darkness, according to the rhetoricians, typifies both moods and both moods have similar symptoms: trembling, sweating, stammering, fainting. On the surface it is a verse ironically cast in the fearful mood, but through suggestion (*dhvani*) it is an erotic verse.

The commentator Kumbha[121] glosses the verse with the standard theoretical classifications of Sanskrit poetics, the technical vocabulary of Sanskrit rhetoric: he indicates that the verse achieves the 'erotic flavour, sentiment or mood' (*śṛṅgāra-rasa*) and that it is 'love-in-separation' (*vipralambha*) about to become 'love-in-enjoyment' (*sambhoga*); the 'hero' (*nāyaka*) and 'heroine' (*nāyikā*) are conventional types, the 'obliging or faithful' (*anukūla*) type and the type 'who has her husband or lover in subjection' (*svādhīnapatikā*); the 'mode' of action (*vṛtti*) is the 'tender' (*kaiśikī*); the 'style' or 'flow' (*rīti*) is the 'southern' (*vaidarbhī*)[122] and the 'quality' (*guṇa*) of the style is 'clarity' or 'grace' (*prasāda*) of word and meaning: the 'metre' (*chandas*) is the 'Tiger's Play' (*śārdūla-vikrīḍita*, i.e., four times: – – – ⏑⏑ – ⏑ – || ⏑ ⏑⏑ – – – ⏑ – – ⏑ –); and so forth. A great deal of the *Gītagovinda* lends itself to this kind of rhetorical exegesis. But while the introductory, connective, narrative, recitative, benedictory and final verses of each of the twelve cantos (*sarga*s) of the *Gītagovinda* are typical of Sanskrit court poetry formally, metrically and in content, the twenty-four songs within the cantos are not. They reflect a popular influence; they suggest the impact of a folk tradition, and this combination of traditions was a unique and original poetic gesture.

[120] Ingalls, *Anthology of Sanskrit Court Poetry*, p. 126.

[121] Author of a fifteenth-century commentary on the *Gītagovinda*, the *Rasika-priyā* (published in the Nirṇaya-Sagar Press edn).

[122] '*Vaidarbhī* [style] avoids compounds and limits alliteration to certain (appropriate) places, it also maintains words in their etymological sense', Rājaśekhara cited by Warder, op. cit., p. 105.

2 *The folk tradition*

The songs are formally characteristic of the vernacular languages, Apabhraṃśa, Early Bhāṣā, Old Bengali: they are composed in rhymed couplets with *moric* metres;[123] each song has a repeated refrain (*dhruva-padam*), a signature line (*bhaṇita*), a given melodic pattern (*rāga*) and rhythmic measure (*tāla*).

The popular influence is clear thematically as well as formally—the mythology of Kṛṣṇa as the cowherd, the lover, particularly as the lover of Rādhā, seems to have originated in the folk milieu.[124] Wandering pilgrims sang the songs of the 'secret love of Mādhava and Rādhā'[125] and it has been suggested that Jayadeva's songs are Sanskrit translations of such vernacular songs.[126]

Jayadeva expresses not only Rādhā's love and longing for Kṛṣṇa, but in juxtaposition to that, his own devotion to Kṛṣṇa. The juxtaposition implies at least an analogy between the sacred and profane dimensions as well as an analogy between the aesthetic, dramatic dimensions of love and the actual, psychological dimensions. Such a juxtaposition is less typical of the literary, courtly tradition than of the tradition of the vernacular literature which was to arise after the twelfth century, a literature which reflected popular ideas and expressed popular sentiments.

[123] That is, metres based on the *gaṇa* system in which the determining principle is the number of *morae* and two short syllables may be substituted for one long or vice versa.

[124] A. B. Keith suggested that Jayadeva was most likely inspired by folk dramas, 'the Yātrās of Bengal, where in honour of Kṛṣṇa in a primitive form of drama dances accompanied by music and song were performed' (*History of Sanskrit Literature*, pp. 191–2).

[125] Ḍimboka cited above, p. 21.

[126] This was first argued by Pischel (*Die Hofdichter des Lakṣmaṇasena* [Gottingen, 1894], p. 22) but it was challenged by Keith (op. cit., p. 197) and others and it remains a generally unaccepted idea. Suniti Kumar Chattarji has, however, taken up the argument more recently: 'It is not unlikely that these [folk] Apabhraṃśa or Old Bengali verses obtained a great popularity and this induced Jayadeva to render them into Sanskrit, to give them a permanent and a pan-Indian form.' He explains that some of the songs 'read [i.e., scan] better as Apabhraṃśa or Old Bhāṣā than as Sanskrit, and fit in better with the scheme of pauses in the line which agrees with Old Bengali very closely' ('Jayadeva Kavi' in *Acarya Dhruva Commemorial*, vol. III, p. 189).

Without records it is impossible to know what the popular conception of love may have been in the twelfth century in north-eastern India, but it is possible to conjecture that unlike the literary notion of love as an art, a cultivated and mannered emotion, a sophisticated pastime, the popular notion would be more of love as a natural, spontaneous, fundamental response, perhaps similar to the popular conception of love in Europe during the same period:

> Live I cannot without you,
> To your beauty I've surrendered.
> Gentle heart, could you love true,
> Heart to whom my love I've tendered?[127]

3 *The religious traditions*

The mythological material, the characterization of the highest Lord incarnated as the cowherd-lover, the love-play with the cowherdesses, the *Rāsa* dance, the motifs of sorrowful love-in-separation from Kṛṣṇa and joyous love-in-union with him—all of this in the *Gītagovinda* seems to be based on the *Viṣṇu* and *Bhāgavata Purāṇas* or at least upon the vast oral literature which found expression in those *Purāṇas*. In those *Purāṇas*, and presumably in that oral folk literature, passionate love had become sacralized as an expression of *bhakti*: the loving-woman's longing became devotion and love-making became worship:

Then the *gopīs*, with face-lotuses wide-open [from joy] saw him coming, Kṛṣṇa the protector of the three worlds who is undefiled in his work. When one girl saw Govinda coming, in her delight she exclaimed, 'Kṛṣṇa! Kṛṣṇa! Kṛṣṇa!' with wide open eyes. Another looked at Hari with a frown, bending her eye-brows, and her eye-bees drank his face-lotus. One girl closed her eyes when she had seen Govinda and meditated on just his beauty; she looked like someone absorbed in *yoga*.[128]

[127] G. Reese, *Music in the Middle Ages* (London, 1940), p. 222.

[128] *Viṣṇu Purāṇa* v. xiv. 42–45 (following Hardy's trans.):

tato dadṛśur āyāntaṃ vikāśi-mukha-paṅkajāḥ;
gopyas trailokya-goptāraṃ kṛṣṇam akliṣṭa-kāriṇam.
kācid ālokya govindam āyāntam atiharṣitā;
kṛṣṇa kṛṣṇeti kṛṣṇeti prāhotphulla-vilocanā.
kācid bhrū-bhaṅguraṃ kṛtvā lalāṭa-phalakaṃ hariṃ;
vilokya netra-bhṛṅgābhyāṃ papau tan-mukha-paṅkajam.

One clever *gopī*, with playful arms, under the pretext of praising his songs, embraced and kissed Madhusūdana.[129]

When he moved backward, they followed him; when he turned round they faced him; in backward and forward movements, the cowherd wives clung to Hari. Madhusūdana had such fun with the *gopī*s, that even a second without him was like a million years. These wives of the cowherds, fond of pleasure, slept with Kṛṣṇa during the nights, although they were hindered by their husbands, parents, and brothers. And Madhusūdana paid honour to his youthful age; of immeasurable self, destroying everything unpleasant, he dallied with them in the nights. [But as] the Lord in the form of his true essence, he pervades their husbands, them, and all beings and he indwells the All; just as ether, fire, earth, water, and wind pervade all elements [or: beings], so he as the *ātmā* pervades and indwells the All.[130]

The *Bhāgavata Purāṇa* homologized the sacred and profane dimensions of love, explicitly theologized sexual love, reconciled human passion as exemplified in the love of the cowherdesses for Kṛṣṇa with *bhakti*, religious devotion valued as the highest means and the highest goal.

How, without the bristling of hair, without the mind dissolving, without being inarticulate on account of joyous tears, without devotion, can the heart be purified? Who stammers his words, whose thought dissolves, who continually weeps and sometimes laughs, who shame-

---

kācid ālokya govindaṃ nimīlita-vilocanā;
tasyaiva rūpaṃ dhyāyantī yogārūḍheva sā babhau.

129 Ibid., v. xiv. 53 (Hardy trans.):
kācit pravilasad-bāhuḥ parirabhya cucumba tam;
gopī gīta-stuti-vyājān nipuṇā madhusūdanam.
(Compare *Gītagovinda* I. 49).

130 Ibid., v. xiv. 56–61 (Hardy trans.):
gate'nugamanaṃ cakrur valane saṃmukhaṃ yayuḥ;
pratilomānulomena bhejur gopāṅganā harim.
sa tathā saha gopībhī rarāma madhusūdanaḥ;
yathābda-koṭi-pratimaḥ kṣaṇas tena vinābhavat.
tā varyamāṇāḥ pitṛbhiḥ patibhir bhrātṛbhis tathā.
kṛṣṇaṃ gopāṅganā rātrau ramayanti rati-priyāḥ.
so'pi kaiśoraka-vayo mānayan madhusūdanaḥ;
reme tābhir ameyātmā kṣapāsu kṣapitāhitaḥ.
tad-bhartṛṣu tathā tāsu sarva-bhūteṣu ceśvaraḥ;
ātma-svarūpa-rūpo'sau vyāpya sarvam avasthitaḥ.
yathā samasta-bhūteṣu nabho'gniḥ pṛthivī jalam;
vāyuś cātmā tathaivāsau vyāpya sarvam avasthitaḥ.

lessly sings and dances, attached by devotion to me, he purifies the world.[131]

But the devotee must turn all his erotic longings away from women, sublimate all carnal desire, and concentrate all his love on Kṛṣṇa who sanctifies the passions when he is their object; the text continues that the devotee must live alone, devoted to Kṛṣṇa alone, abandoning the company of women and also of men who are fond of the company of women.[132]

Various means of worshipping Kṛṣṇa, of expressing devotion to the Lord, are elaborated in the *Bhāgavata Purāṇa.* These include *guṇa-kīrtana*, chanting the names, deeds, and praises of Kṛṣṇa. The simple proclamation of the name, any song about him, leads immediately to liberation. The *Gītagovinda* then is sacred because it is about Kṛṣṇa, regardless of how profanely it depicts him.

Although the *Gītagovinda* is infused with *Purāṇic* attitudes, it diverges from the *Purāṇas* both in the positive value it places upon the delights of human love and in the consideration which it gives to Rādhā. And as Rādhā, the individual cowherdess not mentioned in either the *Viṣṇu* or the *Bhāgavata Purāṇa*, took on special importance, Kṛṣṇa was transformed. The Kṛṣṇa of the *Gītagovinda*, unlike the *Purāṇic* Kṛṣṇa, suffers in his separation from Rādhā; he serves her, bows down in obeisance to her, worships her, and by this suffering Kṛṣṇa is more humanized. He is a man carnally loving a woman. The metamorphosis reflects the literary tradition—Kṛṣṇa is the conventional 'obliging' lover (*anukūla-nāyaka*); but the submissive worship of Rādhā, the passionate devotion to her, reflects also a religious tradition prominent in Bengal, the worship of Devī, the Goddess, 'the great creative power (*mahāmāyā*), which puts the universe through its revolution of appearances'.[133] In the Devī

[131] *Bhāgavata Purāṇa* XI. xiv. 23–24:

kathaṃ vinā roma-harṣaṃ dravatā cetasā vinā;
vinānandāśru-kalayā śudhyed bhaktyā vināśayah.
vāg-gadgadā dravate yasya cittaṃ
rudaty abhīkṣaṇaṃ hasati kvacic ca;
vilajja udgāyati nṛtyate ca
mad-bhakti-yukto bhuvanaṃ punāti.

[132] *Bhāgavata Purāṇa* XI. xiv. 29–30, etc.

[133] W. Norman Brown, introduction to his translation of the *Saundaryalaharī* (Cambridge, Massachusetts, 1958), p. 5.

or Śakti cult the male was subordinate to the female; the man to the woman and the god to the goddess:

The feminine principle, or śakti (power), personified as the goddess Devī, is the first and supreme principle of the universe. It includes both the spiritual and material principles. . . . The feminine principle in conjunction with the masculine principle . . . but with the masculine principle always secondary and subordinate to the feminine, creates the cosmos by exercising its power to produce change (*māyā*). . . . Man's highest goal is to achieve the vision of Devī in her mansion, to become assimilated to Devī, that is, to become identified with the absolute principle.[134]

In the worship of the Goddess the voluptuousness of woman, which had always been a relished theme of *kāvya*, became sacred—praise and adoration of Devī's full breasts and thighs, her flowing hair and lotus-face, her eyes and side-long glances, her ornaments and graceful limbs, became a means to liberation. Her devotee 'dissolves the union of soul and bond and savours the sweetness called supreme brahman'.[135]

Śaktism was a phase of Hindu Tantrism. And the *Gītagovinda* was written in a milieu in which Tantrism had become popularized and proliferated. Between A.D. 700 and 1200 in northeastern India, lyrics in Apabhraṃśa and Old Bengali, *dohā*s and *caryā-pada*s, were composed by Tantric Buddhist *siddhācārya*s. Formally the *caryā-pada*s are very similar to many of Jayadeva's songs—they are in rhymed couplets with moric metres, *rāga*s are given and there is a signature line, a *bhaṇita* or colophon. The close similarity suggests influence, or at least that Jayadeva and the *siddhācārya*s drew influence from a common source. The lyrics are written in a 'twilight language' (*sandhyā bhāṣā*), a symbolic language, a code in which esoteric doctrine was expressed in sexual terms. The sexual experience was equated with the *yogic* experience in the attainment of the state of non-ego, vacuity, the suspension of thought-process and self-consciousness. The internal physiological process by which the state of vacuity was achieved was analogous to the union of the man and the woman and symbolized by the thunder entering the lotus.

[134] Ibid., p. 6.

[135] *Saundaryalaharī* 99: '. . . *jīvanneṣa kṣapita-paśu-pāśavyatikaraḥ parabrahmābhikhyaṃ rasayati rasaṃ* . . .'.

Love is in the heart of the lotus-thunder. Who is there within these three spheres whose desire is not fulfilled by its enjoyment?[136]

Arise, O compassionate master, look at my helplessness. In the union of Great Ecstasy there is the honey of love. Seek it, thou of the nature of the void. Without thee I live not. Do thou arise, O Hevajra. Dispel the stupor of the void. Let the desire of the Śavara girl be fulfilled. O master of love's sport, why dost thou, after inviting guests, remain inactive?[137]

The attitude was anti-ascetical, anti-clerical, anti-doctrinal; the body was said to be the abode of the 'Enlightened One'[138] and the sensual could be enjoyed:

Without meditation and without renunciation, one stays at home in the company of one's wives. Saraha says, 'If one is not released while indulging in the pleasures of sense, how may one talk of complete knowledge?'[139]

Enjoying sensual things, one is not defiled by sensual things; one plucks the lotus, but does not touch the water. So the *yogi* who has gone to the root of things, is not tormented by the senses, although he enjoys them.[140]

The identity of *saṃsāra* and *nirvāṇa*, the profane and the sacred, is asserted. The goal is awareness of that identity, realization of the Innate, the *sahaja*, the experience of the Great Delight, the *mahāsukha*.

The reconciliation of profane and sacred, the anti-ascetical trend, the idealization of delight—all these tendencies are present in the *Gītagovinda*, but the substantial divergence is toward love, for the sake of love and not love as a metaphor for any process of liberation or awakening, but human love as an ultimate reality in itself, celebrated in the body, in the heart, in the soul.

The religious systems aim at release; but human love is bondage and aims at bondage, attachment. In Kṛṣṇa the

[136] Saraha, cited by Sarkar, op. cit., p. 15.

[137] Saraha, cited by Sukumar Sen, *History of Bengali Literature* (New Delhi, 1960), p. 34.

[138] Saraha, *Dohākoṣa* 68.

[139] Ibid., 21 (following the French translation of the Apabhraṃśa by M. Shahidullah, *Les Chants Mystiques de Kāṇha et de Saraha* [Paris, 1928], p. 171).

[140] Ibid., 66 (Shahidullah, p. 176).

paradox of love is realized: bondage is release; the profane dimensions of love are wholly sacred. Spanning *nirvāṇa* and *saṃsāra*, beyond time and space as the Lord, the 'creator of ten forms' (i. 16), and yet entering into time and space out of love for Rādhā, binding himself to *saṃsāra* (iii. 1), Kṛṣṇa homologizes the sacred and the profane.

Jayadeva sings with devotion to Kṛṣṇa (iii. 10), in service of Kṛṣṇa (v. 15; vii. 29), in remembrance of Kṛṣṇa (i. 34; ii. 8; xii. 24), taking refuge at Kṛṣṇa's feet (vii. 10). As devotional songs, songs about love sung with love, the lyrics spread happiness, delight, joy, prosperity (i. 15, 25, 46; ii. 18; iv. 18; vi. 9; ix. 9) and destroy the evil of the current, degenerate, Kali Era (vii. 20, 29):

> Put your compassionate heart in this splendid speech of Śrī Jayadeva, which is an adornment, which dispels the fever and impurity of the Kali [Era] for it is formed with the nectar which is remembrance of the feet of Hari. [xii. 24]

The songs are for those 'whose minds are fixed on Hari' (xi. 9), 'whose minds are passionate, whose minds have *rasa*, in that remembrance of Hari' (i. 3). *Rasa* as both an aesthetic and a devotional term (as well as a physiological term) provides the link between the profane and sacred dimensions and between the literary and religious traditions.

2

THE CONCEPT OF LOVE

My heart! I ask you—
'What is love?' Speak![1]

THE TASTE OF LOVE

They will have no idea of the good and the truly
sweet unless they have tasted it . . . [2]

In the *Ṛg-Veda* the word *rasa* refers to any fluid, but particularly to the sap of plants, juice, the essence of anything; and as the essential quality it came to mean 'taste'.

The meaning 'taste' then develops in two directions: (a) subjective, viz. 'a taste, inclination, desire or relishing of an object'; and (b) objective, viz. 'that to which the inclination, desire, etc., is directed'; 'savor, charm, delight.' Matters would be simple if the subjective and objective meanings were always distinguished. But it is typical of the Indian outlook that often they are not.[3]

Rasa is at once an inner and outer quality as the object of taste, the taste of the object, the capacity of the taster to taste that taste and enjoy it, the enjoyment, the tasting of the taste. The psychophysiological experience of tasting provided a basis for a theory of the aesthetic experience which in turn provided a basis for a systematization of a religious experience, devotion to Kṛṣṇa.

1 *The psychophysiological experience of love*

'Thy lips', Solomon sings to the Shulamite, 'drop as the honeycomb: honey and milk are under thy tongue. . . .'[4] The im-

[1] Friedrich Halm, *Der Sohn der Wildniss*, end of Act II.
[2] Aristotle, *Nicomachean Ethics* 10.9.4 (1179b 15).
[3] Ingalls, 'Words for Beauty . . .', p. 98.
[4] Song of Songs 4: 11.

agery of the sweet taste of the beloved occurs throughout the literature of love both as the literal sweetness of the beloved's kiss and metaphorically as the sweetness of assimilating the beloved, relishing, savouring, devouring the beloved.

In the ecstasy of human love, who is unaware that we eat and devour each other, that we long to become part of each other in every way, and . . . to carry off even with our teeth the thing we love in order to possess it, feed upon it, become one with it, live on it?[5]

Taste is the most intimate of the senses—the taste-object is inside the subject and yet still outside; the lover savours the beloved—the beloved is within and yet without. As *rasa* means 'taste', it may also mean 'love'. The lover relishes the beloved emotionally in loving and physically in kissing; the kiss in Sanskrit poetry is conventionally described as drinking the nectar, honey or mead, from the beloved's lower lip (II. 13; VI. 2; VII. 15; X. 2; XI. 22; XII. 10; etc.):

Bring the *rasa* of nectar from your lower-lip, O passionate-angry-woman! vivify me, your slave, as if I were dead; my mind is fixed upon you, my body is consumed by the fire of love-in-separation. . . . [XII. 6]

The lover draws sustenance and life from the sweetness of the beloved's mouth. Love as ecstasy is a death and a resurrection.

O sweet pleasure, full of sweet thought, when sweetness from the sweet mingling, embraces and unites soul mixed with soul, body sweetly joined with body, O sweet death! O sweet passing! My soul then shaken by great joy, flowing at once from me into you, now high, now low, seeks its ecstasy. When we are on fire with powerful love, Méline, I to be in you, you to take me wholly into you, by that part which goes farthest into you and you receive it, while I remain a lifeless mass; then your mouth comes to return it within my mouth, bringing life back to all my paralyzed limbs.[6]

[5] Jacques Benigne Bossuet, *Méditations sur L'Evangile*, cited and trans. by Perella, op. cit., p. 3. The bishop then compares the sacred with the profane, seeing the meaning of the Eucharist in terms of the human impulse: 'That which is frenzy, that which is impotence in corporeal love is truth, is wisdom in the love of Jesus: "Take, eat, this is my body": devour, swallow up not a part, not a piece but the whole.'

[6] Jean-Antoine Baif, 'O doux plaisir . . .', cited and trans. by Perella, op. cit., p. 221.

Rādhā is dying of love, drowning in a sea of *rasa* (vi. 10)—and she can be saved, the go-between tells Kṛṣṇa, who is 'like a Heavenly Physician', only by his *rasa* (iv.19); *rasa* is love, the pleasure of savouring love and the heavenly elixir, the life-giving honey of love.

The various meanings of the word come into play: lac (x. 7) and sandal unguent (iv. 12) are *sa-rasa*, 'beautiful', 'tasteful', inspiring *rasa* as 'longing', but also *rasa* in the primary sense of 'sap', 'juice of plants'; spring is *sa-rasa* (i. 28 ff.)—'beautiful', 'delectable', 'passionate'—the time when the juices and saps flow, pulsate, throb. Rādhā's breasts are *ati-sa-rasa*, 'very beautiful', 'passionate' but the simile, 'like a coconut', brings in the auxiliary meaning, 'very full of milk', that is, 'large' (ix. 3).

Rasa means 'pleasure', particularly the pleasure of love: a young girl praised Kṛṣṇa in the *pleasure* of the *rāsa* dance (i. 44); Rādhā was languid from the *pleasure* of sexual union with Kṛṣṇa (ii. 17); Rādhā imagines a young girl crying-out with the *pleasure* of love (vii. 17); Rādhā complains that she has been without love's *pleasure* (vii. 28) and the redness of Kṛṣṇa's eyes betrays his enjoyment of that *pleasure* (viii. 2). Pleasure is the taste *of* love; passion is the taste *for* love. *Rasa* is both—the friend urges Rādhā to go to Kṛṣṇa 'passionately', with *rasa*. It is the source of the pleasure and the cause of the passion—Rādhā's hips are *sa-rasa*, 'beautiful' (xii. 28). Rādhā is Kṛṣṇa's *eka-rasa*, his 'one-love', 'one-passion', 'one-pleasure' (xi. 24).

The concept of *rasa* links the psychological and physiological aspects of love—*rasa* is both the emotional pleasure of love and the biological manifestation of that pleasure, i.e. semen. The latter usage occurs in Vedic literature, medical texts,[7] in the *Tantras* and (in connection with Rādhā and Kṛṣṇa) in Vaiṣṇava Sahajiyā doctrine:

To the Sahajiyās, the body is full of *rasa*, of the bliss of union, or, speaking purely physiologically, of semen. The place of *rasa* at the beginning of *sādhana* is in the lowest lotus, the seat of sexual passion. By *sādhana rasa* is raised from lotus to lotus along the spinal column, until it unites with the thousand-petalled lotus in the head; there, in

[7] In the medical texts generally *rasa* refers to the body's constituents and to the 'chyle' which is produced in the digestive process, i.e., to the 'essential part' of the body and to the 'essential part' of the food which nourishes the body.

pure experience and pure consciousness of Rādhā and Kṛṣṇa in union . . . is full and eternal realization of their bliss.[8]

Jayadeva makes the pun: conventionally spring, as a time which inspires love-making, torments both men separated from their beloveds and ascetics trying to practise their austerities; it torments 'travellers who delight in the relish (*rasollāsa*) of union with their beloveds obtained [only] in moments of thinking about them' *or* it torments 'wandering-monks who raise their semen (*rasollāsa*) by the doctrine of breath-control as obtained in moments of meditation' (I. 37).

In the experience of sexual love Baudelaire described his 'insatiable soul' overwhelmed by '*le goût de l'éternel*', a longing for *and* savouring of the eternal, the *rasa* of the eternal.[9] *Rasa* is joy and 'Joy ever desireth an eternity of all things, desireth honey, desireth lees, desireth drunken midnight. . . . Joy would have an eternity of all things—profound, profound Eternity.'[10] The medieval Vaiṣṇavas saw *rasa* as the link between temporal and eternal—the basic emotion of human love, the experience of Kṛṣṇa the cowherd lover, could be transformed into *rasa*, the religious experience of Kṛṣṇa the eternal Lord. The Sahajiyās conceived of *rasa* flowing 'perpetually from the eternal Vṛndāvana to earth, manifested as the stream of *rasa* flowing to and between men and women'.[11] The Vaiṣṇavas based their *rasa*-theology upon the *rasa*-rhetoric of Sanskrit literary theory in which *rasa* was the essence of aesthetic experience. *Rasa* was joy realizing 'profound, profound Eternity'.

2 *The aesthetic experience of love*

Rasa is a technical term in Sanskrit poetics indicating 'mood' or 'sentiment' or 'flavour':

When the literary critics and aestheticians sought a word to sum up the very essence of art, it was this word *rasa* that they chose; it fitted their purpose because it spread over both the inner and outer world. *Rasa* is both the quality in a literary work that enables it to give

[8] Dimock, *Place of the Hidden Moon*, p. 177.

[9] Charles Baudelaire, 'Hymne' in *Les Fleurs du Mal*.

[10] Friedrich Nietzsche, *Thus Spake Zarathustra*, trans. A. Tille, Everyman edn (London, 1933), p. 284.

[11] Dimock, *Place of the Hidden Moon*, p. 168.

delight and the delight, the response, which occurs within the mind and heart of the audience.[12]

Bharata used the gastronomic metaphor to explain the aesthetic experience—just as the basic ingredient in a dish as it is seasoned with secondary ingredients and spices, yields a particular flavour (*rasa*) which the gourmet can savour with pleasure, so the basic emotion (*sthāyi-bhāva*) in *kāvya* as it is seasoned with secondary emotions, with literary spices, verbal herbs, poetic condiments, produces pleasure in the connoisseur (*rasika*) as he experiences the *rasa*, the mood or sentiment, of the work.[13] The taster tastes the taste with taste. The rhetoricians classified the basic emotions, the *sthāyi-bhāva*s, the psychological experiences, in correspondence with the moods or sentiments, the *rasa*s, the aesthetic experiences. The *rasa*s represent the aesthetic fulfilment, the literary realization, of the *sthāyi-bhāva*s:[14]

Basic Emotion (*sthāyi-bhāva*)	Aesthetic Experience (*rasa*)
love (*rati*)	erotic (*śṛṅgāra*)
humour (*hāsa*)	comic (*hāsya*)
grief (*śoka*)	tragic (*karuṇa*)
anger (*krodha*)	furious (*raudra*)
energy (*utsāha*)	heroic (*vīra*)
fear (*bhaya*)	fearful (*bhayānaka*)
disgust (*jugupsā*)	horrific (*bībhatsa*)
astonishment (*vismaya*)	marvellous (*adbhuta*)

The *rasa* arises when the *rasika* experiences pleasure, 'a calm enjoyment, a dispassionate pleasure in the passions',[15] through his apprehension of the basic emotion as it is enhanced, matured, aesthetically realized, through the operation of the various ingredients, e.g., the things, as portrayed in *kāvya*, which in actual experience would be the causes (*vibhāva*) and effects (*anubhāva*) of the respective emotions. The rhetorical classifications of the causes and effects which, when used skilfully by

[12] Ingalls, 'Words for Beauty . . .', p. 98.

[13] *Nāṭyaśāstra* of Bharatamuni, prose following VI. 31 ff.

[14] A ninth *rasa*, the *śānta*, 'calmed' or 'peaceful' was added by writers in the eighth century and the Bengal Vaiṣṇavas added the *bhakti-rasa*.

[15] See above, p. 27, note 101.

the poet to depict love (*rati-bhāva*), convey the erotic mood (*śṛṅgāra-rasa*), transforming the emotional experience of love into the aesthetic one, reveal both the conception of love in Sanskrit literary tradition and the theory of its poetic depiction. There are the fundamental causes (*ālambana-vibhāva*) of love—the lover and the beloved; in the *Gītagovinda*, Rādhā and Kṛṣṇa. And there are secondary, enhancing, 'exciting' causes (*uddīpana-vibhāva*): certain times and places as night and spring, the forest and the banks of the Yamunā in the *Gītagovinda*; thus spring is *sa-rasa* (I. 28 ff.); love is further enhanced by the moon, lotuses, sandalwood, ornaments, bees, cuckoos, breezes, rain-clouds; thus sandal unguent, lac, and musk are *sa-rasa* (IV. 12; X. 7; XII. 21). The effects or external manifestations of the emotion are seductive glances, playful gestures, side-long glances, sweet words, moving the eye-brows, and so forth. The emotion in turn causes other emotions—involuntary or expressive emotions (*sāttvika-bhāva*), *viz.*, trembling, fainting, weeping, paralysis, horripilation, perspiration, change of colour, change of voice. And also various transitory emotions (*vyabhicāri-bhāva*) ranging from shame to pride, joy to despair, and including envy, anxiety, agitation, remembrance, intoxication, sleepiness, madness, etc., but never in the case of love to include terror (*trāsa*), rage (*ugratā*), lassitude (*ālasya*) or the basic emotion of disgust (*jugupsā*). The causes and effects work together to convey the basic emotion:

> . . . a *vibhāva* may be taken as that which makes the permanent mood (*sthāyi-bhāva*) capable of being sensed; an *anubhāva* is that which makes it actually sensed; while a *vyabhicāri-bhāva* is that which acts as an auxiliary or gives a fresh impetus to it.[16]

If the emotion is portrayed with *rasa* and sensed or savoured with *rasa* by the *rasika*, the personal, psychological, emotional experience becomes universalized into the poetic experience.

Beyond its causes and effects, stimulants and manifestations, love was classified further into love-in-separation (*vipralambha-śṛṅgāra-rasa*) and love-in-union or enjoyment (*saṃbhoga-śṛṅgāra-rasa*) and the former was subclassified according to the causes and conditions of the separation. The various types of lovers and beloveds were enumerated and their actions and

[16] S. K. De, *Studies in the History of Sanskrit Poetics*, vol. II (London, 1923), p. 22.

gestures and emotions were categorized. Classifications, sub-classifications, sub-sub-classifications—all this defined love as a very particular effect, a very particular response to particular situations, particular stimuli, in particular circumstances for a particular type of person. And love, or at least the behaviour which expresses it, had been classified in the erotic texts, the *kāmaśāstra*, as well, and those classifications represent an invention of love, of what it should ideally be, an establishment of conventions for love and for its portrayal, for passion and its expression:

Passion and expression are not really separable. Passion comes to birth in that powerful impetus of the mind which also brings language into existence. So soon as passion goes beyond instinct and becomes truly itself, it tends to self-description, either in order to justify or intensify its being, or else simply to keep *going*. . . . The emotions first experienced by an upper class and then through imitation by the masses are literary creation in the sense that a given rhetoric is the sufficient condition for them to be *avowed* and hence for them to be made conscious. In the absence of this rhetoric, the emotions would no doubt still exist, but accidentally, lacking recognition, and they would be treated as unmentionable and contraband peculiarities.[17]

In the context of the main currents of Indian religion, which in practically all of its forms is primarily concerned with release from the world, with cessation of desire and attachment, love is 'contraband'. But the concept of *rasa* justifies and sanctifies love; for *rasa*, which became in classical Sanskrit literature the essence of love, was also equated with the blissful essence of all, 'profound Eternity', *brahman*, the sacred:

Once this man has tasted *rasa* [in all things], he tastes joy. For who could breathe, who could live, if this joy was not throughout space? For this [*rasa*] alone brings joy.[18]

The Vaiṣṇavas of Bengal used this passage to transform an aesthetic system into a theological one—*rasa*, the joyful essence of all, became manifested as Kṛṣṇa. Kṛṣṇa as the Bhagavat, the

[17] de Rougemont, *Love in the Western World*, pp. 183–4.

[18] *Taittirīyopaniṣad* II. vii. 1:

. . . rasaṃ hy evāyaṃ labdhvānandī bhavati; ko hy evāṇyāt kaḥ prāṇyāt yad eṣa ākāśa ānando na syāt; eṣa hy evānandayāti . . .

Lord, is the source of all *rasa*; Kṛṣṇa as Govinda, the cowherd lover, personifies the *śṛṅgāra-rasa* particularly. The *Caitanya-Caritāmṛta* quotes the *Gītagovinda* to show that Kṛṣṇa is the 'embodiment of *Rasas* and amorous love incarnate' and that 'he made his appearance to taste the said *Śṛṅgāra Rasa* and incidentally promulgated all the *rasas*':[19]

Producing the joy of all-creatures by his love, initiating the festival of love by his limbs which are dark and tender like bunches of [blue] lotuses, embraced by the beauties of Vraja of their own free-will, entirely, all-over-his-body, like *śṛṅgāra* incarnate, O friend, Hari plays in spring. [I. 47]

The verse is quoted again to explain that Kṛṣṇa embodied the *śṛṅgāra-rasa* in order to 'attract and enchant the hearts of all'.[20] But Kṛṣṇa is also the carnal manifestation of the erotic sentiment in that Viṣṇu is the presiding deity of that sentiment and the dark hue of Kṛṣṇa's body is, according to the rhetoricians, its appropriate colour.[21] Jayadeva uses the word *śṛṅgāra* in its technical sense, referring to the aesthetic experience of love: Govardhana is praised for his 'compositions which are predominately *śṛṅgāra*' (I. 4) and the *Gītagovinda* itself has 'the truth of the discrimination in *śṛṅgāra*' (XII. 28); 'the clever words of Jayadeva bestow the emotion (*bhāva*) which, like a blessing, has the essence of *śṛṅgāra*' (XII. 31). *Bhāva*, *rasa* and *rasika* are also used technically: 'May this song . . . produce in people of taste (*rasika-jana*) an appreciation of the emotion (*bhāva*) of the sentiment (*rasa*) of love . . . ' (XII. 9). The songs are sung for the joy of *rasika*s, for them to relish: 'May this song . . . spread great joy to *rasika*s' (VI. 9); '. . . may the story-of-the-deeds of Hari cause happiness for *rasika*s' (IX. 9). The ideal *rasika* is described by the aestheticians[22] as being of noble birth, virtuous and honest, learned and cultivated, dispassionate and yet empathetic; he should be *sahṛdaya*—good-hearted, full-of-feeling, sensitive and sensible; he should both *have rasa* as a *rasika* and *know rasa* as a *rasajña*. The aesthetic experience cannot be fulfilled without him—the *rasa* is both in him and in the work and it is

[19] *Caitanya-Caritāmṛta* (N. K. Ray trans.), *Ādi*. IV (vol. I, p. 56), op. cit.
[20] Ibid., *Madhya*. VIII (vol. II, pp. 160–1).
[21] *Nāṭyaśāstra* VI. 42 & 45; *Sāhityadarpaṇa* of Viśvanātha Kavirāja 210.
[22] *Nāṭyaśāstra* XXVII. 49 ff.

the relationship between him and the work. In the context of the *Gītagovinda rasika* also suggests *bhakta*, the devotee: Pūjārī Gosvāmin glosses '*rasika*' as '*bhakta*' as it occurs in the *Gītagovinda*.[23] The equation had been made before Jayadeva, in the *Bhāgavata Purāṇa*—the devotees are addressed as *rasika*s and are urged to relish the *Bhāgavata Purāṇa*:

Fallen from the sacred-wishing-tree which is the Vedic-texts, a fruit having a flow of nectar from the mouth of Śuka, an abode, *rasa*, the *Bhāgavata*—drink [it] at once O *Rasika*s! O you who, on earth, have-a-taste-for-the-beautiful![24]

The biological (*rasa* as the juice of the fruit), the aesthetic (*rasa* as the essence, the beauty, of the text), and the devotional (*rasa* as the transcendental joy of relishing the Lord through the text) all come into play. The word is essentially ambiguous, pointing to various realms of experience, and Jayadeva uses the ambiguity to extend the dimensions of his song: it is for those whose minds are *sa-rasa* in remembrance of Kṛṣṇa (I. 3), those who experience the aesthetic *and* devotional savour of Kṛṣṇa; Kumbha glosses *sa-rasa* as *sa-rāga* and *ekāgra*—those whose minds are 'passionate' and those whose minds are 'intent upon' or 'absorbed in' recollection of Kṛṣṇa. *Ekāgra* ('one-pointed') is a technical term from the *yoga* system for an advanced state during which the mind is concentrated on one object; Kumbha implies, then, an ambiguity which links the erotic and the contemplative. The mind of the devotee has *rasa* as does Rādhā's mind when she encounters Kṛṣṇa (XII. 1)—that which is passionate love in Rādhā is concentrated devotion in the *bhakta* as *rasika*.

Rādhā's friend, the go-between, speaks to Rādhā passionately (*sa-rasam*) (I. 27), with fervour, relish, love, delight, charm and also in such a way as to inspire the emotional experience of

[23] Also called Caitanyadāsa; his commentary, the *Bālabodhinī*, is a Bengal Vaiṣṇava exegesis (seventeenth century?).

[24] *Bhāgavata-Purāṇa* I. i. 3:

nigama-kalpataror galitaṃ phalaṃ śuka-mukhād
amṛta-drava-saṃyutam;
pibata bhāgavataṃ rasam ālayaṃ muhur aho
rasikā bhuvi bhāvukāḥ.

('having a taste for the beautiful' = *bhāvuka*, following M. Monier-Williams, *Sanskrit–English Dictionary* [Oxford, 1899], *s.v.*).

love in Rādhā *and* the universalized poetic experience of love in the reader. The love is sanctified if Kṛṣṇa is its object.

Rādhā experiences the passionate emotion (*sa-rasa-bhāva*) of the sentiment of love (*madana-rasa*) (XI. 19)—the poet's choice of terms acknowledges the rhetorical texts. And he puns on the terms: a husband and wife going off for illicit love-affairs encounter each other accidentally in the darkness and recognize each other by their voices only after making love—'their *rasa* is mixed with embarrassment' (V. 18); their pleasure is mixed, but also the *śṛṅgāra-rasa* is mixed with the *hāsya-rasa*, the comic mood. The comic and erotic sentiments are closely connected: Bharata explains that the comic arises from the erotic, that the comic is a mimicry of the erotic.[25] The erotic and heroic sentiments are juxtaposed—a woman is able to display the heroic *rasa* only by indulging in the pleasure (*rasa*) of a man, by taking the 'man's position' on top of her lover in the 'battle of love' (XII. 13). The erotic *rasa* threatens to become the tragic or pitiful (*karuṇa*) if Rādhā is not united with Kṛṣṇa (VII. 37). In turn, the comic and tragic are juxtaposed in a playful conceit (I. 32).

In the eleventh century Abhinavagupta elaborated upon Bharata's theory of *rasa*, emphasizing the process of the universalization and abstraction of emotion as a transcendental process: the *rasika* is lifted out of himself, above the world, beyond time and space, in an experience of pure *rasa*, the highest joy equated with ultimate beatitude; *rasa* is experienced

> as if quivering in one's presence, as if entering the heart, as if embracing the whole body, as if obscuring everything else, as if causing one to experience the taste of *brahman*.[26]

The aesthetic experience had become a religious experience, '*le goût de l'éternel*'.

3 *The religious experience of love*

Let not the evil, which is produced by the Kali Age, remain here,

[25] *Nāṭyaśāstra* VI. 39–40.

[26] Abhinavagupta cited in Maṃmata's *Kāvyaprakāśa*, prose following IV. 28:
pura iva parisphuran hṛdayam iva praviśan sarvāṅgam
ivāliṅgan anyat sarvam iva tirodadhat brahmāsvādam
ivānubhāvayan . . .

in Jayadevaka, the king of poets—he has a song of *rasa*, he recounts Hari's qualities, he is a servant of the feet of the Enemy-of-Madhu. . . . [VII. 29]

There is a tension: king and servant, poet and devotee, profane and sacred; the reconciliation begins with 'recounting Hari's qualities', an act in which Jayadeva's 'own duty', *sva-dharma*, as both a poet and a devotee is fulfilled; the reconciliation is fully realized in *rasa*, which is at once the aesthetic experience and also the essence of all, the 'joy throughout space', *brahman*, the *rasa* which for the Bengal Vaiṣṇavas became the state of total absorption in, devotion to, enjoyment of Kṛṣṇa, 'the embodiment of the nectar which is the total *rasa*'.[27] The disciples of Caitanya developed a theology of *bhakti-rasa*, a rhetoric of devotional love, based upon the model of Sanskrit poetics.

The mediaeval conception of love is sublimated into a deeply religious sentiment by bringing erotico-religious ideas to bear upon the theme of the literary Rasa, especially to the Erotic Rasa (Śṛṅgāra). . . . The literary Sahṛdaya, as the recipient connoisseur, was replaced by the religious Bhakta, the devotee of nice sensibility; while the love of Kṛṣṇa was installed as the dominant feeling (Sthāyi-bhāva), which by means of its appropriate Excitant (Vibhāva), Ensuant (Anubhāva) and Auxiliary feelings (Vyabhicāri-bhāvas) could be raised to the supreme relishable condition of impersonal enjoyment in his susceptible mind as the Bhakti-Rasa or devotional sentiment.[28]

The basic emotion (*sthāyi-bhāva*) became specifically Kṛṣṇa-*rati*, the fundamental causes (*ālambana-vibhāva*) of which became Kṛṣṇa as the object of love and the devotee as the subject, the lover; the secondary, enhancing causes (*uddīpana-vibhāva*) became Kṛṣṇa's qualities, exploits, ornaments and embellishments; the effects, 'ensuants', manifestations (*anubhāva*) became singing and dancing, sighing and shouting, disregarding the opinion of others, all the symptoms of enraptured devotion; the expressive emotions (*sāttvika-bhāva*) and transitory emotions (*vyabhicāri-bhāva*) as enumerated in orthodox poetics were accepted. Because the *Gītagovinda* had been conceived in terms of traditional aesthetics, *rasa* theory, and because it was about

[27] Rūpa Gosvāmin, *Bhakti-rasāmṛta-sindhu* I. 1: '*akhila-rasāmṛta-mūrtiḥ*'.

[28] S. K. De, introduction to his critical edition of the *Padyāvalī* of Rūpa Gosvāmin (Dacca, 1934), pp. xc–xci.

Kṛṣṇa, it could quite naturally be read as illustrative of this later *bhakti-rasa* theory.[29]

The Bengal Vaiṣṇavas relegated the comic, tragic, furious, heroic, fearful, horrific and marvellous *rasas* of Sanskrit aesthetics to a secondary position and described five primary *rasas* as five modes of *bhakti*: devotion as peace (*śānta*), devotion as servitude (*dāsya*), devotion as friendship (*sakhya*), devotion as parental affection (*vātsalya*), and highest and finest of all devotion as erotic love (*śṛṅgāra*, usually called *madhura* in Vaiṣṇava texts). *Rasa* became a matter of relationship with Kṛṣṇa.

> When one looks upon or listens to a supreme work of art, one's senses become so completely absorbed in that work that they utterly exclude all else. To the Vaiṣṇavas of Bengal, religious devotion, bhakti, is such a state of rasa: the senses and the mind of the worshipper are absorbed in Kṛṣṇa, that personification of rasa, in the most intense experience possible for man.[30]

> . . . this Vaiṣṇava system developed a kind of mystic attitude where Rasa was a mystic experience. The soul being extremely transparent in its nature through dissociation from the Māyā becomes identified, as it were, with the Supreme Reality, God.[31]

In *rasa* subject and object merge—there is no distinction between taster, tasted, tasting, taste. 'You swim', Caṇḍīdāsa sang, 'in a sea of rasa which glitters when you touch it. . . . '[32] Jayadeva had used the image—Rādhā moans and loses consciousness, 'having thoughts of intense passion about you [Kṛṣṇa], the deer-eyed woman is immersed in an ocean of *rasa*, fixed in meditation' (VI. 10). Ascetic meditation is juxtaposed with sensual passion and the two, the sacred and the profane, converge in *rasa*, as *rasa*, the sea of *rasa*, the endless swelling ocean in which the empirical self is dissolved.

[29] Rūpa Gosvāmin, in the *Bhakti-rasāmṛta-sindhu* and the *Ujjvala-nīla-maṇi* (works which really comprise the *śāstra* of *bhakti-rasa*), used the *Gītagovinda* to illustrate these principles of Vaiṣṇava devotional 'aesthetics'; see *Bhakti-rasāmṛta-sindhu* 314, 881, 821, 829 (Murshidabad edn, 1924) and the *Ujjvala-nīla-maṇi* 113, 162, 175, 183, 243, 273, 276, 284, 287, 301, 310, 314, 496) (Kāvyamālā edn, Bombay, 1913) (cited by De, *Vaiṣṇava Faith and Movement*, pp. 152 & 167).

[30] Dimock, *Place of the Hidden Moon*, pp. 21–2.

[31] H. R. Mishra, *Theory of Rasa in Sanskrit Drama* (Chhatarpur, 1964), p. 379.

[32] Cited and trans. by Dimock, op. cit., p. 140.

In the swelling flood
in the echoing sound
in the infinite torment of the World's breath,
be engulfed
sink deep
unknowing
supreme joy![33]

The oceanic symbol occurs throughout mystical literature as an expression of ultimate communion:

As rivers flowing [downwards] find their home
In the ocean, leaving name and form behind,
So does the man who knows, from name and form released,
Drawn near to the divine Person who is beyond the beyond.[34]

The sage becomes 'an ocean, One, without duality'[35] and in that state taster, tasted, tasting, taste, *rasa*, are the same:

Truly when there he does not taste; he is truly tasting although he does not taste, for there is no ceasing of the tasting of the taster on account of imperishableness; but there is no second, nothing else separate which he could taste.[36]

The religious experience of communion, of the 'supreme joy' and ineffable ecstasy of the two merging into one, 'without duality', finds expression in terms of the sexual experience. But there are two modes of sexuality analogous or homologous with the mystical unification: one is the sexual union of lover and beloved—God ravishes the devotee:

Oh night that joined the lover
To the beloved bride
Transfiguring them each into the other.

[33] Last lines of Wagner's libretto for *Tristan and Isolde*, cited by de Rougemont, *Love Declared*, p. 159.

[34] *Muṇḍaka Upaniṣad* III. ii. 8 (Zaehner trans.):
yathā nadyas syandamānās samudre astam gacchanti
nāma-rūpe vihāya;
tathā vidvān nāma-rūpād vimuktaḥ parāt-param puruṣam
upaiti divyam.

[35] *Bṛhadāraṇyaka Upaniṣad* IV. iii. 32 (Zaehner trans.):
'*salila eko draṣṭādvaito bhavati. . .*'.

[36] Ibid., IV. iii. 25:
yad vai tan na rasayati rasayan vai tan na rasayati na hi
rasayitū rasayater viparilopo vidyate avināśitvāt na tu
tad dvitīyam asti tato' nyad vibhaktam yad rasayet.

Within my flowering breast
Which only for himself entire I save
He sank into his rest
And all my gifts I gave . . .[37]

The other is union of infant and mother—Freud suggested that the 'oceanic feeling', the experience of 'one-ness', the loss of distinction between self and other, subject and object, was concomitant with regression to the primal experience of the pre-natal unity, or the experience at the mother's breast when gratification was total and immediate and when there were no ego-boundaries; the experience is libidinal, 'orgastic', but the sexuality is oral.[38]

I alone am inert, like a child that has not yet given
a sign [by stretching its hand towards some object];
Like an infant that has not yet smiled. . . .
. . . wherein I most am different from men
Is that I prize no sustenance that comes not from the
Mother's [i.e., the Way's] breast.[39]

In this state the soul is like a little child still at the breast, whose mother, to caress him whilst he is still in her arms, makes her distill into his mouth without his even moving his lips. So it is here . . . Our Lord desires that our will should be satisfied with sucking the milk which His Majesty pours into our mouth, and that we should *relish the sweetness* without even knowing that it cometh from the Lord.[40]

Rasa as the enjoyment of Kṛṣṇa through devotion was based on the former mode, the joy experienced by lovers; *rasa* as 'taste', 'liquid' (and derivatively 'milk') suggests the latter mode; '*sa-rasa*' Rādhā was venerated by the medieval Vaiṣṇavas of Bengal as the Goddess, the Mother:

Sri Radhika is Devi, Krishnamayee, Paradevata, Sarva-Lakshmimayee, Sarvakanti and Para Sanmohini. . . . Lord Krishna enchants

[37] Saint John of the Cross, 'Canciones del alma . . .', op cit. (Campbell trans.), pp. 27–9.

[38] See Freud's *Civilization and Its Discontents*, Chapter 1.

[39] *Tao Tê Ching* xx, Arthur Waley trans., *The Way and Its Power* (London, 1934), pp. 168–9.

[40] Saint Francis de Sales, *Chemin de la Perfection*, xxxi, cited by William James, *Varieties of Religious Experience*, Mentor paperback edn (New York, 1958), pp. 27–8. [Italics mine.]

the world but Sri Radha enchants Him. Therefore Sri Radharani is the crown-jewel of all: Sri Radha is the full power. . . .[41]

And by their conception of *rasa* the two modes, the worship of Rādhā and the worship of Kṛṣṇa, were unified:

Radha and Krishna are one and the same. And they have assumed two bodies and sport eternally to taste their mutual sweetness and *Rasa.*[42]

THE VOCABULARY OF LOVE

Love—*love,* affection, friendship, charity, Eros; agapism; true love, real thing; . . . *feeling*; kindness, tenderness; fondness, liking, predilection, inclination . . . fancy, *caprice*; attachment, devotion . . . sentimentality, susceptibility, amorousness . . . Cupid's sting, yearning, longing, *desire*; eroticism, prurience, lust, *libido*; admiration . . . infatuation; worship . . . passion . . . rapture, transport, *excitable state.* . . .[43]

Thought and language are inseparable, as language categorizes experience, feeling about experience, understanding of experience.

. . . discrepancies in the categorizing of basic notions are frequent between the Indian languages on the one hand and those of Europe on the other. The notions of love, law and religion, to mention but a few, follow different patterns in the two cultural traditions.[44]

An Indian concept of love is inherent in the vocabulary used in Sanskrit literature to express love, that is, in the words which might be translated from Sanskrit into English as 'love'. The words with this denotative meaning established connotative meanings in *kāvya.* When the vocabulary of human love was adopted by the medieval Vaiṣṇavas the denotative meanings remained constant but the connotative meanings changed. The distinction between sacred and profane was a matter of connotation.

[41] *Caitanya-Caritāmṛta*.(N. K. Ray trans.), *Ādi.* IV (pp. 46–7).

[42] Ibid. (p. 43).

[43] Roget's *Thesauraus*, 'love', *s.v.*, Penguin edn (Harmondsworth, 1966).

[44] Ingalls, 'Words for Beauty . . .', p. 87.

In the *Gītagovinda* the change is occurring—the words for love point in both directions, connote human longing and sensual enjoyment as well as religious devotion and beatitude.

1 *Rati*

Rati is the basic emotion which in literature crystallizes into the aesthetic experience of love, the *śṛṅgāra-rasa*. It is the feeling of love that Rādhā experiences in relation to Kṛṣṇa; the *rasika*'s potential for that feeling enables him to empathize with Rādhā (or Kṛṣṇa) and through that empathy to experience *rasa* as a literary connoisseur or as a Vaiṣṇava devotee or as both. The *rasika*'s own experience of love, or *rati*, enables him to perceive the *rasa* in the literary or devotional work and thereby to move from the immanent delight of his own experience, Rādhā's or Kṛṣṇa's experience, to the transcendent joy of the universal experience. The aesthetic theory of universalization and the *bhakti-rasa* theology sanctify, give meaning and significance to *rati* which as an individual, emotional, sexual experience perpetuates entanglement in the empirical world, the world of pain and pleasure, but which through art and/or devotion is a means of transcendence—the profane is transformed into the sacred by the poetic and/or devotional act.

As the basic emotion (*sthāyi-bhāva*) described by the rhetoricians, *rati* 'has the nature of delight on account of the enjoyment of lovely places, the [amatory] arts, [lovely] occasions, clothes, pleasures, and so forth'.[45] It is expressed by smiles, sweet words, playful glances, etc.[46] Smiles and glances also cause *rati* (xi. 27, 28, etc.). Jayadeva uses the term '*rati*' to indicate the emotion (*bhāva*)—*rati* is the emotion which 'delights the mind' (*manorama*) (xii. 9); '*bhāva*' is used synonymously with *rati* as *the* emotion—the go-between commands Rādhā to 'enjoy the emotion-of-love' (*bhaja bhāvam*) (xi. 4). *Rati* is the enjoyable emotion of love, love's pleasure rather than its passion; *rati* is both the emotion of love and the pleasure of the act which engenders the emotion; *rati* is the feeling of pleasure and the pleasure of feeling, the pleasure of love-making. Sexual union is the game of *rati* (i. 2; xi. 30; xii. 12) or the battle of *rati* (vii. 19; xi. 7) or

[45] The *Daśarūpa* of Dhanaṃjaya iv. 56: '*ramya-deśa-kalā-kāla-veṣa-bhogādi-sevanaiḥ pramodātmā ratiḥ. . . .*'

[46] *Nāṭyaśāstra* vii. 9.

the arts or skill of *rati* (v. 4). *Rati* is sexual pleasure (II. 17; v. 8) and the pleasurable sexual act—Rādhā has a violent-desire for *rati* (II. 11; IX. 1; XI. 14). A young girl cries out with the pleasure (*rasa*) of *rati* (VII. 17). The vulva is the 'abode of *rati*' (VII. 26).

As amorous, sexual enjoyment, *rati* became personified as a wife (along with Prīti) of the love-god, Kāma-deva. Thus Jayadeva refers to the love-god as the master, lord or lover of Rati (v. 7; VII. 23; XII. 18).

Vātsyāyana distinguished between *rati* and *rata* or *surata* and the commentator Yaśodhara indicates that the latter is the cause (*hetvavasthā*), coition, and the former is the effect (*phalāvasthā*), pleasure.[47] *Surata* refers specifically to the act of copulation (II. 16; XII. 10, 16) whereas *rati* is all the pleasurable activity in which the lover and beloved indulge. Both come from the root √*ram-*, 'to delight, please, gladden', a verb used in the sexual sense—to erotically delight, carnally please, to gladden through love-play (I. 45; II. 11 ff.; III. 6; VII. 22 ff., 30; XI. 10; etc.) —Kṛṣṇa is sexually-pleased (*ramita*) by a cowherdess (VII. 12) and in turn some young girl is likewise pleased by him (VII. 31).

When the followers of Caitanya adapted the aesthetic theory of *rasa* to their system of devotion, *rati* remained the 'basic emotion'. *Rati* was the universal and natural emotion which, if it became Kṛṣṇa-*rati*, love-pleasure concentrated upon Kṛṣṇa, could lead to the experience of *rasa*, an infinitely and eternally joyous absorption in Kṛṣṇa. Rūpa Gosvāmin defined *rati* in the clearly sexual imagery of love's 'warm and sweet flow': '*Rati* truly has the form of a very mighty river of joy which is naturally warm and incessant; but despite emitting heat it is sweeter than the cusp of the nectarous moon.'[48] In his consideration of *rati* as the 'basic emotion' (*sthāyi-bhāva*) of *madhura-bhakti-rasa* he classified it into three types: (a) 'universal' (*sādhāraṇī*) *rati*—yearning for (or delighting in) union with Kṛṣṇa for self-gratification, for one's own pleasure; (b) 'proper' (*samañjasā*) *rati*—yearning for (or delighting in) union with Kṛṣṇa for the sake of mutual gratification; (c) *rati* having a

[47] *Kāmasūtra* II. i. 64. Vātsyāyana gives as synonyms for *rati*: *rasa*, *prīti*, *bhāva*, *vega* and *samāpti*.

[48] *Bhakti-rasāmṛta-sindhu* I. iii. 61:

ratir aniśa-nisargoṣṇa-prabalatarānanda-pūra-rūpaiva;
uṣmāṇam api vamantī sudhāṃśu-koṭer api svādvī.

'suitable aim' (*samarthā*)—yearning for (or delighting in) union with Kṛṣṇa for *his* gratification.[49] While the selfless *samarthā rati* was considered better than the self-centred *sādhāraṇī rati*, the latter was nevertheless accepted as a genuine and fulfilling form of *bhakti*; it was exemplified by the courtesan Kubjā or Sairandhrī:

> . . . her desire for sporting with Kṛṣṇa is not deprecated in itself; for whatever may have been the character of her desire, she did not long for inferior worldly objects but for the Bhagavat himself, and there is no doubt from her words about the intensity of her feeling. It is deprecated only in comparison with the feeling of the Gopīs, for her desire for sport was entirely for her own sensual pleasure, while that of the Gopīs was exclusively intended for Kṛṣṇa.[50]

The medieval Vaiṣṇavas of Bengal esteemed Rādhā as the paragon of *samarthā rati*, but in the *Gītagovinda* she longs for Kṛṣṇa for self-gratification—she is jealous when she imagines another woman being sexually-pleased (*ramita*) by Kṛṣṇa (vii. 31 ff.), jealous when she feels she is not his favourite-beloved (ii. 1). But the justification, again, is in terms of the object and the intensity of her love. Naming self-centred love '*sādhāraṇī rati*', the 'universal' or 'common' *rati*, is revealing—the clinging to a sense of 'I' and 'mine' and the longing to be pleased were accepted by the medieval Vaiṣṇavas as 'universal', natural, human propensities, and such acceptance must have been necessary to a religious movement which had its roots and its fulfilment in the people, the masses, in contrast to more ascetic movements which demanded overcoming the sense of 'I' and 'mine' for the sake of liberation: 'The man who puts away all desires and roams around from longing freed, who does not think, "This I am" or "This is mine", draws near to peace.'[51]

[49] *Ujjvala-nīla-maṇi* cited by S. C. Chakravarti, *Philosophical Foundation of Bengal Vaiṣṇavism* (Calcutta, 1969), pp. 253–4.

[50] De, *Vaiṣṇava Faith and Movement*, p. 287.

[51] *Bhagavad-Gītā* ii. 71 (Zaehner trans.):
vihāya kāmān yaḥ sarvān pumāṃś carati niḥspṛhaḥ
nirmamo nirahaṃkāraḥ sa śāntim adhigacchati.
Commenting on the verse Zaehner suggests that the ideal of total detachment and selflessness was essentially a Buddhist one and that ultimately the *Bhagavad-Gītā* seeks to reconcile this with 'a scheme of things which also makes room for a personal God'.

But in the highly emotional and erotic *bhakti* movement in medieval Bengal the sense of 'I' and 'mine' could be sanctified if one could say '*I* worship and love Kṛṣṇa, longing for him to be *mine*'. Liberation took an ironic second place to love as an end in itself and 'peace' took a second place to the weeping, trembling, running about, all the *un*-peaceful symptoms (the 'expressive emotions' [*sāttvika-bhāvas*]) of that love, *rati*, in its fullest intensity.

2 *Rāga*

In the Sāṃkhya system, in the orthodox systems generally, *rāga*, passion, is something binding, something to be overcome. Patañjali lists *rāga* among the five afflictions or defilements (*kleśa*), the hindrances to liberation, integration (*samādhi*).[52] This *rāga* is the result of *rajas*, passion as one of the 'three constituents of nature', one of the 'three strands' out of which everything in the phenomenal world is woven.[53] The teacher in the *Maitrī Upaniṣad* expounds the nature of *rajas*:

[The following] belong to [the constituent of Nature called] 'Passion' [*rajas*]: inward craving, cloying love [*sneha*], passion [*rāga*], greed, wishing others ill, sexual pleasure [*rati*], hatred, secretiveness, envy, desire [*kāma*], instability, fickleness, distraction, ambition, acquisitiveness, favouritism, reliance on worldly wealth, aversion to objects repellent to the senses and attachment to [objects] attractive [to them], churlish speech and gluttony.[54]

[52] Patañjali, *Yoga-sūtra* II. 3 ff.—the five '*kleśas* are *rāga*, *avidyā* (ignorance), *asmitā* ('egotism', 'I-am-ness'), *dveṣa* (aversion, hatred), *abhiniveśa* (tenacity of mundane existence)'.

[53] *Bhagavad-Gītā* XIV. 5 (Zaehner trans.):

Goodness [*sattva*]—Passion [*rajas*]—Darkness [*tamas*]:
these are the [three] constituents from Nature sprung
that bind the embodied [self] in the body though [the
self itself] is changeless.
(sattvaṃ rajas tama iti gunāḥ prakṛti-saṃbhavāḥ
nibadhnanti mahābāho dehe dehinam avyayam.)

[54] *Maitrī Upaniṣad* III. 5 (Zaehner trans.):

. . . antas-tṛṣṇa sneho rāgo lobho hiṃsā ratir dviṣṭir
vyāvṛtatvam īrṣyā kāmam asthiratvam calatvam vyagratvam
jigīṣārthopārjanaṃ mitrānugrahaṇaṃ parigrahāvalambo
niṣṭeṣvindriyārtheṣu dviṣṭiriṣṭeṣvabhiṣvaṅgaḥ
śuktasvaro'nnatamastv iti rājasāny. . . .

But in the medieval Vaiṣṇava tradition *sneha*, *rati*, *kāma*, *rāga* became, as expressions of devotion, more important than liberation; *bhakti* transcended *mokṣa*, the Bhagavat transcended *brahman*;[55] the revelation of love became more important than Vedic injunction or Brahmanical precept. Rūpa Gosvāmin defined *rāga* as the subject being 'completely engrossed in the desired-object',[56] and similarly Jīva Gosvāmin described *rāga* as 'love (*preman*) spontaneous and consisting of an abundance of desire for union with the sense-object'.[57] Kṛṣṇa was *the* desired sense-object, the Lord who, by incarnating as a love 'manifested the *rāga-mārga* [the way of passion] . . . he taught it by his *līlā* [his sport or love-play]'.[58] The way of *rāga*, 'following *rāga*' was a type of devotion (*rāgānuga-bhakti*) in which the devotee imagined himself to be female, to be Rādhā, or one of the other cowherdesses, and imagined himself then as a lover of Kṛṣṇa, at play with him, united with him in love. By virtue of this type of devotion the *Gītagovinda* was particularly sacred—the *rasika* could sing, with intense empathy and full vicarious ecstasy, Rādhā's songs ('O friend! make him make-love to me passionately, I am engrossed with desire for his love' [II. 11 ff., etc.]). *Rāgānuga-bhakti* was a

> . . . concentrated imaginative process . . . for a mystic union with the beloved object. It is indeed not achieved by the direct injunction of the Śāstras, but it does not also arise spontaneously in one's own self. It is engendered by external effort, by elaborately imitating the action and feeling of those connected with Kṛṣṇa in Vraja. . . . The devotee by his ardent meditation not only seeks to visualize and make the whole Vṛndāvana-līlā of Kṛṣṇa live before him, but he enters into it imaginatively, and by playing the part of a beloved of Kṛṣṇa, he experiences vicariously the passionate feelings which are so vividly pictured in the literature.[59]

For the Sahajiyās, the path of *rāga* was considered more effective than the path of the *śāstras*, passion and spontaneous love were

[55] The highest reality conceived of in the *Upaniṣads*, *brahman*, was considered but an imperfect manifestation of Kṛṣṇa, the Bhagavat.

[56] *Bhakti-rasāmṛta-sindhu* I. ii. 272: '*iṣṭe svārasikī rāgaḥ paramāviṣṭatā bhavet . . .*'.

[57] *Bhaktisandarbha* 310 (Sanskrit cited by S. C. Chakravarti, op. cit., p. 214): '*. . . svābhāviko viṣaya-saṃsargecchātiśaya-mayaḥ premarāgaḥ.*'

[58] *Caitanya-Caritāmṛta*, *Adi.* IV. 220 (cited by Dimock, *Place of the Hidden Moon*, p. 194).

[59] De, *Vaiṣṇava Faith and Movement*, p. 131.

more potent than knowledge and ritual; the initiate was commanded to '. . . leave the Vaidhi path [the path following the Vedic injunctions] and worship only in *rāga*. If there is no *rāga* there can be no union.'[60]

Jayadeva foreshadows the later Vaiṣṇava trends—*rāga* as a defect, defilement or affliction (*kleśa*) to be absolved by austerities, by *yoga*, does not exist as in the Sāṃkhya system; Kṛṣṇa, who is both 'without-*kleśa*' (II. title) and yet passionate (*rāgin*) (IX. 10), passionate for love's pleasure (*rati-rāga*) (XI. 27), whose passion (*anurāga*) is visibly manifested on his chest in the sweat caused by his vigorous love-making (I. 26), by his grace takes away all defilements and afflictions: 'May young Keśava take away your *kleśa*' (II. 21). Kṛṣṇa purifies passion—the devotee may feel passion because all that is defiling or afflicting in it will be absolved by Kṛṣṇa. Kṛṣṇa justifies *rāga* by inspiring it: a cowherdess embraces Kṛṣṇa passionately (*sa-rāgam*) (I. 40); Rādhā is passionate (*anurakta* [from the same root as *rāga*]) in separation from him (VI. 1) and as he delights and impassions the women of Vraja so too he is the delighter of all people (*jana-rañjana*) (I. 19), 'producing the joy of all-creatures by his passionate-love (*anurañjana*)' (I. 47).

As well as meaning 'love, passion, affection', *rāga* can mean 'red' and its root form √*rañj*- may mean both 'to be impassioned, enamoured, delighted' and 'to redden or be reddened' (and the participle *rakta* then means 'excited with passion' or 'red'). Rādhā's lower-lip is red and impassioned (*rāgavān*), red with the increased blood flow of sexual arousal and also, perhaps, red with a cosmetic (III. 14). Jayadeva plays on the ambiguity of the word: Kṛṣṇa's eye is red with passion and the redness is his passion made visible (VIII. 2); 'Rādhā impassions Kṛṣṇa' = 'Rādhā reddens what is black' (X. 5); his chest, sprinkled with red lac from enjoying a certain coital posture, displays his passion (VIII. 10). Jayadeva combines the ambiguity red/passion with the ambiguity chest/heart, *hṛdaya* as both the outer breast and the inner seat of feeling and emotion: Rādhā's feet, painted with red lac, redden his chest in a certain sexual position[61] and delight or impassion his heart (X. 7); the ornaments upon

[60] *Premānanda-laharī* cited by Dimock, op. cit., p. 195.

[61] The *krodha-bandha*, an 'inverse' posture in which the woman's feet are upon the man's chest (noted by Śaṅkara Miśra in his commentary on vs. VIII. 5).

Rādhā's breast redden or brighten her chest and impassion her heart (x. 6). The ambiguity of the words *rāga* and *hṛdaya* emphasizes the connection between the physical and the psychological, the inner and the outer. The body is the physical manifestation of the psyche, the outer manifestation of the inner life.

3 *Kāma*

In the *Vedas kāma* designates 'desire, wish, drive, urge' generally but prevalently with the notion of carnality and sensuality inherent in the usage.[62] *Kāma* was, furthermore, revealed in the *Ṛg-Veda* as the very bond of the cosmos, the basic force of cohesion, the source of becoming:

> Desire [*kāma*] in the beginning came upon [took possession of] that [unevolved universe], (desire) that was the first seed of mind. Sages seeking in their hearts with wisdom found out the bond of the existent in the non-existent.[63]

The human experience of wanting something, desiring someone, is then the manifestation in the individual of the primal force, the very energy which evolves the universe, 'the bond of the existent in the non-existent'. *Kāma* is the ontogenetic and phylogenetic energy, a psychogenetic and cosmogenetic power, a particular and universal force.

By the Epic period *kāma* meant 'pleasure' as well as 'desire', the desire for pleasure and the pleasure itself, and in the various *śāstras* the concept of the *tri-varga*, the three ends of life, *kāma*, *artha*, and *dharma*, was developed with the general consensus that the three should be reconciled to work together:

> It is said that *dharma* (religious duty) and *artha* (material gain) [together] are best and that *kāma* and *artha* [together are the best] and

[62] *Kāma*, like the English word 'desire', indicates both sensual desire and desire for *any* object, but, unlike the English word, the Sanskrit word can include the *object* of (the wider form of) desire.

[63] *Ṛg-Veda* x. 129. 4 (trans. A. A. Macdonell, *A Vedic Reader For Students* [Oxford, 1917], p. 209):

kāmas tad agre sam avartatādhi,
manaso retaḥ prathamaṃ yad āsīt
sato bandhum asati nir avindan
hṛdi pratīṣyā kavayo manīṣā.

that *dharma* alone [is best] or that *artha* alone [is best] but the best is keeping to the *tri-varga* [i.e., heeding the three together].[64]

When the sons of Pāṇḍu debate the issue in the *Mahābhārata* Bhīma argues the supremacy of *kāma*:

Without *kāma* there is no desire for *artha* and without *kāma* there is no wish for *dharma*, for without *kāma* there is none who desires—therefore *kāma* is the best. The sages are impelled by *kāma* when they are immersed in asceticism. . . . All is held together by *kāma*. . . . *kāma* is more excellent than *dharma* and *artha*; *kāma* is a nectar-essence (*rasa*) like the honey of flowers—from *kāma* happiness arises. [Therefore:] Delight with women who have lovely voices, who are drunken, beautifully dressed and ornamented, having chosen *kāma* [over *artha* and *dharma*], for *kāma*, O king, assails with might![65]

And Kṛṣṇa says essentially the same thing to Yudhiṣṭhira later on in the Epic:

In [this] world men do not commend a man whose very self is desire, and [yet] there can be no progress without desire; for the gift of alms, study of the Veda, ascetic practices, and the Vedic sacrificial arts [are motivated] by desire. Whoever knowingly undertakes a religious vow, performs sacrifice or any other religious duty, or engages in the spiritual exercise of meditation without desire [does all this in vain (?)]. Whatever a man desires, that is [to him his] duty (*dharma*): it cannot be sound to curb one's duty.[66]

[64] *Manu-smṛti* II. 224:
dharmārthāv ucyate śreyaḥ kāmārthau dharma eva ca
artha eveha vā śreyas trivarga iti tu sthitaḥ.

[65] *Mahābhārata* (crit. edn) XII. 161. 28–36:
nākāmaḥ kāmayaty arthaṃ nākāmo dharmam icchati;
nākāmaḥ kāmayāno'sti tasmāt kāmo viśiṣyate.
kāmena yuktā ṛṣayas tapasy eva samāhitāḥ
. . . .
. . . sarvaṃ kāmena saṃtataṃ
. . . .
. . . kāmo dharmārthayor varaḥ;
puṣpato madhv iva rasaḥ kāmāt saṃjāyate sukham.
sucāru-veṣābhir alaṃkṛtābhir
madotkaṭabhiḥ priya-vādinībhiḥ;
ramasva yoṣābhir upetya kāmaṃ
kāmo hi rājaṃs tarasābhipātī.

[66] *Mahābhārata* XIV. 13–91–0 (Zaehner trans.):
kāmātmānam na praśaṃsanti loke
na cākāmāt kācid asti pravṛttiḥ;

In the *Bhagavad-Gītā* Kṛṣṇa rails against *kāma* as the enemy, an 'evil thing':

> Desire it is: Anger it is,—arising from the constituent of Passion,—all devouring, mightily wicked, know that this is [your] enemy on earth. . . . this [world] is obscured by that [desire]. This [desire] is the wise man's eternal foe; by this is wisdom overcast, whatever form it takes, a fire insatiable. Sense, mind, and soul, they say, are the places where it lurks; through these it smothers wisdom, fooling the embodied [self]. Therefore restrain the senses first: strike down this evil thing! . . .[67]

Desire is the enemy, the great evil because it magnifies the ego, the sense of 'I and mine', because it is the 'bond of the existent', perpetuating and sustaining *saṃsāra*. Love as desire is attachment and whatever joys there are in attachment to anyone or anything ultimately turn to sorrow and pain. But the sorrow of attachment, the pain of the lover separated from the beloved, was given meaning as an essential phase of loving, idealized, in Sanskrit poetry; and that sorrow, that love, was given further meaning in Vaiṣṇava literature:

> The pain of separation . . . and the resultant constant dwelling of the minds of the *gopīs* on Krishna is their salvation. . . . Their *viraha*, their pain of separation, draws their interests away from worldly concerns and leads to meditation on Krishna which is the essence of *bhakti* and leads to attainment of him.[68]

---

dānaṃ hi vedādhyayanaṃ tapaś ca
kāmena karmāṇi ca vaidikāni.
vrataṃ yajñāni yamān dhyāna-yogān
kāmena yo nārabhate viditvā;
yad yad dhy ayaṃ kāmayate sa dharmo
na yo dharmo niyamas tasya mūlam.

[67] *Bhagavad-Gītā* III. 36–41 (Zaehner trans.):
kāma eṣa krodha eṣa rajo-guṇa-samudbhavaḥ;
mahāśano mahā-pāpmā viddhy enam iha vairiṇam.
. . . . tathā tenedaṃ āvṛtam.
āvṛtam jñānam etena jñānino nitya-vairiṇā;
kāma-rūpeṇa (kaunteya) duṣpūreṇānalena ca.
indriyāṇi mano buddhir asyādhiṣṭhānam ucyate;
etair vimohayaty eṣa jñānam āvṛtya dehinam.
tasmāt tvam indriyāṇy ādau niyamya . . . ;
pāpmānaṃ prajahi hy enaṃ . . . .

[68] Dimock, 'Doctrine and Practice among the Vaiṣṇavas . . . ', op. cit., pp. 56–7 (cf. *Bhāgavata-Purāṇa* x. 46–47).

Kāma is used to transcend *kāma.*

Jayadeva, writing within both the literary and the religious traditions, transformed the Kṛṣṇa of the *Bhagavad-Gītā*, the opponent of desire, into the subject and object of desire; Kṛṣṇa longs passionately, loves fervently, desires without restraint; he is pierced by the arrows of *kāma* (his own desire personified), impaled through his desire for Rādhā (xii. 13). Desire is *not* to be 'struck down': the go-between orders Rādhā to 'fulfil the desire (*kāma*) of (and/or for) the Enemy-of-Madhu' (v. 14) and this use of the word as sensual desire is immediately juxtaposed to desire as an expression of *bhakti*—Kṛṣṇa is 'desirable for his virtues' (*sukṛta-kamanīya*) (v. 15). Rādhā and the woman she imagines to be trysting with Kṛṣṇa are *kāminīs*, loving-women, desiring and desirable (vii. 6, 11; xii. 2). Rādhā mutters (*japati*) 'Hari! Hari!' with *kāma*, passionately, according to her desire: to make a *japa* of the name 'Hari' is devotional activity—the loving woman invokes the name of her beloved, the devotee invokes the name of his Lord. This reflects the general tenor of the *Gītagovinda*: the sexualization of the devotional, the sanctification of the erotic.

Repeatedly Jayadeva qualifies *kāma* as being *vāma,*[69] paradoxical, contrary: 'the way of love (*kāma*) is paradoxical (*vāma*)' (xii. 11)—Kṛṣṇa obtains delight and pleasure from being scratched, bitten, crushed and crazed in love-making; Rādhā's mind is paradoxical, perverse, in loving Kṛṣṇa despite his faithlessness—she loves him in spite of herself and love (*kāma*) is cruel (*vāma*) (ii. 10); 'even when he is cruel again my heart is attached to him by force—perverse (*vāma*) is the lotus-eyed-women's love (*kāma*), for it makes the cool breeze seem like a fire, the nectarous rays of the moon seem like poison' (vii. 40). Love is perverse, paradoxical, cruel, but also 'lovely, desired'—the homonymity itself reflects the paradox of love. Love 'by force', '*amors par force*',[70] passion is 'all devouring . . . a fire

[69] *Vāma* means 'lovely, beautiful', often applied to the eye-brows or eyes (i. 49); it also (homonymously) means 'paradoxical, contrary, froward, cruel, perverse' or 'left, left-hand' and the various meanings are combined: Rādhā tells the wind which is *dakṣiṇa* (southern, right) to stop being *vāma* (left, cruel, perverse) (vii. 39); Kṛṣṇa chases after a woman who is *vāma*—'beautiful', but that he must follow her suggests that she is also being 'cruel' by running from him and 'perverse' by running from him although she wants him (i. 45).

[70] A hermit in Beroul's *Tristan* says to Tristan and Iseult, '*Amors par force vos*

insatiable', the wicked enemy. But in the *Gītagovinda* the paradox of love is celebrated: the wicked is the good, the enemy is beloved (in the 'battle of love'), the profane is the sacred, pain is pleasure, and attachment is liberation.

The repeated qualification of *kāma* with *vāma* suggests *vāma* as a technical term from the *tantra*s and Sahajiyā texts—the *vāmācāra* or left-hand practice of worship involving sexual intercourse, the identification of sexual impulse in the individual man with that 'desire [which] in the beginning took possession of the universe . . . the first seed of mind':

The central *sādhanā* of tantrism, Buddhist and Hindu alike, is the exercise of sexual contact. . . . The Hindu tantric tradition makes some distinction between the *sādhanā* that is performed on a purely mental plane and that which involves actual handling of the ritualistic 'ingredients' including woman, meat, wine. The Hindu schools usually refer to the latter as the left-handed (*vāmācāra*), to the former as the right-handed (*dakṣiṇācāra*).[71]

In the sphere of practical culture, the Sahajiyās say that the worshipper should not follow the *dakṣiṇa* course but should stick to the *vāma* mode. In the *Tantra*s it is also said that the *vāma* is better than the *dakṣiṇa,* for the latter is practically based on Vedic principles and hence falls within the sphere of *Vaidhi* culture, which is also denounced by the Sahajiyās because they prefer the *rāgānuga* mode. . . . [72]

But the Sahajiyās, like the followers of Caitanya, distinguished between *kāma* and *prema* as the profane and sacred dimensions of love.

4 *Prema(n)*

The distinction made by the disciples of Caitanya between *kāma* and *prema* is analogous to the distinction between the Platonic *Eros* and the Pauline *Agape*:

---

demeine!' ('love by force dominates you'); Denis de Rougemont considers the phrase the most poignant description of passion ever penned by a poet. . . . the whole of passion is summed up . . .' (*Love in the Western World*, p. 42).

[71] Agehananda Bharati, *The Tantric Tradition*, Anchor Books edn (New York, 1970), p. 228.

[72] M. M. Bose, *The Post-Chaitanya Sahajia Cult of Bengal* (Calcutta, 1930), p. 132.

The *prema* of the Gopīs is . . . a pure and spotless *prema* with no element of *kāma* in it. The manifestations of *kāma* and *prema* are as different as iron and gold. . . . *Kāma* is the desire for the satisfaction of the self, but *prema* is the desire for the satisfaction of the senses of Kṛṣṇa. The sole object of *kāma* is the pleasure of the self, but *prema* has as its only object the pleasure of Kṛṣṇa. . . . So abandon all else and worship Kṛṣṇa; serve him in *prema*, with his pleasure as your sole object.[73]

Jayadeva uses the metaphor of *prema* as good—the 'night manifests [itself] as a touchstone for the gold which is love for him' (xi. 12). The correlation, *prema* = gold and *kāma* = lead, suggests the alchemical process—the goal of the Vaiṣṇavas was the transmutation of the baser emotion into the pure emotion. The Sahajiyā Vaiṣṇavas stressed the mutability of *kāma* into *prema*; their distinction between the two is analogous to Augustine's distinction between *cupiditas* and *caritas*, Plato's lower and higher *eros*:

Prema is derived from *kāma*, but the motives of the two are different. . . . In whom there is a suggestion of *kāma*, *prema* is born.[74]

. . . the Sahajiyās [say that] the same flow of emotion . . . that becomes *kāma* in association with the selfish desires, transforms itself into *prema* when dissociated from such desires through physical and psychological discipline. *Prema* is but the purified form of *kāma*, and as such the former has its origin in the latter. There cannot be *prema* without *kāma*, and hence *prema* cannot be attained through the absolute negation of *kāma*; it is to be attained through the transformation of *kāma*.[75]

Distinctions between *kāma* and *prema* were not made in classical *kāvya*—Rādhā is 'blind with love' (*premāndha*) (i. 49) and Kumbha glosses *prema* as *kāma*. Kṛṣṇa is 'distracted by the burden of his *prema*' (iv. 1)—the phrase suggests longing and the longing of lovers in *kāvya*, the longing of Rādhā and Kṛṣṇa in the *Gītagovinda*, is rarely selfless. The lover longs to be pleased by the beloved and to please the beloved—mutual enjoyment is the object of both *kāma* and *prema*.

In the medieval Vaiṣṇava texts *prīti* is used synonymously with *prema* as the sacred expression of love (both are derived

[73] *Caitanya-Caritāmṛta, Ādi.* iv.139 ff. (cited by Dimock, *Place of the Hidden Moon*, p. 162).

[74] *Vivarta-vilāsa*, cited by Dimock, ibid., p. 163.

[75] S. B. Das Gupta, *Obscure Religious Cults*, p. 135.

from the same root √*prī-*, 'to please, gladden, delight'):[76]

> *Prīti* or *Prema-bhakti*, wish as devotional love for the Bhagavat is the highest type of *Bhakti*, is the *summum bonum* of life. . . . *Prīti* for the Bhagavat is considered as the highest end. . . . [bringing] happiness which is unalloyed and imperishable and consequently causes the absolute and permanent cessation of misery. . . . *Prīti* is the highest form of *Mukti*. . . . *Bhakti* is *Prīti* directed toward Kṛṣṇa.[77]

Jayadeva uses *prīti* not as 'love' but as the enjoyment of love —*prīti* is delight; Kṛṣṇa squeezes Rādhā on account of his *prīti* (XI. 34); in love-making lovers are delighted (*prīta*) (v. 18). *Prīti* is sexual joy, the sensual delight of love—Kṛṣṇa imagines that in uniting with him Rādhā will attain *prīti* (XI. 10). And he is the source of *prīti* not only for Rādhā, but also for the *rasika*, the devotee: 'May Hari grant you *prīti*!' (x. 16).

5 *love and Love and LOVE*

A list of words for love is given in the *Rasaratnākara*:

> *Premā*, *abhilāṣa*, *rāga*, *sneha*, *preman*, *rati* and *śṛṅgāra* are said to be seven types of love-in-enjoyment (*saṃbhoga*); *premā* is desire to see [the beloved] in a lovely-abode; *abhilaṣaka* is also thought about him (or her); *rāga* awareness of attachment to him (or her); *sneha* is activity devoted to him (or her); *preman* is being unable to bear separation from him (or her); *rati* is abiding with him (or her); *śṛṅgāra* is love-play together with him (or her); thus *saṃbhoga* is arranged in seven ways.[78]

The vocabulary of love encompasses the pain of separation and the joy of union, and the longing of separated lovers for union. As the vocabulary became adapted to Vaiṣṇava theology, the motif of separation and union, the course from the awakening

[76] From the same root, *priya*(*ā*) is used adjectively as 'beloved, dear' or 'pleased' or, as a noun, 'the beloved' or 'love, love's pleasure'.

[77] De, *Vaiṣṇava Faith and Movement*, pp. 204, 288, 296–7.

[78] Cited by Richard Schmidt, *Beiträge zur Indischen Erotik* (Leipzig, 1902), p. 93 (as a citation by Mallinātha given in several commentaries):

premābhilāṣo rāgaś ca snehaḥ prema ratis tatha;
śṛṅgāraś ceti saṃbhogaḥ saptavasthaḥ prakīrtitaḥ.
prema didṛkṣa ramyeṣu taccintapy abhilaṣakaḥ;
rāgas tatsaṅga-buddhiḥ syāt snehas tatpravaṇa-kriya.
tadviyogāsaham prema ratis tat-saha-vartanaṃ;
śṛṅgāras tat-samaṃ kriḍa saṃbhogaḥ saptadhākramaḥ.

of love through the painfulness of craving to the ultimate joy of union, remained. Love, the joy of loving attachment encompassing both the pain of separation and the pleasure of union, was classified into stages resembling the literary types:

> (1) *Rati* which produces delight in the mind, (2) *preman* which makes the devotee regard the Bhagavat as his own, (3) *praṇaya* which begets confidence, (4) *māna* which through excess of affection generates a peculiar sensitiveness resulting in a diversity of feelings, (5) *sneha* which causes a melting of the heart, (6) *rāga* which excites an extreme form of eager longing for the object of love, (7) *anurāga* which makes the object of love appear ever and ever new and (8) *mahābhāva* which maddens by the wonderful display of unequalled and unsurpassed ecstasy.[79]

Love, *prema*, for the medieval Vaiṣṇavas, the passionate desire which is love (*kāma* and *rāga*) and the deep joy which is love (*rati* and *prīti*), became the ultimate and universal *dharma* (virtue, religious duty)—'*Prema hi dharma hai*', sang Caṇḍīdāsa.[80]

Jayadeva frequently uses the epithets of the love-god, *L*ove as a proper noun, to indicate *l*ove as a common noun, usually in the sense of 'desire'. The lack of distinction between capital and lower-case letters in Indian scripts provides an ambiguity: LOVE may refer either to the god or the human experience or both at once and this ambiguity provides a link between the psyche and the cosmos—desire is the emotional and biological manifestation of the same force that is divinely manifested in the Hindu pantheon as Kāma-deva.

THE MYTHOLOGY OF LOVE

> Love, my children, is a god, young and beautiful and winged. That's why he delights in youth and pursues beauty and gives wings to the soul. And he can do greater things than Zeus himself. He has power over the elements, he has power over the stars, he has power over his fellow gods. . . . The flowers are all Love's handiwork. The trees are his creations. He is the reason why the rivers run and winds blow.[81]

[79] S. C. Chakravarti, op. cit., p. 252.

[80] Cited by Lal, op. cit., p. 80.

[81] Longus' *Daphnis and Chloe*, II. 7, trans. Paul Turner, Penguin edn (Harmondsworth, 1968), p. 48.

The Indian god of love, Kāma-deva, was worshipped more by the poet than by the priest—his place is limited in religious texts and yet he appears throughout *kāvya*. The poet's 'worship' is sensual activity, a ritual that mocks ascetic activity. Vallaṇa sang:

> I praise the consecration of the god of love,
> which wins the victory from the best of other rites.
> No priest need be invited;
> the sacred image is one's mistress;
> the acts of worship: looks, embraces, kisses;
> and the gift, oneself.[82]

Belief in the god requires aesthetic faith—belief in the validity of the metaphor, in the meaning of an image which presents emotional experience as mythological structure. The image makes the concept of love specific, concrete.

In the *Bṛhadāraṇyaka Upaniṣad*[83] the sacrificial ritual had been compared with sexual union, described in erotic terms. The structure of the ritual was based on sexual experience and both were invested with the same magical power. Vallaṇa's poem makes the same comparison, but in jest—sexual union becomes a parody of the ritual, a 'humorous blasphemy' (typical of courtly literature) which suggests the 'antagonism of the two ideals', *kāma* and *tapas*, sensuality and asceticism. Essentially the same type of poems were written in twelfth-century France, *jeu d'esprit* celebrating the love religion of the god Amor:

> . . . erotic religion arises as a rival or parody of the real religion and emphasizes the antagonism of the two ideals. . . . The worship of the god Amor had been a mock-religion in Ovid's *Art of Love*. The French poet [of the *Concilium in Monte Romarici*] has taken over this conception of an erotic religion with a full understanding of its flippancy, and proceeded to elaborate the joke in terms of the only religion he knows

[82] *Subhāṣitaratnakoṣa* 333 (Ingalls trans.):
yācyo na kaścana guruḥ pratimā ca kāntā
pūjā vilokana-vigūhana-cumbanāni;
ātmā nivedyam itara-vrata-sāra-jetrīṃ
vandāmahe makara-ketana-deva-dīkṣām.

[83] *B.U.* vi. ii.13 (see above, p. 16).

—medieval Christianity. The result is a close and impudent parody of the practices of the Church, in which Ovid becomes a *doctor egregius* and the *Ars Amatoria* a gospel . . . and the god of Love is equipped with cardinals and exercises the power of excommunication. The Ovidian tradition, operated upon by the medieval taste for humorous blasphemy, is apparently quite sufficient to produce a love religion, and even in a sense a Christianized love religion, . . . the love religion often begins as a parody of the real religion.[84]

As the *Gītagovinda* exemplifies the Vaiṣṇava traditions, Jayadeva's use of the vocabulary of the orthodox religious systems is a serious sanctification of human love as an emotion readily mutable into *bhakti*; but as the work exemplifies the courtly, literary tradition, the use of the sacred vocabulary represents a playful profanization—the poet makes a parody of the practice of ascetics, *kāma* mocks *tapas*, Kāma derides Śiva. When Jayadeva describes the perfections (*siddhi*) of the love-god (*rati-pati*) as attained in the love-god's great pilgrimage place (*mahātīrtha*), the ascetics who go into the forest to meditate are caricatured in the lovers going into the grove to tryst; and as Kṛṣṇa meditates ($\sqrt{dhyai}$-) there on Rādhā, chanting ($\sqrt{jap}$-) *mantras* as an invocation to her, ascetic activity is taunted (v. 7). Kāma-deva, 'the presiding deity of wanton sports',[85] 'the family priest of womankind, who consecrates them for sports',[86] prevails:

> Bodiless he is and with a bow of flowers,
> with flower arrows, too, that do not touch;
> but such is his art at will to work great wonders
> that at one blow he strikes the world to heart.
> Long live the mind-born god, with fiat
> honored by all creatures.[87]

[84] C. S. Lewis, *The Allegory of Love* (London, 1936), pp. 18 ff.

[85] *Utpalarāja*, '*lalita-surata-līlā-daivata*' (Ingalls trans., *Subhāṣitaratnakoṣa* 332).

[86] Rājaśekhara, '*kula-gurur abalānāṃ keli-dīkṣa-pradāne . . .*' (Ingalls trans., *Subhāṣitaratnakoṣa* 327).

[87] Manovinoda, *Subhāṣitaratnakoṣa* 325 (Ingalls trans.):
manasi kusuma-bāṇair eka-kalaṃ trilokiṃ
kusuma-dhanur anaṅgas tāḍayaty aspṛśadbhiḥ;
iti vitata-vicitrāścarya-saṃkalpa-śilpo
jayati manasi-janmā janmibhir mānitājñaḥ.

The court poet used the mythology of the love-god as it had been developed in Epic and *Purāṇic* literature, transforming mythological symbols into poetic conceits. Similarly the European poets moulded a rhetoric of love out of the classical mythology of the 'great *daimon*'. Dante explained that to 'speak of love *as if* it were a bodily thing, even *as if* it were a man' is justified by the precedents of the 'ancient poets [who] spoke of inanimate things *as if* they had sense and reason . . . '.[88] The 'as-if-ness' is at the heart of belief in the love-god—the divinity of love was a poetic device by which the internal could be expressed *as if* it were external, the psychological could be revealed *as if* it were physical:

The God of Love who, with bended bow, had been constantly intent on following and watching me . . . he straightway took an arrow and when the string was in the notch, he drew it back to his ear and loosed the arrow from his mighty bow with such skill that he shot it fiercely through my eye and lodged it in my heart. . . .[89]

Vergil's '*Omnia vincit amor*' echoed through the lyrics of the Middle Ages and the Renaissance. Likewise the omnipotence of Kāma-deva was recognized by the Sanskrit poets: Bhartṛhari salutes Kāma-deva as more powerful than the Creator, the Preserver and the Destroyer;

By whom Śambhu [Śiva], Svayambhu [Brahmā] and Hari [Viṣṇu] are constantly made house pot-servants of deer-eyed-women . . . Hail the *makara*-bannered Lord![90]

Kṛṣṇa acknowledges the supremacy of the love-god: 'O you who conquer all [the world] for sport!' (III. 12); '[love's] weapons . . . conquer the world' (III. 13); 'the flower-weaponed [god of love] conquers all' (X. 14).

In the Vedic period Kāma appeared as desire personified,

[88] Dante Alighieri, *La Vita Nuova*, XXV, trans. Barbara Reynolds, Penguin edn (Harmondsworth, 1969), pp. 72 ff. [Italics mine.]

[89] Guillaume de Lorris, *Roman de la Rose* (First Part), extracts in *The Penguin Book of French Verse I*, p. 150.

[90] Benedictory verse to the *Śṛṅgāraśataka* 1:

śambhu-svayambhu-harayo hariṇekṣaṇānāṃ
yenākriyanta satataṃ gṛha-kumbha-dāsāḥ;
[vācāmagocara-caritra-vicitritāya]
tasmai namo bhagavate makara-dhvajāya.

not yet a god of love, but 'a deity who fulfils all desires'.[91] Kāma, chanted in the *Atharva-Veda* as the 'first-born',[92] was a cosmogonic force much like Hesiod's Eros:

First of all, the Void came into being, next broad bosomed Earth, the solid and eternal home of all, and Eros, the most beautiful of the immortal gods, who in every man and every god softens the sinews and overpowers the prudent purpose of the mind.[93]

This Eros, Jung has called a '*kosmogonos*, a creator and father-mother of all higher consciousness' and he adds, '. . . we are in the deepest sense the victims and the instruments of cosmogonic love'.[94] Rādhā and Kṛṣṇa are shown to be both the targets and sources of the arrows of the love-god.

The image of the 'arrow of Desire' emerged; the 'fulfiller of all desire' was called upon in magical incantation to fulfil sexual desire in particular:

Let the up-thruster thrust thee up, do not
stick to thy own bed; love's arrow that is
terrible, with that I pierce thee in the heart;
The arrow, feathered with longing, tipped with desire,
necked with resolve, may Kāma make it well-
directed and pierce thee in the heart;
The well-directed arrow of love that dries the spleen,
forward-winged, consuming, with that I pierce
thee in the heart.[95]

[91] A. A. Macdonell, *Vedic Mythology* (Strasbourg, 1897), p. 120.

[92] *Atharva-Veda* IX. ii. 19. The hymn is a magical incantation to Kāma as a powerful personal god: 'Kāma the Bull, I worship with molten butter, sacrifice, oblation. Beneath my feet cast down mine adversaries with thy great manly power, when I have praised thee. . . . May Kāma, mighty one, my potent warder, give me full freedom from my adversaries. . . . be joyful, ye Gods whose chief is Kāma. . . . First before all sprang Kāma into being. Gods, Fathers, mortal men have never matched him. Stronger than these art thou, and great for ever, Kāma, to thee, to thee I offer worship . . .' (trans. R. T. H. Griffith [Benares, 1916]).

[93] Hesiod, *Theogony* II. 116 ff., trans. N. O. Brown (New York, 1953), p. 56.

[94] C. G. Jung, *Memories, Dreams, Reflections*, trans. Richard and Clara Winston, Fontana Library edn (London, 1967), pp. 386–7.

[95] *Atharva-Veda* III. 25 (Sten Konow trans.):
uttudas tvottudatu ma dhṛtās śayane sve;
iṣuḥ kāmasya yā bhīmā tayā vidhyāmi tvā hṛdi.
ādhiparṇāṃ kāmaśalyām iṣuṃ saṃkalpakulmalām;
tāṃ susamnatāṃ kṛtvā kāmo vidhyatu tvā hṛdi.

The mythology, iconography, the biography of the love-god developed in the Epics and *Purāṇas,* and it was used, modified, embellished, invigorated, in the literary tradition. The archer's flower-tipped arrows are fired from a sugar-cane bow; he traps men and gods with his hook and noose; his emblem and vehicle is the *makara,* a sea-monster or crocodile; his wives are Pleasure and Delight (Rati and Prīti) and his friends are Spring and the Malayan wind; he is surrounded by heavenly nymphs. But always the mythology remained open, flexible, adaptable to the context.

The love-god has various epithets which are revealing about the Indian conception of love: Love is Desire (Kāma), Remembrance (Smara), the Bodiless (Anaṅga, Atanu), the Intoxicater (Madana), the Mind-Churner (Manmatha), the Five-arrowed-one (Pañca-bāṇa, Asama-bāṇa), Death (Māra), Kandarpa (sometimes etymologized as 'inflamer of even a god'), and so forth. As the concept became personified into a god, so also the epithets referred back to the concept, the god became re-abstracted; but most often they are used ambiguously with the distinction between the god and the concept blurred. Love is called the Mind-born or Heart-Born (Manoja, Manasija, Svāntaja, etc.): in the *Ṛg-Veda kāma* had been said to be the 'first seed of mind'[96] and in later mythology Kāma was said to be born from the mind of Brahmā—again the ambiguity between love and Love provides a link between the human realm and that of the gods, the internal and external, psychological and cosmological; 'the arrows of Love' are the 'arrows produced in the mind' (iv. 2 ff.; vi. 32), the 'joy of Love' is the 'joy produced in the mind' (vii. 39), the 'heated-pain of Love' is the 'heated-pain produced in the mind' (xii. 5), and so forth. The individual gives birth to the *daimon* just as Brahmā did. And as soon as he is born he begins to enchant.

Brahmā created all the gods and all the *prajāpatis,* including Dakṣa.

---

yā plīhanaṃ śoṣayati kāmasyeṣuḥ susaṃnatā;
prācīnapakṣā vyoṣā tayā vidhyāmi tvā hṛdi.

Konow suggests that the image of Love's arrow developed from the use of arrows in magic rites: '. . . we learn from the *Kauśika Sūtra* 35. 22 ff. that . . . to win a woman's love . . . an arrow was shot against a clay image of the woman' ('Anaṅga, the bodiless Cupid', in *Festschrift Jacob Wackernagel* [Gottingen, 1923], p. 5).

96 '*manaso retaḥ prathamaṃ* . . . ', *Ṛg-Veda* x. 194. 4 (cited above, note 63).

Then a beautiful woman named Sandhyā was born from his mind. . . . Then from his mind was born Kāma, with his five marvellous flower arrows. . . . [Brahmā] said to Kāma, 'Enchant men and women with your five flower arrows and your own beauty, maintaining creation eternally. No one will be able to withstand you—not even Viṣṇu and Śiva and I.' Then Kāma decided to begin his work right then and there, starting with Brahmā, and he excited Brahmā and all the sages with his arrows. As Brahmā gazed upon Sandhyā, his senses became aroused, and she too showed the signs of desire.[97]

Brahmā sends Kāma to assault Śiva 'out of spite and in revenge against Śiva (for opposing Brahmā's incestuous behaviour) as well as against Kāma (for causing this behaviour)'.[98] *Kāma* is pitted against *tapas*, sensual-desire against asceticism. 'Śiva is the natural enemy of Kāma because he is the epitome of chastity, the eternal *brahmacārin*, the very incarnation of chastity.'[99] To arouse Śiva from his deep meditation and fill him with desire for Pārvatī, Kāma tried to shoot one of his arrows, but Śiva released the fires of his asceticism from the eye in his forehead and reduced Kāma to ashes. But Kāma persists, 'bodiless', in revenge, in tormenting gods and men.

While Śiva, Hara, is the eternal ascetic, chastity personified, Kṛṣṇa, Hari, is the eternal lover, sensual playfulness personified. Kṛṣṇa asks the love-god not to mistake signs of the lover's longing, the articles worn to allay the heat of love's fever, for the signs of asceticism, the heat of *tapas*, the emblems of Śiva;

This is a lotus-tendril necklace on my chest, not the Lord-of-Serpents; this is a row of [blue] lotus petals on my neck, not the radiance of [the blue *kālakūṭa*] poison; this is sandal dust, not ashes, on me [for I am] deprived of my beloved; do not attack me, mistaking me for Hara; O Bodiless-love-god! why do you chase me angrily? [III. 11]

There is an analogy between the mythological description of Kāma and the psychoanalytic description of the *id* as pure desire; the *id* produces 'a striving to bring out the satisfaction of the instinctual needs subject to the observance of the pleasure principle . . . the id stands for the untamed passions'.[100] What

[97] *Śiva Purāṇa* 2. 2. 2. 15–42 (cited and trans. by Wendy O'Flaherty, *Asceticism and Eroticism in the Mythology of Śiva* [London, 1973], p. 118).

[98] O'Flaherty, ibid., p. 141. [99] Ibid.

[100] Sigmund Freud, *New Introductory Lectures on Psychoanalysis*, trans. and ed. James Strachey, College edn (New York, 1965), pp. 73 and 76.

Freud called the '*id*' is 'the first seed of mind', primal and 'mind churning'. The *ego* develops as a mediator between the *id* and the world, the external reality. If the *ego* attempts to fulfil this function by denying the presence of the instinctual desires, by repressing them, the result is greater conflict, anxiety, guilt—it is like trying to destroy 'mind-born' Kāma—it is made 'bodiless', invisible, but still present, still tormenting. Freud saw 'sublimation' as 'the way out, a way by which the claims of the ego can be met without involving repression';[101] through sublimation the energy of the *id*, and particularly the sexual energy, could be made constructive:

The sexual are amongst the most important of the instinctive forces thus utilized; they are in this way sublimated, that is to say, their energy is turned aside from its sexual goal and diverted towards other ends, no longer sexual. . . .[102]

Śiva 'permanently ithyphallic, yet perpetually chaste'[103] exemplifies the path of sublimation: through chastity the ascetic is able to generate *tapas* as a great creative force; he controls and transforms desire, the heat of *kāma*, 'diverting it towards other ends'. Kṛṣṇa exemplifies another path—desire quenched by its fulfilment. The *ego* seeks to satisfy the demands of the *id*, to achieve union with the world in pleasure. It is *bhakti* as passionate love as opposed to *yoga* as ascetic practice. Kṛṣṇa mocks that *yogic* practice when he proclaims that his mind is fixed in a state of *samādhi* (III. 15), a technical term in classical *yoga:*

With the advance of this state the sage ceases to have inclinations even towards the processes of concentration, and there is only discriminative knowledge; this state of *samādhi* is called *dharma-megha*. At this stage all the roots of ignorance and other afflictions become absolutely destroyed, and in such a state the sage, though living (*jīvann eva*), becomes emancipated (*vimukta*).[104]

'How then', Kṛṣṇa asks, 'can the sickness of love-in-separation

[101] Idem, *Collected Papers* IV. 52, ed. and trans. Joan Rivière and James Strachey (New York, London, 1924–50).

[102] Idem, *A General Introduction to Psychoanalysis*, trans. Joan Rivière, rev. edn (New York, 1952), p. 27.

[103] R. C. Zaehner, *Hinduism*, OUP paperback edn (Oxford, 1966), p. 86.

[104] S. N. Das Gupta, *A History of Indian Philosophy*, vol. II (Cambridge, 1952), p. 251.

increase?'; and he has answered his own question—because the objects of his meditation are Rādhā's features (III. 15). Kāma-deva does not need to arouse him from his meditation in order to fill him with desire for a consort—his meditation is the result of that desire—he has already been pierced by the arrows:

Do not hold that mango-arrow in your hand, do not string that bow! O you who conquer all [the world] for sport! What kind of bravery is it to strike people who are stupefied [already]? O mind-born-love-god! Torn apart by the rows of arrows which are the trembling, side-long glances of the deer-eyed-woman, of her alone, not even now does my mind even slightly recover. [III. 12]

Rādhā's heart is also impaled by the flower-arrows (IV. 2 ff.; VII. 4, 8; VIII. 1; etc.). For Śiva this would mean defeat by Kāma and for the Śaivite ascetic, taking the god's behaviour as his model, pursuing *samādhi* through *yoga*, it would mean defeat by *kāma*; but for Kṛṣṇa or for Rādhā or for the Vaiṣṇava *bhakta*, taking the god's behaviour and that of Rādhā as his model, the defeat would be a victory. The *samādhi* of love is attained by the five arrows.

Kāma, like Longus' sylvan Love, 'the reason why the rivers run and winds blow', was predominantly associated with the renewal of the world, the spring:

The most popular festival in early times was the Festival of Spring, in honour of Kāma, the Love-god, who, though he played only a small part in the thought of theologians, was evidently a very popular deity.[105]

The mythopoetic image expresses the sentimental psychology —the flower-tipped arrow causes pain because it causes attachment to what is pleasurable; it is the beauty of spring that torments, joy that pierces the heart. Simultaneously a source of misery and delight, the five arrows of Kāma are like those of Amor:

On Love's mighty quiver in terrible characters of fate the legend shines: 'I bear the sweetest arrows of joy; I contain the bitterest arrows of pain.'[106]

[105] A. L. Basham, *The Wonder That Was India*, Evergreen edn (New York, 1959), p. 207.

[106] Friedrich Müller, 'Aus Amors Köcher' in the *Penguin Book of German Verse*, ed. Leonard Forster (Harmondsworth, 1959), p. 233.

Love is the inner manifestation of the same vital rhythm which is outwardly manifested as nature coming into full bloom:

> As the mango puts forth shoot and leaf,
> puts forth bud and flower,
> so in our hearts does Kāma shoot
> and leaf and bud and flower.[107]

Love lurks in every spring sound, in the warm breeze, in every 'leaf and bud and flower'—the cuckoos sing the 'behests of Love' (XI. 14); the wind, warm and sandal-scented, is like the 'breath of Love' (I. 36), 'bringing Love near' (V. 2), the 'joy of Love' (VII. 39).

> When the opening-up of the *keśara* flower has the appearance of the golden parasol of the emperor of passion, when the love-god's quiver makes its appearance as *pāṭala* full of bees-arrows, Hari plays, now, in the amorous springtime, endless for separated lovers, he dances with the young-girls O friend! [I. 31]

Crimson flowers look like the finger-nails of the Love-god, red with blood from lacerating the hearts of the young in the 'amorous springtime' (I. 30). The new blossom, an image of life reaffirming itself, looks like a blood-smeared nail or, elsewhere, like an arrow, or spear—images of death. Kāma as lord of spring is the personification of the 'life-instinct', *Eros*, the force which 'seeks to force together and hold together portions of living substance',[108] the personification of 'the bond of the existent in the non-existent'.[109] But by causing attachment to the world, Kāma is also death:

> . . . in the *Gītā* 'death' is equivalent to the ever-dying world of material Nature and 'immortality' to the changeless category of Ātman-Brahman.[110]

Kāma was called Māra, Love was called Death. Kāma and Māra were identified throughout the Buddhist tradition:

> The one who they call Kāma-deva here-on-earth, he who has varie-

[107] *Subhāṣitaratnakoṣa* 188 (Ingalls trans.):
aṅkurite pallavite korakite vikasite ca sahakāre;
aṅkuritaḥ pallavitaḥ korakito vikasitaś ca hrdi madanaḥ.

[108] Freud, as cited above, p. 10, note 34.

[109] *Ṛg-Veda*, as cited above, p. 64, note 63.

[110] R. C. Zaehner, commentary on the *Bhagavad-Gītā*, op. cit., p. 128.

gated weapons, flower-tipped-arrows, likewise they call *him* Māra, the ruler of the way of desire, the enemy of liberation.[111]

When the Buddha, resolved upon fulfilling his quest for the highest wisdom, sat down in meditation, Kāma/Māra explained to his sons, Vibhrama (Confusion), Harṣa (Joy), Darpa (Pride) and to his daughters, Rati (Sexual-delight), Prīti (Loveliness), Tṛṣṇā (Thirst), that the Buddha must not gain enlightenment or the world, the abode of love and death, will become empty. Māra was Kāma, death became love, as he seized his flower-arrows and attempted to shake the Buddha with the delights of the senses, to tempt the Buddha with Rati. When this failed, Kāma was Māra, love became death, as he summoned a grotesque army of demons to terrify the Buddha. But the Buddha was impervious to the weapons; he felt neither fear nor desire.[112]

The equation of Kāma and Māra gives mythological expression to the central revelation of Buddhism—thirst, craving, desire, produces suffering, produces the 're-existence' which is this world, the world of death. Māra rules over the Kāmaloka, this world, the world of desire.

The identification of Love and Death, Eros and Thanatos, also occurred in European literature:

> Though I am young, and cannot tell
> Either what Death or Love is well,
> Yet I have heard they both bear darts,
> And both do aim at human hearts;
> And then again I have been told
> Love wounds with heat, as Death with cold;
> So that I fear they do but bring
> Extremes to touch, and mean one thing.[113]

Both the Buddhist legend and the Elizabethan poem suggest the same psychological axiom: to become impervious to death one must be impervious to love as desire, that is, fear of death is

[111] Aśvaghoṣa's *Buddha-carita* XIII. 2:
yaṃ kāma-devaṃ pravadanti loke citrāyudhaṃ puṣpa-śaraṃ
tathaiva; kāma-pracārādhipatiṃ tam eva mokṣa-dviṣaṃ
māram udāharanti.

[112] Aśvaghoṣa's *Buddha-carita* XIII. 1–73.

[113] Ben Jonson, 'Death and Love' from *The Sad Shepherd* in *The New Oxford Book of English Verse*, ed. Helen Gardner (Oxford, 1972), p. 211.

overcome only when love's longing and attachment is overcome. The axiom was recognized by the medieval European poets but their attitude was that neither love nor death was to be overcome—death was worth love:

By your arrows, Love, I swear, and by your mighty, sacred torch, that, although the one burn me and waste my heart, and the others wound, I do not mind. . . .[114]

> Loving torments me, I die from the wound
> in which I glory. Ah! if she would cure me
> by a single kiss, how my heart would rejoice
> to be wounded by such a happy dart![115]

And the same attitude is expressed by Kṛṣṇa:

May the arrow of your side-long glance, placed on the bow of your brow cause pain where I am vulnerable, also may the mass of your braids, dark natured and curly, make the strenuous-effort [to perform the function] of Māra; meanwhile may this [your] impassioned-reddened lower-lip, which is like the *bimba* fruit, spread [my] infatuation! . . . [III. 14]

That is, 'may your hair kill me by making me love you, by making me attached to you'.

The Buddha pays no attention to the arrows of love; he embodies the ideal of the cessation of desire; his path is non-attachment. But Kṛṣṇa is pierced to the heart by the arrows (III. 2, 12; V. 3; X. 11), he burns with passion (III. 9); he is inflamed by Love (with love) (Kandarpa) (XI. 22), he surrenders to Love and calls for more; he embodies the ideal of the fulfilment of desire; his path is total-attachment. For Jayadeva, Kṛṣṇa encompasses the Buddha, took on the form of the Buddha, not to teach non-attachment, but out of compassion for tethered-animals (I. 13).

The love-making of Rādhā and Kṛṣṇa has 'the mark of Māra' (XII. 12)— it is a battle of love and love-play, the commentator Mānāṅka[116] explains: as a battle sexual union is

[114] Gaspara Stampa, 'Per le saette tue, Amor . . . ' in the *Penguin Book of Italian Verse*, ed. George R. Kay, rev. edn (Harmondsworth, 1965), p. 177.

[115] *Carmina Burana*, cited and trans. Perella, op. cit., p. 106 (see above, p. 52, note 26).

[116] Author of the fourteenth-century (?) commentary on the *Gītagovinda* (published in the V. M. Kulkarni edn).

marked by Māra as Death and as play it is marked by Māra as Love. In the union of Rādhā and Kṛṣṇa love and death are reconciled.

The love-god is also identified with the god of death, Yama:

Her dwelling is like the forest and the garland of her dear friends is like the snare and her heated sorrow with her sighing-breath is like the sheet of flames in a burning forest; and, alas, through your absence, she is like the doe [caught in the snare, in the burning forest]; Ah! how, moreover, is the love-god (Kandarpa) like the death-god (Yama) performing tiger's play? [IV. 10]

In separation from Kṛṣṇa she is easy prey for the murderous love-god, as a deer caught in a snare is easy prey for a tiger. The go-between then asks Kṛṣṇa, 'how can she live through long separation, having seen the mango branch, its tip in flower?' (IV. 22) and she tells Kṛṣṇa that only 'the nectar of contact with his body' can cure Rādhā (IV. 20). The implications for the followers of Caitanya were clear—the wounds from the arrows of Love were good if they made one long for the soothing 'nectar of contact' with the Lord. The Sahajiyā initiate was instructed 'always to worship by means of the arrows of the love-god'.[117]

There is an alliance between Viṣṇu-Kṛṣṇa and Kāma-deva[118] —according to the *Matsya Purāṇa* 'there is no difference between Kāma and Viṣṇu', Viṣṇu is to be worshipped with the names of the love-god.[119] Kṛṣṇa, the dwelling place of love, the love-god (Anaṅga) (XI. 24), has the appearance of the love-god (Madana) (V. 8); he initiates the festival of the love-god (Anaṅga) (I. 47); he wears *makara* ear-rings (II. 7) and his eyes glance about like the restless love-god (Anaṅga) (II. 8). The go-between tells him of Rādhā's suffering:

[117] Attributed to Caṇḍīdāsa, cited by Dimock, *Place of the Hidden Moon*, p. 59.

[118] The spring festival honouring the love-god evolved into the Kṛṣṇaite festival, *holī*; in the *Brahmāṇḍa Purāṇa* (IV. viii. 24–29; xi. 7) Kṛṣṇa blesses the love-god with a boon to ensure his eternal omnipotence (cited by V. R. R. Dikshitar, *The Purāṇa Index* [Madras, 1955], *s.v.* 'Kāmadeva'); in the *Viṣṇu Purāṇa* (V. 26–27) Kāma, having been immolated by Śiva, is reborn as the son of Kṛṣṇa and Rukminī, i.e., as Pradyumna who slew the demon Śambara—Jayadeva's use of the epithet Śambara-dāraṇa for Kāma (XII. 23) acknowledges the episode.

[119] *Matsya Purāṇa* 70, 52 and 70, 34 ff., cited by J. Gonda, *Aspects of Early Viṣṇuism*, 2nd edn (Delhi, 1969), p. 17.

Secretly she draws you with musk as the love-god—she bows down [to you] placing the *makara* beneath you and the fresh mango arrow in your hand! [IV. 6]

Drawing the beloved as the love-god was conventional:

May the god of the five arrows bring you fortune who have painted him incarnate as your absent lord, for while your fingers [the bouquet of branches of your hand] trembled on his body the bow and crocodile are drawn in steady lines.[120]

But in the Vaiṣṇava context the motif has broader implications—for the followers of Caitanya Kṛṣṇa was considered the 'new' god of love, 'Kṛṣṇa is known as Kāma or Kandarpa, as he attracts the mind of all creatures towards Him'.[121] 'A new invisible Madana is in Vṛndāvana,' Kṛṣṇadāsa Kavirāja declared, 'and his worship is by the *kāma-gāyatrī* [*mantra*]. . . .'[122] One example of this *mantra* is: 'We meditate on the god of love, whose arrows are flowers, so that the bodiless one may compel it [*rasa*/semen].'[123] And the god was worshipped with the *mantra*: 'O Kandarpa, love of my life, ruler of *rasa*, delighting in *rasa*, may I be your slave!'[124] Kandarpa was Kṛṣṇa.

And here on earth the Lord completely over-powered the minds of the Gopis with His sacred beauty. He charmed Kandarpa, the God of Love and He is called Madana-mohana, the enchanter of the God of Love himself. His is love that is transcendental. It is far above sexual desire and all its power. This makes Him the new God of Love and as such He jubilates in transcendental glory with His beloved Gopis.[125]

A similar evolution took place in the Christian tradition—Vergil's '*Omnia vincit amor*' was elaborated by Paulinus of Nola:

[120] *Subhāṣitaratnakoṣa* 740 (Ingalls trans.):

diśatu sakhi sukhaṃ te pañca-bāṇaḥ sa sākṣād
anayana-patha-vartī yas tvayālekhi nāthaḥ;
taralita-kara-śākhā-mañjarīkaḥ śarīre
dhanuṣi ca makare ca svastharekhā-niveśāḥ.

[121] S. B. Das Gupta, op. cit., p. 133.

[122] *Caitanya-Caritāmṛta, Madhya.* VIII. 109, cited by Dimock, *Place of the Hidden Moon*, p. 229.

[123] *Vivarta-vilāsa* (cited by Dimock, loc. cit.): '*kāmadevāya vidmahe puṣpa-bāṇāya dhīmahi tan no'naṅgaḥ pracodayāt.*'

[124] *Nāyikā-sādhana-ṭīkā* (cited and trans. Dimock, ibid., p. 239).

[125] *Caitanya-Caritāmṛta, Madhya.* XXI. (N. K. Ray trans., p. 530).

'*Amor omnia Christi vincit*';[126] the pagan love-god was identified with the 'new love-god':

Jesus, Thou mighty God of love, come near to me for I am languishing almost to death in desire of love. Take up Thine arms and swiftly pierce my heart with Thine arrow—oh, wound me![127]

The identification of Kāma with Kṛṣṇa had been made in the *Mahābhārata* by Kṛṣṇa himself. But there the identification was based not on love but on power; Kṛṣṇa sings to Yudhiṣṭhira the 'song which knowers of ancient lore celebrate as having been sung by Desire':

I cannot be slain by any being whatever since he is wholly without the means. If a man should seek to slay me, putting his trust in the strength of a weapon, then so I appear again in the very weapon he uses. If a man should seek to slay me by offering sacrifices and paying all manner of fees, then do I appear again as the 'self that dwells in all action' in moving things. If a man should seek to slay me by means of the Vedas and the ways of perfection [prescribed] in the Vedas' end, then do I appear as the 'stilled, quiet self' in unmoving things. If a man should seek to slay me by steadfastness, a very paladin of truth, then do I become his very nature, unaware of me though he is. If a man should seek to slay me by ascetic practice, strict in his vows, then do I appear again in his very ascetic practice. If a man should seek to slay me wise and bent on liberation, then do I dance and laugh before him as he abides in the bliss of liberation. Of all beings I alone cannot be slain, eternal [as I am].[128]

[126] Cited from the *Corpus Scriptorum* by Perella, op. cit., p. 267.

[127] Angelus Silesius, '*Jesu, du mächtiger Liebesgott*' in *The Penguin Book of German Verse*, p. 145.

[128] *Mahābhārata* xɪv. xiii. 12–17 (Zaehner trans.):

nāhaṃ śakyo'nupāyena hantuṃ bhūtena kena-cit;
yo māṃ prayatate hantuṃ jñātvā praharaṇe balam;
tasya tasmin praharaṇe punaḥ prādur bhavāmy aham.
yo māṃ prayatate hantuṃ yajñair vividha-dakṣiṇaiḥ;
jaṅgameṣv iva karmātmā punaḥ prādur bhavāmy aham.
yo māṃ prayatate hantuṃ vedair vedānta-sādhanaiḥ;
sthāvareṣv iva śāntātmā tasya prādur bhavāmy aham.
yo māṃ prayatate hantuṃ dhṛtyā satya-parākramaḥ;
bhāvo bhavāmi tasyāhaṃ sa ca māṃ nāvabudhyate.
yo māṃ prayatate hantuṃ tapasā saṃśita-vrataḥ;
tatas tapasi tasyātha punaḥ prādur bhavāmy aham.
yo māṃ prayatate hantuṃ mokṣam āsthāya paṇḍitaḥ;
tasya mokṣar atisthasya nṛtyāmi ca hasāmi ca;
avadhyāḥ sarva-bhūtānām aham ekaḥ sanātanaḥ.

This unslayable Epic god was transformed into a vulnerable lover in the *Gītagovinda*, open to the arrows of Love as they are fired from Rādhā's eyes. The bodiless love-god attains a body in Rādhā (following Kumbha's gloss of xi. 34); his assault is made through her:

The bow is the sprig of her brow, the arrows are her side-long glances, the bow-string is the tip of her ear: have [love's] weapons, which conquer the world, been transferred by love onto her, the living goddess who is the triumph of the Bodiless-love-god? [iii. 13]

Her finger-nails are like arrows of the love-god (x. 3; xi. 8); her breasts are pitchers of Love's festival (xii. 17); her eye releases Love's arrows (xii. 18); her ear is like Love's snare (xii. 19); her hair is the fly-whisk and banner of Love (xii. 22); her thighs are the cave-dwelling of the elephant of Love (xii. 23).

This lower-lip is akin in splendour to the *bandhūka* [flower]; your shiny cheek has the complexion of the *madhūka* [flower], O fierce-passionate-woman! Your eye, which emits the lustre of a blue lotus, shines; your nose resembles the sesame flower; your teeth are like jasmine; O beloved, above all, by employment of [the features of] your face the flower-weaponed [love-god] conquers all. [x. 14]

Kumbha correlates each flower-like feature with one of the flower arrows of Kāmadeva: her lip is the arrow that impassions (*raktākarṣaṇa-bāṇa*); her cheek is the arrow that subjugates (*vaśikāra-bāṇa*); her eye is the arrow that intoxicates (*unmādana-bāṇa*); her nose is the arrow that puts to flight (*drāvaṇa-bāṇa*);[129] her teeth are the arrow that desiccates. And again, after love-making, five of Rādhā's features become like the five arrows of love—her breasts, her eyes, her lower-lip, her hair, and her thighs (exposed by her loosened girdle)—'by these arrows of love, fixed in his eyes, the mind of her lord was impaled . . . ' (xii. 13).

The same image had become conventional in European lyric poetry by the twelfth century:

. . . the lady has arrows and javelins and other warlike equipment constantly at her command. She looks like an angel; but this angel is in permanent ambush, there is a bent bow in her eyes, and woe unto

[129] '*drāvaṇa*' is a term used in the *Ratirahasya* for 'bringing about orgasm'; cited by Schmidt, op. cit., p. 883.

him who stops one of her glances. . . . The lover is left more dead than alive, with the dart of love festering in his heart.[130]

The sixteenth-century poet Joshua Sylvester sang of his beloved's 'locks more than golden . . . where in close ambush wanton Cupid lurks . . . '[131] and Sir Charles Sedley in the seventeenth century addressed his Phillis:

> The God of Love, in thy bright eyes,
> Does like a tyrant reign![132]

The image, occurring throughout the history of the romantic love lyric, represents a change from the classical attitude toward love:

The romantic lover thus offers the greatest possible contrast to the classic hero in his amatory aspect. In the classic view love is an appetite; in the romantic it is a thunderbolt. The heroic lover pleases himself; the romantic strives to please his beloved. . . . Achilles is served by women. Troilus serves.[133]

The same contrast might be made between the Indian Epic hero and the lover in Sanskrit court poetry. The Epic hero demands obedience and dependence from his women (who are, after all, his property); the lover in Sanskrit poetry makes obeisance to his mistress, and if separated from her he laments:

> Time and again the flood of tears prevents my sight;
> my body numbed by thoughts of her is paralyzed;
> my hand here tries to paint,
> but see how it breaks forth in sweat; see how its fingers tremble.
> How can I paint my love?[134]

[130] Maurice Valency, *In Praise of Love* (New York, 1958), p. 18.

[131] 'Sweet mouth, that send'st a musky-rosed breath . . . ' in J. Betjeman and G. Taylor (eds), *English Love Poems* (London, 1957), p. 36.

[132] 'Song', ibid., p. 82.

[133] Valency, loc. cit.

[134] *Subhāṣitaratnakoṣa* 753 (Ingalls trans.):

> vāraṃ vāraṃ tirayati dṛśor udgamaṃ bāṣpa-pūras
> tat-saṃkalpopahita-jaḍima stambham abhyeti gātram;
> sadyaḥ-svidyann ayam aviratotkampa-lolāṅgulīkaḥ
> pāṇir lekhāvidhiṣu nitarāṃ vartate kiṃ karomi.

In the Rāmāyaṇa, the hero, Rāma, laments when he discovers that Sītā has been abducted, but Lakṣmaṇa and Sugrīva, tell him that he must not grieve, that his sorrow will weaken him and put his life in jeopardy; love's grief does not suit the Epic hero.

Kṛṣṇa, the Epic hero, the heroic lover of the *Purāṇas*, the killer of Madhu, Keśin, Mura, Kaṃsa, countless demons, became the 'romantic lover' in the *Gītagovinda*. The Bhagavat became human, deeply human, in love.

3

THE LOVER AND BELOVED

As Beloved dwells in Lover
Each in other did reside,
And that same love that unites Them
Did in both of Them abide.[1]

KṚṢṆA

His mouth is most sweet: yea, he is altogether lovely. This is my beloved. . . . I am my beloved's and my beloved is mine. . . .[2]

It has been usual to speak historically of three Krishnas, although each melts into the others. . . . There is Krishna the chief of the Yādavas, who served as Arjuna's charioteer in the Bhārata epic. There is Krishna the god incarnate, the instructor of Arjuna and through him of all mankind, who appears not in the old epic, but in that larger religious work, the *Mahābhārata*, into which it was expanded. Then there is Krishna of Gokula, the god brought up among the cowherds, the mischievous child, the endearing lover, the eternal paradox of flesh and spirit.[3]

Kṛṣṇa the hero, Kṛṣṇa the lover, Kṛṣṇa the lord, he embodies the heroic sentiment (*vīra-rasa*), the erotic sentiment (*śṛṅgāra-rasa*), and the peaceful sentiment (*śānta-rasa*), and exemplifies three modes of being: power, love and transcendence.

The three are brought together in a tight juxtaposition, a compact interlinking, in the second song of the *Gītagovinda* (I. 17-25), a series of vocative descriptions of Kṛṣṇa, a victory chant to Hari 'which causes happiness and joy':

[1] Saint John of the Cross, 'Romance I', trans. E. Allison Peers in *The Complete Works of Saint John of the Cross*, vol. II, 2nd edn (London, 1953), p. 433.

[2] Song of Songs 5: 16–6: 3.

[3] Ingalls, 'Foreword' to Singer (ed.), *Krishna: Myths, Rites and Attitudes*, p. v.

Kṛṣṇa the lover:

You lay upon the roundness of Kamalā's breasts, wore ear-rings, wore a fair forest-garland (17). . . . O delighter of men (19). . . . Your eye is a spotless lotus-petal (21). . . . You made ornaments for Janaka's daughter (22). . . . You are beautiful like a young rain-cloud . . . O *cakora* [bird] to the moon which is the face of Śrī (23). . . .

Kṛṣṇa the hero:

O queller of the venom-bearing serpent Kāliya . . . O sun to lotus which is the Yadu family (19). . . . O annihilator of Madhu, Mura and Naraka, O rider of Garuḍa (20). . . . you slew Dūṣaṇa, you destroyed the ten-necked-one in battle (22). . . . you supported [Mount] Mandara (23). . . .

Kṛṣṇa the god:

you break-up the world, O Mānasa gander of the sages (18). . . . O liberation from phenomenal-existence, O support of the mansions of the three worlds (21). . . . We bow down to your foot—know this is so! Cause prosperity among us who are bent down [in obeisance] (24). . . .

In one compact couplet he is both liberation from the world and has a lovely lotus-petal eye (*amala-kamala-dala-locana bhava-mocana e*) (21)—that is the essence of the conception of Kṛṣṇa—the particular, the delicate, evocative of the erotic sentiment, is immediately juxtaposed to the universal, beyond time and space, beyond all quality, all sentiment. The juxtaposition, the counterbalance, is itself the meaning; Kṛṣṇa embodies a metaphysical paradox; he is 'what IS and what is not'.[4] And through *bhakti* the devotee could experience the paradox, could be both in and beyond the world: 'the man who loves-and-worships Me . . . becomes fit [to share in] my own mode of being.'[5]

1 *The lover*

Sweet, sweet is the body of this Lord, sweet, sweet is the face, sweet; this gentle smile, fragrant as honey, oh, sweet, sweet, sweet, sweet.[6]

[4] Or 'Being and Not-Being' (*sad asat*), *Bhagavad-Gītā* IX. 19; XI. 37, etc.

[5] Ibid., XIII. 18 (Zaehner trans.): '*mad-bhakta etad . . . mad-bhāvāy' opapadyate.*'

[6] *Kṛṣṇakarṇāmṛta* of Līlāśuka Bilvamaṅgala I. 92:

madhuraṃ madhuraṃ vapur asya vibhor
madhuraṃ madhuraṃ vadanaṃ madhuram.
madhu-gandhi mṛdu-smitam etad aho
madhuraṃ madhuraṃ madhuraṃ madhuram.

Kṛṣṇa, the lover, the embodiment of erotic sentiment, emerged in the first few centuries A.D. in the folk tradition:

The clever *gopī*, going to the side of her friend, and pretending to praise her dance, kisses Kṛṣṇa in his reflection on her cheek.[7]

While you, Kṛṣṇa, remove with the breath of your mouth a particle of dust from Rādhikā, you take away the pride of your other beloveds.[8]

And by the twelfth century the loves of Kṛṣṇa, a theme developed in the *Harivaṃśa*, the *Viṣṇu Purāṇa* and the *Bhāgavata Purāṇa*, had become a conventional subject for *kāvya*:

> Victorious is Hari
> who, thinking it the black border of her garment,
> tries to wipe away
> the reflection of his face as dark as a raincloud
> from the golden globe of Rādhā's breast [i.e., bright as a touchstone with gold]
> whence being laughed at by his mistress,
> he drops his head in shame.[9]

Jayadeva expanded the theme into a 'canto-composition' (*sarga-bandha*), a *mahākāvya*, maintaining the conventions of the literary tradition, the song-forms of the folk tradition, and the sentiments of both traditions.

After the introductory eulogies of Kṛṣṇa in his various aspects, the god is presented in a completely erotic benediction—his chest is flooded with sweat from his love-making, 'may it fulfil your pleasure!' (I. 26). And then he is depicted 'delighting in ... the excitement of the embraces of many women' (I. 38),

[7] *Sattasaī* of Hāla 114 (Friedhelm Hardy trans.):
ṇaccaṇa salāhaṇaṇihe
ṇa pāsaparisaṃṭhiā ṇiuṇagovī;
sarigoviāṇa cuṃvai
kavolapaḍimāgaaṃ Kaṇhaṃ.

[8] Ibid., 89
muhamārueṇa taṃ Kaṇ-
ha! goraaṃ Rāhiāi avaṇeṃto;
eāṇa vallavīṇaṃ
aṇṇāṇa vi goraaṃ harasi.

[9] *Subhāṣitaratnakoṣa* 147 (Ingalls trans.):
kanaka-nikaṣa-svacche rādhā-payodhara-maṇḍale
nava-jala-dhara-śyāmām ātma-dyutiṃ prati-bimbitām
asita-sicaya-prānta-bhrāntyā muhur muhur utkṣipañ
jayati janita-vrīḍānamra-priyā-hasito hariḥ.

playing and dancing 'in the coquettish flock of artless women' (I. 39 ff.); the depiction is redolent of the *Viṣṇu* and *Bhāgavata Purāṇas*—Kṛṣṇa enchants and enraptures nameless cowherd-women one after another and all together, in frenzy, ecstasy and joy. When he has a favourite it is temporary:

He embraces one, kisses one, sexually-pleases some sexually-pleasing one, he sees yet another beauty more charming still on account of her smiles and he chases her. . . . [I. 45]

He is the very essence of love, the embodiment of love and loving, hungrily 'embraced by the beauties of Vraja of their own free-will, entirely, all-over-his-body' (I. 47) and that his love produces 'the joy of all-creatures' (in the same verse) suggests that the cowherd-women are those who immediately and carnally enact and enjoy in the mythopoetic sphere the sentiments and impulses which are generally felt in the sphere of everyday existence by 'all-creatures'. Kṛṣṇa loves all the women equally (II. 1), desires to kiss all their mouths (II. 4), 'a thousand cowherd girls were encircled by the shoots which are his very bristled arms' (II. 5), 'fickle Kṛṣṇa delights among the girls' (II. 10).

But then the beloved god is transformed into the human lover, the *Purāṇic* god is transfigured into the literary hero: Kṛṣṇa watches the coquetries of the cowherdesses, 'for a long time thinking within-himself, [and] his desire for them was dispelled' (II. 21); he then concentrated all of his love upon Rādhā 'as the chain binding him with desire for the world', [and he] abandoned 'the beauties of Vraja' (III. 1); Rādhā became Kṛṣṇa's 'one love', his 'one *rasa*' (XI. 24), his one and only passion and pleasure; 'there is no room for another' in his heart (X. 10). In European terms the 'heroic lover' had been transmuted into the 'romantic lover':

For the heroic lover no woman is unique. The hero's interest in the other sex is purely sporadic, and he is generally indifferent to the individual lady as such. He is able therefore to maintain a manly independence through the variety of his amatory experiences. What he desires, ideally, is pleasure without passion. . . . The romantic lover . . . is psychically destined for a single mistress. She alone completes him, and nobody else in the world, as he thinks, will fit his preconceived ideal of perfection.[10]

[10] Valency, op. cit., p. 30.

The transformation represents the evolution of *libido* into the *eros* of love and it reflects not only the literary influence, the imposition of the conventions, conceits and themes of court poetry onto the Kṛṣṇa legend, but also it reflects the Bengali–Orissan milieu in which the cult of the Goddess, the Mother, has always been significant—'the romantic hero assumes the posture of a suppliant, a child, one might say, at his mother's knee'.[11] Rādhā's attitude toward Kṛṣṇa is typical of the attitude of the cowherd-women in the *Purāṇas*; her joy and union with him and torment in separation from him, her descriptions of him, her longing—all this had been defined and explored in the *Purāṇic* depiction of the women of Vraja. But Kṛṣṇa's attitude toward Rādhā in the *Gītagovinda* is a completely new development in Kṛṣṇaism; Kṛṣṇa became even more humanized (and thereby more personal), subject to the pain of love which in prior literature he had inflicted but not himself felt; he became like the European *'fin aman'*, the *gentle*man who 'desired not to conquer and to dominate, but to serve and adore'.[12] He is repentant about his love-play with the flock of herdswomen, he feels guilty and despondent, he searches for Rādhā, laments and begs for forgiveness (III); he suffers, seems to die, falls ill, moans, longs for her (v. 2 ff.), undergoes all the symptoms of love-in-separation as defined in the literary texts:

At one moment he scatters sighs, then looks towards the heavens, then resorts to the grove humming, then gasps-for-breath, then prepares the bed, then looks around bewilderedly—O lovely-woman, your beloved is wearied by the suffering of love. [v. 16]

Kṛṣṇa is a suppliant, stammering before Rādhā: 'you are my life!' (x. 4). The vast power that supports the three worlds, Viṣṇu the Preserver, makes obeisance at Rādhā's feet—he massages her feet, paints them with lac, puts his head on them, worships them (x. 7, 8; XI. 2; XII. 3). Kṛṣṇa worships Rādhā's lotus-feet 'like a slave' (XI. 22); God incarnate declares himself to be Rādhā's servant:

Bring the elixir of nectar from your lower-lip, O passionate-angry woman! vivify me, *your slave*, as if I were dead; my mind is fixed upon you, my body is consumed by the fires of love-in-separation; I am

[11] Ibid. [12] Ibid., p. 26.

without the pleasure-of-play! For a moment now, follow Nārāyaṇa, [me, as I have] followed [you], O Rādhikā! [XII. 6]

And then the 'Primal God, the Primeval Person, the last prop-and-resting-place of the universe, the knower and what is to be known, . . . whose forms are infinite',[13] in complete subservience to Rādhā, his mind having been impaled by the arrows of her beauty (XII. 13), assumes the role of servant and sees to Rādhā's toilet:

'Put a pattern on my breasts, make a design on my cheeks, fasten a girdle on my hips, fix the mass of my braids with artless garlands, put rows of bracelets on my arms and jewelled anklets on my feet'—thus directed, the yellow-robed-one was pleased, and he did so. [XII. 25]

Kṛṣṇa's subordination to Rādhā reflects a coalescence of the Vaiṣṇava tradition with a Tantric śakti-cult tradition, a merger which foreshadowed the development of a Rādhā-theology, the growth of a Rādhā-cult which placed Rādhā above Kṛṣṇa as the Supreme Goddess, Devī. The *Tantratattva* declares that 'To worship Her, even the Lord of the Universe descended to earth . . . and became Her servant.'[14]

Kṛṣṇa, in the *Gītagovinda*, (not as he is hailed in the various benedictory verses, but) as he is described in relationship to Rādhā is the 'gentleman' of erotic *kāvya*, the hero (*nāyaka*) as a 'fundamental ingredient' in a literary work which has and which evokes the *śṛṅgāra-rasa*. The rhetoricians described the hero as he ought to be portrayed in a *kāvya*:

generous, accomplished, high-born, handsome, energetic and youthful in body, adroit, loved by the world, possessing splendour, intelligence and character, a leader.[15]

And as a lover the hero was subclassified into four types;

[13] *Bhagavad-Gītā* XI. 38 (Zaehner trans.):
'. . . *ādi-devaḥ puruṣaḥ purāṇas*
. . . *asya viśvasya paraṃ nidhānam*
vettā'si vedyaṃ ca . . . ananta-rūpa.'

[14] *Tantratattva*, a nineteenth-century Bengali treatise, trans. Sir John Woodroffe, *Principles of Tantra*, 2nd edn (Madras, 1955), p. 399.

[15] *Sāhityadarpaṇa* 64: '*tyāgī kṛtī kulīnaḥ suśrīko*
rūpa-yauvanotsāhī; dakṣo'nurakta-lokas
tejo-vaidagdhya-sīlavān netā.'

Jayadeva's Kṛṣṇa assumes each of the four roles at various stages in the poem:

(a) the *dhṛṣṭa nāyaka*—the faithless, bold, audacious lover who lets the evidence of his infidelity, the marks of love-making on his body, show:

Although he is guilty, he is not afraid; although he is threatened, he is not ashamed; although his crime is visible, he lies about it.[16]

And so Kṛṣṇa appears before Rādhā with red eyes, mascara on his lips, nail-marks on his body, lac on his chest, teeth-marks on his lower-lip (VIII. 2-6) and Rādhā is both tormented by jealousy and by the humiliation caused to her when her friends see that evidence of his love-making.

(b) the *śaṭha nāyaka*—is the false, sly, deceitful lover, the rogue, the cheat, who pretends to have affection for one woman while actually loving another. Rādhā asks her friend, 'if my pitiless false-lover [*śaṭha*] has not come, O messenger, why should you burn-with-sorrow?' She accuses Kṛṣṇa of making love to another cowherdess after pledging a tryst with her (VII; VIII).

(c) the *dakṣiṇa nāyaka*—is clever and has an equal passion for many women—this is the Kṛṣṇa of the *Purāṇa*s, the Kṛṣṇa of the first two cantos of the *Gītagovinda*, the Kṛṣṇa who is transformed into:

(d) the *anukūla nāyaka*—the faithful, kind, obliging lover, who has only one beloved, whose love is constant; the rhetoricians usually give Rāma as the primary example of this type of lover.

Jayadeva's depiction of Kṛṣṇa in each role anticipated the theological poetics of the Gosvāmins of Vṛndāvana, the followers of Caitanya:

The classification of the hero in orthodox poetics . . . as a lover into *Anukūla*, *Dakṣiṇa*, *Dhṛṣṭa* and *Śaṭha* is applied to Kṛṣṇa [by Jīva Gosvāmin]. . . . Even if some of these qualities are apparently inconsistent with each other they can reside without conflict in Kṛṣṇa as a deity.[17]

Feminine beauty is described in the rhetorical texts in physi-

[16] Ibid., 72: '*kṛtāgā api niḥśaṅkas tarjito'pi na lajjitaḥ; dṛṣṭa-doṣo'pi mithyāvāk . . .*'.

[17] De, *Vaiṣṇava Faith and Movement*, p. 137. Kṛṣṇa was said to have 'all qualities' and those attributes which were defects in other heroes become virtues in him.

cal terms; it is closely related to the woman's capacity for love, whereas masculine beauty is a matter of character, of the man's capacity for heroism and virtue; the beauty of the *nāyaka* is in 'compassion for his inferiors, emulation of his superiors, heroism and adroitness'.[18] Rādhā in the *Gītagovinda* is described completely conventionally, with the same adjectives as might be applied to any heroine, with the same features which all beautiful women in *kāvya* share. Kṛṣṇa, however, had an established iconography; the unique conventions for describing his beauty had been developed in the *Purāṇic* literature—his blue-black complexion, his yellow-robe, his forest-garland, his particular ornaments, his lotus-eyes, lotus-mouth, lotus-feet. That his ravishing beauty is described in detail is significant in the context of the *bhakti* cult, for *bhakta*s were to be won over by his beauty, converted not by being convinced of the truth of doctrines but by being enraptured by Kṛṣṇa's charms; the *bhakta* was to fall passionately in love with Kṛṣṇa:

As when a girl [*nāyikā*] sees an attractive man [*nāyaka*] and has a passionate desire for him—such should be the emotion with which one calls constantly for Kṛṣṇa. In this passion all thought of self disappears, together with the darkness of the mind.[19]

Beauty is the very essence of Kṛṣṇa; he was considered the ultimate 'fulfilment of loveliness':

It is admitted that there are thousands of eminent
  aestheticians,
and there are those thousands who have vowed to completely
  understand the subtleties of beauty.
We have no quarrel with these people, nor do we flatter
  you, O Lord,
but, truly, the fulfillment of loveliness is realized only
  in you.[20]

Kṛṣṇa is 'mind-stealing' with his smile (I. 49), a conventional

[18] *Daśarūpa* II. 16: '*nīce ghṛṇā'dhike spardhā* (*śobhāyāṃ*) *śaurya-dakṣate*.'

[19] The *Karacā* of Govinda-dāsa cited by Dimock, *Place of the Hidden Moon*, p. 212.

[20] *Kṛṣṇakarṇāmṛta* I. 100 (trans. Frances Wilson, op. cit.):
kāmaṃ santu sahasraśaḥ katipaye sārasya-dhaureyakāḥ;
kāmaṃ vā kamanīyatā-parimala-svārajya-baddha-vratāḥ;
nai'vai'tair vivadāmahe na ca vayaṃ deva priyaṃ brūmahe
yat satyaṃ ramaṇīyatā-pariṇatis tvayy eva pāraṃ gatā.

word for beauty, but one which has devotional ramifications when applied to Kṛṣṇa—his beauty inspires a passion in which 'all thought of self disappears, together with the darkness of the mind'. A particular mode of *bhakti*, a chief expression of devotion, was 'remembrance' (*smaraṇa*) and when Rādhā remembers Kṛṣṇa it is recollection of his beauty; his lips, his tremulous ear-ornaments, his hair, his robe, his arms, chest, hands, feet, cheeks, forehead-mark, side-long glances (II. 2-8). And after the enraptured catalogue of his charms, Jayadeva adds that his song has the 'beauty of the Enemy-of-Madhu, infatuating, so beautiful . . . suitable for the virtuous for remembrance of Hari' (II. 9). The impassioned activity of sexual recollection of the beloved's beauty is homologized with devotional recollection, equally impassioned, of the beauty of god. Rādhā again enumerates his graces as she imagines another cowherdess being sexually-pleased

> by the one whose eyes are like lotuses moving in the breeze. . . . whose lovely mouth is an opened lotus. . . . whose sweet speech is more gentle than the nectar-of-immortality. . . . whose feet and hands resemble lotuses on the bank. . . . who is beautiful like a mass of rain-clouds. . . . who in his shining robe resembles a touchstone of gold. . . . [VII. 31–36]

And then again the sacred and profane dimensions of love are homologized: 'Through this speech as sung by Jayadeva, may Hari too enter [your] heart!' (VII. 38). Again, it is the beauty of the god which is the basis of his appeal for the devotee, the basis and crux of his divinity. Yet again his beauty is described: the pearls on his chest are like the masses of foam floating on the Yamunā's dark waters; his yellow-robe is like a veil of pollen encircling a blue lotus; his eyes are like wagtail-birds at play; his ear-rings are like suns; his hair, with flowers in it, is like a cloud streaked with moon-beams; his forehead mark is like the moon; 'his body was beautiful on account of his ornaments radiant with a mass of light-beams from the multitude of his jewels . . .;' (XI. 24-30). And once again the erotic vision of the lover is juxtaposed to the devotional: 'placing him in your heart, bow to Hari . . .' (XI. 31).

Kṛṣṇa is like a blue-jewel (V. 20), an image which emphasizes the particular characteristic of his beauty—his dark radiance.

'Although dark,' Līlāśuka Bilvamaṅgala sang, 'you are a light in the deep darkness.' He is the resolution of all paradox; inconsistent qualities 'can reside without conflict in Kṛṣṇa as a deity':

> Although but a boy, you hold a mountain aloft
> with a finger tip.
> Although dark, you are a light in the deep darkness.
> Although immovable, you are drawn by the eyes of Rādhā.
> Although an adulterer, you destroy the inevitability of
> rebirth. How are you all this?[21]

For Rādhā, the literal 'deep darkness' is the night of the tryst, for the devotee it is the figurative 'deep darkness' of the present, degenerate age, the *Kali Yuga*. It is the heroic Kṛṣṇa who is the light in that night.

2 *The hero*

> In the early morning to quiet the evils and great pain
> of existence I call to mind
> Nārāyaṇa, who is the colour of black mascara, whose
> chariot is Garuḍa,
> who caused the release of the great elephant overcome by
> the crocodile,
> whose weapon is the discus, whose lotus eyes sparkle.[22]

Jayadeva combines the heroic sentiment with the erotic sentiment in his depiction of Kṛṣṇa: there is a benedictory verse in which Kṛṣṇa's arm is described lifting up Mount Govardhana as an umbrella over Gokula and being kissed by the cowherdwomen—they kiss it as if in gratitude for the heroic act, as if they are proud of the arm, but actually their kisses are an expression of their passionate love, 'their supreme joy' (IV. 23).

[21] *Kṛṣṇakarṇāmṛta* II. 73 (Wilson trans.):
bālo'pi śailoddharaṇāgra-pāṇir
nīlo'pi nīrandhra-tamaḥ-pradīpaḥ;
dhīro'pi rādhā-nayanāvabaddho
jāro'pi saṃsāra-haraḥ kutas tvam.

[22] Ibid., additional verse cited by Wilson, p. 228:
prātaḥ smarāmi bhava-pāpa-mahārti-śāntyai
nārāyaṇaṃ garuḍa-vāhanam añjanābham;
grāhābhibhūta-vara-vāraṇa-mukti-hetuṃ
cakrāyudhaṃ tarala-vārija-patra-netram.

The Govardhana myth[23] is used also in an erotic context in a poem attributed to Jayadeva:

> 'O artless-woman!'
> 'Lord, what you are saying?'
> 'My arm is bent from the weight of this mountain.'
> 'How, dear, can I give help?'
> 'O beautiful-one, exert the creeper which is your arm!'
> [the glances] which fell upon Rādhā's breasts which
> were exposed by the unsteady border-of-her-garment as
> it fell from her shoulder when she raised her arm,
> may the glances of the Enemy-of-Kaṃsa prevail![24]

Jayadeva uses Kṛṣṇa's dark complexion as the unifying feature in a cluster of metaphors—he is presented as a lover in relation to Rādhā and the other cowherd-women, a heroic warrior in relation to Kaṃsa[25] and the other demons, and the highest god in relation to the three worlds:

> [He is] the bee on the lotus which is Rādhā's artless face, the appropriate blue jewel for the adornment of the region which crowns the three worlds, the death of those [demons] whose descent is a burden to the world, spontaneously the beginning of a night of pleasure for the minds of the beauties of Vraja, a smoke-bannered-comet for the destruction of Kaṃsa—may he, the son-of-Devakī, protect you! [v. 20]

Similarly his flute delights the women who love him and the gods who are aided by him, and it is invoked to remove the misfortunes of the *rasika* (VIII. 11).

The heroic and erotic are counterbalanced in a description of Kṛṣṇa battling Kuvalayāpīḍa, the war-elephant of Kaṃsa (X. 16; see also XI. 35)—the lobes on the forehead of the ele-

[23] Indra, jealous of Kṛṣṇa, and angry with the cowherds for adoring him, caused storms over Gokula; Kṛṣṇa lifted up the mountain to shelter his devotees. See *Bhāgavata-Purāṇa* x. 25 and *Viṣṇu Purāṇa* v. 10.

[24] *Saduktikarṇāmṛta* I. 60.5 (Śarma edn):
mugdhe tanvi nātha kim āttha śikhari-prāg-bhāra-bhugno bhujaḥ
sāhāyyaṃ priya kiṃ bhajāmi subhage dor-vallim āyāsaya
ity ullāsita-bāhu-mūla-vicalac-cālāñcala-vyaktayo
rādhāyāḥ kucayor jayanti calitāḥ kaṃsa-dviṣo dṛṣṭayaḥ.

[25] Kaṃsa, the cousin of Devakī (Kṛṣṇa's mother), was a tyrannical ruler of Mathurā; in response to a prophecy that he would be killed by a son of Devakī, he tried to kill all her children. Kaṃsa made many efforts to slay Kṛṣṇa but was ultimately slain by him.

phant remind Kṛṣṇa of Rādhā's breasts and he begins to show the conventional involuntary signs (the *sattvika-bhāvas*) of love—sweating and closing his eyes; Kaṃsa mistakes these for the symptoms of fear and fatigue and cries out 'He is conquered!' That the symptoms of love and of losing a battle are the same emphasizes the link between sexuality and aggression, love and war, the procreative and the destructive act. The link is psychological—'The aggressive instincts,' Freud wrote, 'are never alone but always alloyed with the erotic ones.'[26] And it is seemingly universal—'Every lover,' Ovid declared, 'is a soldier.'[27] In the European traditions of the Middle Ages and Renaissance the knight was idealized as a lover and a soldier (and as a Christian)—each role was fulfilled in service, service to the lady and the lord (and as a Christian, in service to *the* Lady and *the* Lord). As sexual language was used to describe religious experience, and religious language to describe sexual experience, likewise, the language of war was used in the erotic context:

Do not believe that the battle of love is like other fights in which the clamour and fury of an appalling war prevail on either side; for love fights only by means of caresses and threatens only with tender words. Its arrows and blows are gifts and blessings. Its encounter is a most effective promise. Sighs make up its artillery. Its taking possession is an embrace. Its slaughter consists in giving one's life for the beloved.[28]

In the Indian traditions the link is implicit in the mythology of the love-god and in the frequent use of the expression 'the battle of love' (*rati-raṇa*, *rati-kalaha*, *rati-vimarda*, etc.). Vātsyāyana compared sexual intercourse with a battle (*kalaha*), explaining that it consists of quarrels (*vivāda*), prescribing striking (*prahaṇana*) to arouse and maintain passion,[29] and warning of dangers of these violent acts with anecdotes of lovers actually killed in the battle of love.[30] All erotic activity—embracing, kissing, hair-pulling, biting, scratching, and so forth—is described in violent terms in the textbooks for the sexual life and in *kāvya*. Rādhā chides Kṛṣṇa for his faithlessness:

[26] Freud, *New Introductory Lectures on Psychoanalysis*, p. 111.

[27] Ovid, The *Amores* I. ix. 1.

[28] *Ley de Amor* of Francisco de Ossuna, cited by de Rougemont, *Love in the Western World*, p. 259.

[29] *Kāmasūtra* II. vii. 1.

[30] Ibid., II. vii. 27–29.

Your body, which has lines of wounds from hard curved-nails [inflicted] in the battle of love, resembles a record of [her] victory in love-pleasure. . . . [VIII. 4]

Kṛṣṇa begs Rādhā to scratch, bite and bind him (X. 3, 11) and the go-between urges her to go to Kṛṣṇa in martial language:

. . . your body too is ready for the battle of love-pleasure; O fierce-passionate-woman, go to him noisily with the battle-drum-uproar of your jangling girdle. . . . [XI. 7]

It is in the battle of love that the woman can display the heroic sentiment, riding on top of her lover for victory over him, biting, scratching, squeezing him, pulling his hair and holding him captive in her arms (XII. 11, 12).

In a poem attributed to Jayadeva in the *Saduktikarṇāmṛta* the heroic and the erotic, the aggressive and the sexual, are coalesced in a series of puns—the poem is presumably an encomium of his patron (most likely King Lakṣmaṇasena of Bengal):

You make a sport of {letting-loose [women's] bodices / stirring-up the Colas}, you engage in the {pulling of [women's] hair / tormenting of the Kuntalas}, you have power over the {curve of [women's] girdles / hiding-place of the Kāñcis}, you make contact with the {bodies [of women] passionately [in love] / Aṅgas violently [in battle]} . . .[31]

Jayadeva uses the heroic epithets of Kṛṣṇa to indicate the link between the warrior and the lover: 'Make the noble Slayer-of-Keśin make-love to me passionately' (II. 11 ff.); the lotus face of the Killer-of-Madhu inspires love (I. 40); the Enemy-of-Mura delights in love-play (I. 38); the Enemy-of-Kaṃsa infatuates the cowherd-women with his flute (VIII. 11).

Kṛṣṇa fulfils his heroic role in his various incarnations or 'descents' (*avatāras*) (I. 5–16):

For whenever the law of righteousness withers away and lawlessness

[31] *Saduktikarṇāmṛta* III. 15.5:
tvaṃ colollola-līlāṃ kalayasi kuruṣe karṣaṇaṃ kuntalānāṃ
tvaṃ kāñci-nyañcanāya prabhavasi rabhasād aṅga-saṅgaṃ karoṣi.

arises, then do I generate Myself [on earth]. For the protection of the good, for the destruction of evil-doers, for the setting up of the law of righteousness I come into being age after age.[32]

The mythology of the incarnations of Viṣṇu, the earthly manifestations of the Lord, the *avatāra* scheme, which by the twelfth century had become a conventional subject for *kāvya*[33] and a theological doctrine which served to incorporate other cults and movements into Vaiṣṇavism,[34] is utilized by Jayadeva to sanctify the erotic—for Kṛṣṇa, the cowherd lover, is considered in the opening paean not as *one* of the incarnations of Viṣṇu (as he was generally considered), but as the very source of *all* the incarnations; he is not an *avatāra*, but the *avatārin*. Kṛṣṇa is the 'Lord of the world', the 'creator and supporter of the ten forms'; as in the *Bhāgavata Purāṇa* 'Kṛṣṇa is the Bhagavat himself'.[35] The sacralization of the erotic is achieved through an irony: the god, who in various ages generates himself 'for the destruction of evil-doers, for the setting up of the law of righteousness', generates himself as a lover for the sake of Rādhā; she keeps him in the world (III. 1); the Preserver, the Pervader, the god of infinite power, is subdued by love.

The doctrine of the incarnations served a devotional function by establishing various forms in which the Lord could be worshipped, and Jayadeva hails each form; but by presenting Kṛṣṇa primarily as the beloved lover he stresses love as the primary devotional activity.

[32] *Bhagavad-Gītā* IV. 7–8 (Zaehner trans.):
yadā yadā hi dharmasya glānir bhavati (Bhārata),
abhyutthānam adharmasya tadā'tmānaṃ sṛjāmy aham.
paritrāṇāya sādhūnāṃ vināśāya ca duṣkṛtām
dharma-saṃsthāpanārthāya saṃbhavāmi yuge yuge.

[33] There is an *avatāra* hymn in Māgha's *Śiśupālavadha* (XIV. 78–86); Kṣemendra based his *Daśāvatāra-caritra* on the scheme (some of the songs are remarkably similar to Jayadeva's, e.g. VIII. 173, cited by Sarkar, op. cit., p. 21); see also *Subhāṣitaratnakoṣa* VI.

[34] Ingalls, *Anthology of Sanskrit Court Poetry*, p. 95: 'The scheme of *avatāras* was more than a convenient method for brahmin theologians to strengthen their religion by attaching to it local cults of totem animals, culture heroes and non-Aryan gods. It opened a path by which Vaiṣṇava thought could reach one of its ultimate goals . . . where the incarnations of God are considered to be universal. Every man is Krishna if he but knew it and every woman Rādhā. Religious realization is but the discovery and practice of this truth.'

[35] *Bhāgavata Purāṇa* I. iii. 28: '. . . *kṛṣṇas . . . bhagavān svayam*.'

The consideration of Kṛṣṇa as the *avatārin* continued the doctrine of the *Bhāgavata Purāṇa* and foreshadowed the conception of the incarnations held by the followers of Caitanya. They conceived of the one Reality as having three phases: *Brahman*, *Paramātman*, *Bhagavat*. *Brahman* is the absolute, impersonal and without quality and as such cannot love nor be loved; the *Bhagavat* is the Lord, personal and with qualities, the Lord who loves and is loved; the *Paramātman* is a partial manifestation of *Bhagavat* which acts as an intermediary between the individual being and the *Bhagavat*; the *Paramātman* is the source of the *avatāras*. They held, furthermore, that the *Bhagavat* was the highest aspect of the three, incorporating the other two.

> As Kṛṣṇa, in the opinion of this school, is the Bhagavat himself, the Avatāras proceeding from the Paramātman stand relatively on a lower level, and never possess the perfection of the highest deity.[36]

Each of the *avatāras* comes into the world to perform a particular function, to rid the world of a particular evil: the Fish saves the *Vedas* from the great flood (I. 5), the Tortoise supports the world (I. 6), the Boar rescues the earth from the sea (I. 7), the Manlion kills the demon Hiraṇyakaśipu (I. 8), the Dwarf rescues the three worlds from the demon Bali (I. 9), Paraśu-rāma restores the priestly-caste to power (I. 10), Rāma kills Rāvaṇa (I. 11), Bala-rāma diverts the Yamunā with his plough (I. 12), Buddha censures the sacrifice of cattle (I. 13), Kalki will purge India of barbarian invaders (I. 14):

> Lifting up the *Vedas*, supporting the world, raising-up the globe, tearing the demon [Hiraṇyakaśipu] to pieces, outwitting Bali, destroying the warrior-caste, conquering Rāvaṇa, bearing a plough, extending compassion, deranging barbarians, creator of ten forms, homage to you, Kṛṣṇa. [I. 16]

The commentator Vanamālibhaṭṭa (the *Sañjīvanī*) glosses the hymn as an interfusion of the heroic and erotic aspects of Kṛṣṇa, an interlacing of references to his 'power-and-glory, good-fortune-and-success, etc.' (*pratāpa-saubhāgyādi*) and to the 'love-making which is the play of the Lord' (*bhagavat-līlā-vilāsa*). In

[36] S. K. De, 'The Doctrine of Avatāra (Incarnation) in Bengal Vaiṣṇavism', p. 150 in his *Bengal's Contribution to Sanskrit Literature and Studies in Bengal Vaisnavism*, reprint (Calcutta, 1960), pp. 143–53.

the erotic exegesis the *Veda* signifies the Kāmasūtra, the fish and tortoise forms indicate coital postures, the tusk of the boar and the nails of the man-lion refer to the biting and scratching in which the lover indulges; the heroic exploits of the anthropomorphous *avatāras* suggest erotic exploits.

The difference between the heroic and erotic exploits, modes, aspects, for the Gosvāmins of Vṛndāvana was a matter of motive: the heroic activity was purposeful, tasks performed *for* the world; the erotic activity was purposeless, the unmotivated 'play of the Lord'. And the relationship between Kṛṣṇa and Rādhā was the earthly expression of the eternal sport of the Supreme Deity with his *Śakti*, his own energy and power, related to him as the flame is to the fire.

3 *The lord*

> In the early morning I bow with heart, word, and head
> to the pair of lotus feet of the supreme soul,
> Nārāyaṇa, who saves from the ocean of hell,
> who is the chief object of brahmans engaged in
>   meditation.[37]

In the *Bhagavad-Gītā* Kṛṣṇa was identified with *brahman*, the highest godhead, and with the highest Self (*paramātman*), completely immanent, completely transcendent, the infinite wholly contained in the finite, the eternal wholly contained in the temporal. He revealed himself as both *brahman* and as a personal god, loving and accessible to those who return his love:

In the region of the heart of all contingent beings dwells the Lord. ... In Him alone seek refuge with all your being, all your love; and by his grace you will attain an eternal state, the highest peace. ... And now again give ear to this my highest Word, of all the most mysterious: 'I love you well.' Therefore will I tell you your salvation. Bear Me in mind, love Me and worship Me, sacrifice, prostrate yourself to Me: so will you come to Me, I promise you truly, for you are

[37] *Kṛṣṇakarṇāmṛta*, additional verse cited and trans. Wilson, op. cit., p. 228:
prātar namāmi manasā vacasā ca mūrdhnā
pādāravinda-yugalaṃ paramasya puṃsaḥ;
nārāyaṇasya narakārṇavatāraṇasya
pārāyaṇa-pravaṇa-vipra-parāyaṇasya.

dear to Me. Give up all things of law, turn to Me, your only refuge, [for] I will deliver you from all evils; have no care.[38]

Jayadeva is cognisant of Kṛṣṇa the lover as this Lord: he declares himself to be a servant of Kṛṣṇa (vii. 29) who goes for 'refuge at the feet of Hari' (vii. 10); the *Gītagovinda* is 'a meditation consecrated to Viṣṇu' (xii. 28), in sacred 'remembrance' of Him (xii. 24), a veneration of Him for the destruction of evil (ix. 11); 'bow to Hari' the poet sings (v. 15; xi. 31) that He who play*ed* in the historical Vṛndāvana, who play*s* in the eternal Vṛndāvana, will appear and play in the region of the heart (v. 6; xi. 31); Kṛṣṇa and songs about him dispel the evil of the present age of degeneration, the *Kali Yuga* (ii. 8; vii. 20); the 'Lord of the World' (*jagad-īsa*) is invoked to allay distress (ii. 21) and misfortune (viii. 11), to protect (v. 20; xii. 26, 27) and to give tranquillity (iii. 16), prosperity (iv. 23) and joy (vii. 42; xi. 36).

Mythologically Kṛṣṇa or the lord Bhagavān is described in the *Purāṇas* [e.g. the *Bhāgavata Purāṇa*] as occupying His throne in the transcendent Heaven (*Vaikuṇṭha*) in His resplendent robes, surrounded by his associates. . . . Since it [*Vaikuṇṭha*] is non-spatial and non-temporal, it is as true to say that God exists in *Vaikuṇṭha* as to say that He Himself is *Vaikuṇṭha*.[39]

Kṛṣṇa is called Vaikuṇṭha in the *Gītagovinda* (vi. title) and described in his yellow robe surrounded by a 'retinue of sages, men, spirits and gods' (ii. 7); he is also called Puruṣottama, the Highest Being, the Supreme Person, and this Supreme Being plays in love with Rādhā on the banks of the Yamunā: 'May the actions of Puruṣottama's hands [upon the breasts of Rādhā] give much joy and success!' (xii. 32).

[38] *Bhagavad-Gītā* xviii. 61–66 (Zaehner trans.):
īśvaraḥ sarva-bhūtānāṃ hṛd-deśe ('rjuna) tiṣṭhati
. . . .
tam eva śaraṇaṃ gaccha sarva-bhāvena (Bhārata);
tat-prasādāt parāṃ sāntiṃ sthānaṃ prāpsyasi śāśvatam.
. . . .
sarva-guhyatamaṃ bhūyaḥ śṛṇu me paramaṃ vacaḥ;
iṣṭo'si me dṛḍham iti tato vakṣyāmi te hitam.
man-manā bhava mad-bhakto mad-yājī māṃ namas-kuru;
mām evaiṣyasi satyaṃ te pratijāne priyo'si me.
sarva-dharmān parityajya mām ekaṃ śaraṇaṃ vraja;
ahaṃ tvā sarva-pāpebhyo mokṣayiṣyāmi mā śucaḥ.

[39] S. N. Das Gupta, *History of Indian Philosophy*, vol. iv, p. 15.

Jayadeva calls Kṛṣṇa an 'ornament of the sun, you break-up the world, O Mānasa *haṃsa* of the sages!' (I. 18). The *haṃsa* bird is a wild gander that migrates to Lake Mānasa, a pilgrimage place for sages, but it is also the Supreme Spirit (Kumbha gives the gloss '*param-brahman*') in the minds of the sages; he is the goal of meditation. In the *Śvetāśvatara Upaniṣad* the *haṃsa* is considered a symbol of the embodied soul that abides in the centre of the world[40] and in the *Maitrī Upaniṣad*:

A golden bird abides in the heart and in the sun; it is a diver bird, a *haṃsa*, the best of fiery energies—let us worship him in this fire.[41]

In the *Matsya Purāṇa* the specific identification of the *haṃsa* and Viṣṇu is made; Nārāyaṇa, at the time of the flood, declares, 'I who am truly the Lord . . . am called the *haṃsa* while I roam in the world of blood in the course of time.'[42]

The macrocosmic gander, the divine self in the body of the universe, manifests itself through a song. The melody of inhaling and exhaling, which the Indian yogī hears when he controls through exercises the rhythm of his breath, is regarded as a manifestation of the 'inner gander'. The inhalation is said to make the sound 'ham', the exhalation 'sa'. Thus, by constantly humming its own name, 'haṃsa, haṃsa', the inner presence reveals itself to the yogi-initiate. . . . The song of the inner gander has a final secret to disclose. 'Haṃsa, haṃsa' it sings, but at the same time 'so'ham, so'ham' . . . 'This am I'. I, the human individual, of limited consciousness, steeped in delusion, spellbound by Māyā, actually and fundamentally am This, or He, namely, the Ātman, the Self, the Highest Being, of unlimited consciousness and existence.[43]

The symbol is used to link the sacred and the profane dimensions: Kṛṣṇa delighting upon Rādhā's breasts is like the *haṃsa* delighting amidst the lotuses on Lake Mānasa and Kṛṣṇa is in Rādhā's heart as the gander is in the heart or mind of the contemplative, the sage, the devotee (XI. 36).

[40] *Śvetāśvatara Upaniṣad* III. 18 and VI. 15.

[41] *Maitrī Upaniṣad* VI. 36:
hiraṇya-varṇaḥ śakuno hṛdyāditye pratiṣṭhitaḥ
madgur haṃsas tejo-vṛṣaḥ so'sminn agnau yajāmahe.

[42] *Matsya Purāṇa* 167.67: '*yo'ham eva . . . prabhur api haṃsa-saṃjñito'sṛj-jagad viharati kāla-paryaye.*'

[43] Heinrich Zimmer, *Myths and Symbols in Indian Art and Civilization*, Harper Torchbook edn (New York, 1962), pp. 49–50.

There is a purely devotional song attributed to Jayadeva in the *Granth Sāhib* (or *Ādi-Granth*), the sacred book of the Sikhs:[44]

Gujarī Rāga:

The highest Primal Spirit is incomparable—He enjoys the conditions of Being [Awareness and Bliss]; He is the highest wonder, beyond matter, incomprehensible and omnipresent!

Speak only the beautiful name of Rāma which consists of the immortal truth! By remembrance of Him fear of birth, old-age and death does not consume you!

You wish the defeat of Yama [death] and the others! O Glory! So be it! Do good deeds! He is past, present and future equally, imperishable, the ultimate, pure!

Speak only the beautiful name. . . .

Abandon looking at another's house [his wife and property] with covetousness and so forth; [abandon] all behavior which is not prescribed, bad action and bad mental disposition! Seek the refuge of the Discus-bearer [Viṣṇu]!

Speak only the beautiful name. . . .

With heart, deed and word be devoted to Hari alone! What's the use of *yoga*, sacrifice, alms-giving, asceticism?

Speak only the beautiful name. . . .

O Man! repeat [the prayer], 'Govinda, Govinda'—the phrase which is all perfection [*or*: His foot is all perfection]; Jayadeva has gone to the one who is clearly omnipresent in the past and present!

Speak only the beautiful name. . . .[45]

[44] The book, compiled in the sixteenth century by Guru Arjun in the Punjab, contains two songs attributed to Jayadeva and they are considered by the Sikhs to be by the Jayadeva who wrote the *Gītagovinda*. He is the oldest poet represented in the book and presumably earned his place there by his popular reputation as an ardent devotee.

[45] The text of the song along with a Sanskrit *chāyā* is given by Chattarji, op. cit., p. 192. He comments on the language of the poem: 'The poem is in Sanskrit corrupted by scribes who read it in a vernacular Eastern Indian pronunciation, with a number of Apabhraṃśa and vernacular forms: to start with it may have been wholly in Apabhraṃśa, and then badly Sanskritized, with a vernacular Bengali or Eastern Indian pronunciation showing through the spelling which was further

Although the authenticity of the verse is highly questionable, its ascription to Jayadeva recognizes its affinity to the songs of the *Gītagovinda* both formally (indicated *rāga*, rhyme scheme, moric metres, signature line) and in the sentiments expressed: utterance of the name, remembrance and the seeking of refuge in Hari, the power of devotion over '*yoga*, sacrifice, alms-giving, asceticism'.

Jayadeva's conception of Kṛṣṇa as at once the lover, hero and lord, at once Lakṣmī-pati, the Enemy-of-Madhu and Puruṣottama, at once playing in love with cowherd-women, battling with demons, and liberating devotees from phenomenal-existence, a conception which had been expressed in legend and discourse in the *Bhāgavata Purāṇa*, became elaborated in the theology of the followers of Caitanya. They developed a soteriological system based on devotion to the cowherd lover as the perfect manifestation of the Lord, the one truth, the one reality, the supreme joy.

> Bhagavān, or Hari, is the name given to the Supreme [by the followers of Caitanya]. He is infinite in nature, power and attributes. The creative, destructive and sustaining aspects which appear in Hindu theology as Brahmā, Śiva and Vishṇu, are manifestations of his nature. All the forms in which the Supreme has been conceived of and worshipped in Hindu thought are included in Bhagavān. He is the source of infinite forms. Of all these forms, that of Kṛishṇa is the most perfect. He is deity in his most entrancing aspect; the Supreme at his best. . . . Nothing can be conceived of the beauty and blissfulness of the

---

modified in the Gurmukhī script of the *Granth*.'

Srī Jaidewajīu-kā Pāda (*Rāga Gujarī*)

Paramādi purukha manopimaṃ sati ādi bhāwa-rataṃ
paramadbhutaṃ parakritiparaṃ jadi cinti saraba-gataṃ.
Rahāū-
Kewala Rāma-nāma manoramaṃ badi amrita-tata-maïaṃ;
na danoti jasamaraṇena janama-jarādhi-maraṇa-bhaïaṃ.
ichasi Jamādi-parābhawaṃ jasu swasti sukriti-kritaṃ;
bhawa-bhūta-bhāwa samabyiaṃ paramaṃ prasannamidaṃ.
lobhādi-drisaṭi paragrihaṃ jadi bidhi ācaraṇaṃ;
taji sakala duhakrita duramatī bhaju Cakradhara-saraṇaṃ.
Hari-bhagata nija niha kewalā rida karamaṇā bacasā;
jogena kiṃ jāgena kiṃ dānena kim tapasā.
Gobinda Gobindeti japi nara sakala-sidhi-padaṃ;
Jaidewa āiu tasa saphuṭaṃ bhawa-bhūta-saraba-gataṃ.

My translation follows Chattarji's *chāyā* and the English version of M. A. Macauliffe, *The Sikh Religion* (Oxford, 1909), vol. vi, pp. 15–16.

Eternal Reality that transcends Śrī Kṛishṇa. . . . He seeks the loving devotion of his worshippers, and graciously gives himself to them. He is himself the essence of love, the home of all blissfulness and delight. His supreme delight is in love. Only by love and adoration can he be attained.[46]

The medieval Vaiṣṇavas of Bengal elaborated a Kṛṣṇaite metaphysic, a description of reality based upon the idea of Kṛṣṇa's powers (*śaktis*). 'As a beam of light is part of the sun, or a flame part of the fire, so his *śaktis* are of the nature of [the Bhagavat].'[47] These infinite powers were held to be of three basic types:

(a) the *jīva-śakti*—the source of the individual soul (*jīva*), the power by which the individual soul relates to the Lord. The Bhagavat is a fire and individual souls are like infinite sparks;[48] according to Jīva Gosvāmin, 'as Bhagavat is the ground of the *jīva-śakti* the *jīva* is a part, but an infinitesimal part of Bhagavat'.[49] The relationship between the *jīva* and the Bhagavat is *bhedābheda*, they are the same and yet different, a dualistic monism which is inscrutable (*acintya*), a mystery.

(b) the *māyā-śakti*—the power of illusion, the source of the creation of the empirical world. It is external and extrinsic to the Bhagavat. He controls it and the *jīvas* are subject to it until they free themselves from the influence of *māyā* through devotion.

(c) the *cit-* or *svarūpa-śakti*—the internal, intrinsic power of the Bhagavat, inseparable from him, the power by which his essential nature is manifested. And of this power the highest phase was called the *hlādinī-śakti*, the power of joy, the Bhagavat's 'infinite capacity for imparting [and enjoying] ravishing sweetness':[50]

Hlādinī is so named because of giving delight to Kṛṣṇa, who tastes delight through that power. Kṛṣṇa himself is delight and tastes delight. *Hlādinī* is the cause of the *Bhakta*'s delight, the essence of *Hlādinī* is called *prema* (love).[51]

[46] Melville T. Kennedy, *The Caitanya Movement* (Calcutta, 1925), pp. 92–3.

[47] *Caitanya-Caritāmṛta, Madhya*. 20: 102, cited and trans. by Dimock, *Place of the Hidden Moon*, p. 127.

[48] Ibid., *Ādi*. 7.111 cited by A. K. Majumdar, *Caitanya: His Life and Doctrine* (Bombay, 1969), p. 278.

[49] *Paramātmā-Sandarbha* cited by Majumdar, loc. cit.

[50] See above, p. 24.

[51] *Caitanya-Caritāmṛta, Madhya*. 8 (cited by Kennedy, op. cit., p. 93).

The power of joy was identified:

Rādhā is the modification of Kṛṣṇa's love. Her name is the very essence of the delight-giving power [the *hlādinī-śakti*]. *Hlādinī* makes Kṛṣṇa taste delight. Through *hlādinī* the *bhaktas* are nursed.[52]

Rādhā is the *hlādinī-śakti* of Kṛṣṇa, the means by which he tastes and imparts joy, always one with Kṛṣṇa and yet always separate.

RADHA

> Thou hast ravished my heart. . . . How fair is thy love . . . how much better is thy love than wine and the smell of thine ointments than all spices! Thy lips . . . drop as the honeycomb. . . .[53]

The Lady (*nāyikā*) was classified in the erotic texts according to her physical characteristics, the dimensions of her vulva, her temperament, her age, disposition, and nationality.[54] And she was classified in the rhetorical texts according to her marital status, her experience in love, and her relationship with her lover.[55] Each category was subclassified again and again; the categories perpetuated ideals, the classifications promulgated literary conventions. The woman, as she was depicted in *kāvya*, was 'glorified by joining her to the appropriate type of heroine. In that guise she became eternal. Everything about her became

[52] Ibid., *Ādi*. 4 (cited by idem, pp. 93–4).

[53] Song of Songs 4: 9–11.

[54] Physical characteristics: lotus-lady (*padminī*), picture-lady (*citriṇī*), shell-lady (*śaṅkhinī*), elephant-lady (*hastinī*). Vulva dimensions: deer-lady (*mṛgī*), horse-lady (*vaḍavā*), elephant-lady (*hastinī*). Temperament: phlegmatic (*kapha-prāya*), bilious (*pittalā*), flatulent (*vātalā*). Age: girl (*bālā*), young-woman (*taruṇī*), mature-woman (*prauḍhā*), old-woman (*vṛddhā*). Disposition: divine disposition (*deva-sattvā*), human disposition (*nara-sattvā*), serpentine disposition (*nāga-sattvā*), nymphish disposition (*yakṣa-sattvā*), Gandharva disposition (*gandharva-sattvā*), the disposition of various animals, the monkey (*vānara*), crow (*kāka*), or ass (*khara*). Following the *Kāmasūtra*, *Anaṅgaraṅga*, *Ratirahasya*, and other texts as collated in Schmidt, op. cit., pp. 205–56.

[55] Marital status: another man's woman (*parakīyā*) or one's own (*svakīyā*). Experience in love: innocent (*mugdhā*), experienced (*pragalbhā*), average (*madhyā*). Relationship to her lover: having him in subjection to her (*svādhīnabhartṛkā*), or being separated from him (*virahiṇī* or *viyoginī*). See Schmidt, pp. 256–338.

lovable . . .'.[56] But the glorification also meant abstraction; becoming 'eternal' also meant becoming impersonal, stereotypical. Classical Sanskrit poetry 'reflects rather the ideals of men than the men themselves'.[57] And Sanskrit love poetry presents love idealized; an ideal hero and an ideal heroine love each other ideally, suffering the ideal of separation, enjoying the ideal of union. In the erotic and rhetorical texts and in the erotic literature itself love was formalized and conventionalized into that ideal. The same thing happened in Europe: in the poetry of of the *Stilnovisti* of Florence the Lady was transformed from flesh and blood into ethereal abstraction and angelic ideality:

> The Lady as an individual lost her significance for the lover; she merely symbolized on the material plane the ideal beauty which was the true object of his desire.[58]

All the women of *kāvya* (like all the Ladies of the *canzone*) seem identically beautiful—all have the same hyperbolically described charms, the same personalities, all fulfil the same ideals. The poet's response was not so much to nature as to art itself, not so much to a woman as to the projection of an aesthetic prototype of Woman.

> The over-idealized woman invariably evokes a religious sense of awe to which she is sacrificed as the fatal object. Being love's martyr and scapegoat, she is ultimately depersonalized; only the ideal espouses her.[59]

In Europe the idealized Lady became the Madonna, Sophia, Beauty; Beatrice, the girl, became Beatrice, the symbol. And within the Indian traditions Rādhā, cast in the idealized mould of literary conventions, also came to 'evoke a religious sense of awe'; she became a symbol of Woman, of the devotee, of the *hlādinī-śakti*, of the female principle, of the Mother, the Goddess.

[56] Ingalls, 'Sanskrit Poetry and Poetics' (an introduction to his *Anthology of Sanskrit Court Poetry*), p. 26.

[57] Ibid., p. 25.

[58] Valency, op. cit., p. 227. In this the lover fulfilled the Platonic ideal in his quest for the Beautiful. This is in contrast to the Indian traditions: '. . . the Indians never developed a Platonic division of the universe into beautiful and non-beautiful. Sanskrit . . . has no word for spiritual beauty. . . . Beauty is conceived by the Sanskrit poet far more subjectively than in the west' (Ingalls, 'Words for Beauty . . .', pp. 106–7).

[59] Edward Honig, *Dark Conceit* (London, 1959), p. 34.

Within the profane dimensions of love Rādhā was the ideal, beauty, the ideal loving-beloved; and for the medieval Vaiṣṇavas that ideality persisted and expanded in the sacred dimensions of love.

The ideal of feminine beauty[60] had been described in the Epics, in the *Mahābhārata*:

> Draupadī has black curling hair . . . her face is like the lotus flower . . . . her thighs are firm and hard . . . . her brows and eyes are round-arched, red as the bimba-fruit are her lips . . . her face is like the full-moon. . . .[61]

and in the *Rāmāyaṇa*:

> . . . a beautiful woman was expected to be delicate (*sukumāra*); she is slim by nature (*svabhāvatanu*) and has well-formed limbs. . . . Hair had to be black (*nīla* or *asita*), long and curly, while the face was round and charming like a lotus or the full-moon . . . . the breasts . . . were expected to be round (*vṛtta*) and full grown (*pīna*) . . . and comparable to 'the fruit of the palm' . . . the hip and loins were considered beautiful if they were large . . . . the thighs were to be round and full like the trunk of an elephant. . . .[62]

This ideal became codified in the erotic texts as the 'lotus-lady' (*padminī*)—she is fragrant and lovely, as delicate as a lotus-blossom; her eyes, hands and feet are also lotuses; her nose is like a sesame-flower and her lips are like the *bimba*-fruit; her face is like the moon and her breasts are heavy, swelling, like coconuts; her arms are lotus-fibres or creepers; she smells like a lotus and her complexion is golden; her hips are large and her thighs are said to surpass the plantain. Personality traits are linked with physical traits—the lotus-lady is modest, loving and sweet though prone to pique and jealousy.[63]

The ideal is sustained in Rādhā—she is the slender loving-woman with limbs as delicate as spring blossoms (I. 29), lotus-feet

[60] Heinrich Zimmer suggests that the ideal extends back to the pre-Aryan, Indus Valley civilization—clay figurines, presumed to represent a Mother Goddess, excavated at Harappa and Mohenjo-Daro, show the same large breasts and hips, the small waist, the same ornamentation, that typify Indian sculpture of women at least until the Muslim conquest of India. See his 'Indian Ideals of Beauty', *The Art of Indian Asia* (New York, 1955), pp. 68–158.

[61] Meyer, op. cit., pp. 430–1.

[62] Ananda Guruge, *The Society of the Rāmāyaṇa* (Maharagama, Ceylon, 1960), p. 185.

[63] A collation of the Sanskrit texts given by Schmidt, op. cit., pp. 220–4.

(XII. 2; etc.), golden complexion (V. 12); her thighs are like elephant trunks (XI. 5) or the plantain (X. 15) and her breasts are round, swelling, heavy (IX. 3; etc.) usually like pitchers (XII. 5; etc.); her face is like a lotus (III. 5; etc.) or the round, resplendent, nectarous moon (III. 16; etc.); her lower-lip is the sweet red *bimba* (III. 14; etc.) and her hair is dark, curly, massive (III. 14; etc.), her brows are curved like serpents (X. 11) or most often like creepers or like the bow of the love-god over lotus-eyes or deer-eyes.

A woman's loveliness (*kānti*), Dhanaṃjaya says, is the 'lustre supplied by love (*manmathāvāpitacchāya*)'.[64] Feminine beauty had its origin and fulfilment in love. Her love made her beautiful and her beauty made her loved.

The ideals were made by asserting the aesthetic reality over the biological reality:

> Truly the deer-marked-moon is not like her face, nor are her eyes like a pair of lotuses, neither is her slender form made of gold—nevertheless even one knowing the truth, his mind thus misled by the poets, having considered that the body of the deer-eyed women consists of skin, flesh and bone, serves-and-worships it.[65]

Ideals are made of poets' lies. Art did not reflect or express the world, nor was it meant to—it reflected and expressed itself, its own formalized ideals; its own values.

1 *Another man's woman*

In the twelfth century in Provence, Marie the Comtesse of Champagne proclaimed that love could not exist in marriage:

[64] *Daśarūpa* II. 54.

[65] Bhartṛhari 108:

> no satyena mṛgāṅka eṣa vadanī-bhūto na cendīvara-
> dvandvaṃ locanatāṃ gataṃ na kanakair apy aṅga-yaṣṭiḥ kṛtā;
> kiṃ tv evaṃ kavibhiḥ pratārita-manās tattvaṃ vijānann api
> tvan-māṃsāsthi-mayaṃ vapur mṛga-dṛśaṃ matvā janaḥ sevate.

And the same poet (vs. 159) can be really vituperous in making the same point:

> The bulbs of flesh which are her breasts are compared to golden pitchers; her face, that receptacle of spit, is compared to the hare-marked-moon; putrid with dripping urine are the thighs which [are said to] rival elephant trunks. Oh, this disgraceful form is given importance by splendid poets!

> stanau māṃsa-granthī kanaka-kalaśāv ity upamitau
> mukhaṃ śleṣmāgāraṃ tad api ca śaśāṅkena tulitam;
> sravan-mūtra-klinnaṃ karivara-kara-spardhi jaghanam
> aho nindyaṃ rūpaṃ kavi-jana-viśeṣair guru kṛtam.

We declare and we hold as firmly established that love cannot exert its powers between two people who are married to each other. For lovers give each other everything freely under no compulsion of necessity, but married people are in duty bound to give in to each other's desires and deny themselves to each other in nothing.[66]

Marriage was accepted as a social ideal, socially useful, while love was celebrated as a personal ideal (*the* aesthetic ideal), personally and emotionally elevating. The social and the personal, the ordered and the passionate, the dutiful and the free, the outer and the inner, were in continual conflict.

A good marriage—if such a thing exists—rejects the company and conditions of love. . . . Marriage has for its portion utility, justice, honor and constancy. . . . Love is founded upon pleasure only, and in truth its pleasure is more exciting, livelier and sharper, a pleasure inflamed by difficulty; it requires stinging and burning. It is no longer love when it is without arrows and without fire.[67]

Passion, 'the arrows and the fire', defended itself against 'utility, justice, honour', the scales and the tablets, by becoming a separate world of value:

. . . in its wilful disobedience of social and religious law, adulterous love states or implies as its rationale and justification that it has its own society and its own religion and its own laws, all of which it holds superior to any other. . . . Adultery . . . is a way of making the love affair of a man and a woman into an independent world of value, a competitor of that other, established world.[68]

Marriage

In India, as in feudal Europe, marriage was founded on the ideal of duty (*dharma*), upon social utility and 'any idealization of sexual love in a society where marriage is purely utilitarian, must begin by being an idealization of adultery'.[69] Marriage had nothing to do with love.

We may hold the view which prevails throughout the Epic to be the usual Indian one: the daughter shall live in complete chastity and

[66] Quoted by Andreas Capellanus, *Tractatus de Amore*, I. vi. 7 (trans. John J. Parry [New York, 1949]). Andreas adds: 'Love cannot exist without jealousy. Jealousy between lovers is commended by every man who is experienced in love, while between husband and wife, it is condemned throughout the world' (loc. cit.).

[67] Michel de Montaigne, *Essais* III. 5, cited by Valency, op. cit., p. 81.

[68] Jonathan Saville, *The Medieval Erotic Alba* (New York, 1972), p. 40.

[69] Lewis, op. cit., p. 13.

implicit obedience towards her father, mother, and other kinsfolk and await from them her husband.[70]

And then she shall live in complete obedience to that husband and provide him with a happy after-life by giving him progeny. Marriage meant the fulfilment of *dharma* in the performance of household sacrifices.

It should be observed that from remote antiquity Indian opinion never made a confusion between love and marriage; the god of love is conceived as different from the deities who preside over marriage and fertility. . . . a girl [very often pre-pubertal] left her father's home to enter the home, not of her husband, but of her father-in-law, and the husband is often merely one of the factors of the bigger joint-family. . . . Wedded life was, of course, highly prized for its comfort and security, but ordinary marriages in the regular form were generally prompted and arranged by motives of convenience. . . .[71]

Of the various types of marriage described in the ordinances for social life,[72] love plays an initial part only in the *Gāndharva* marriage, 'marriage by mutual consent'. But this form was not considered a 'religious' marriage, not appropriate for the priestly caste and not beyond annulment.

[The *Gāndharva* marriage] which might be solemnized merely by plighting troth . . . often clandestine . . . . which might often amount to no more than a liaison . . . . forms the basis of many romantic stories, and has given rise to one of the stock figures of later poetic convention—the *abhisārikā*, the girl who secretly leaves her father's home by night to meet her lover at the appointed trysting place.[73]

It was not only the father's house that was left in secret; the

[70] Meyer, op. cit., p. 54.

[71] S. K. De, 'Ancient Indian Erotics' in *Ancient Indian Erotics and Erotic Literature* (Calcutta, 1959), pp. 100–1.

[72] The *Brahmā* marriage—a dowered girl given by her father to a man of the same caste in a religious marriage ceremony; the *Daiva*—the girl given to a sacrificial priest by her father in exchange for services rendered; the *Ārṣa*—a cow and a bull are given by the father in place of the dowry; *Prājāpatya*—the father gives the girl without a dowry; the *Gāndharva* marriage; the *Āsura*—the husband purchases the bride; *Rākṣasa*—the husband takes the girl by force; *Paiśāca*—the man rapes the girl while she is drunk or asleep. Only the first four were acceptable for the priestly class.

[73] Basham, op. cit., p. 168.

adulterous nocturnal tryst was as conventional in *kāvya* (see v. 18) as in European literature:

> My husband is no easy fool,
> the moon is bright, the way is mire
> and people love a scandal;
> yet it is hard to break a lover's promise. . . .[74]

The *abhisārikā* moves in secrecy and darkness (xi. 11, 12); she wears a dark cloak and silences her ornaments; 'Abandon the noisy, capricious anklet . . . go, O friend, to the dense, dark grove; wear a dark-blue cloak . . .' (v. 11). The convention of the woman going to the man reflects perhaps the cosmological notion of the female as the active principle and the social precept that the man was free from the moral stain of adultery if the woman came to him;[75] even the *yogin* was permitted to break his vow of chastity if the woman made the advances:

> . . . it is said in *Śruti*, '*talpāgatāṃ na pariharet*' (she who comes to your bed is not to be refused), for the rule of chastity which is binding to him [the *yogin*] yields to such an advance on the part of a woman.[76]

The rhetorical texts classified the heroine in terms of her relationship with the hero as one's-own-woman (*svakīyā*), another-man's-woman (*parakīyā*) and every-man's-woman (*sādhāraṇa-strī*), that is, the prostitute (who was not to be depicted except in farces).[77] The *parakīyā nāyikā* was subclassified further into another-man's-wife (*paroḍhā*) and the maiden (*kanyakā*) who is another-man's daughter.

The question as to whether Rādhā was *svakīyā* or *parakīyā* and if *parakīyā*, whether she was *paroḍhā* or *kanyakā*, became an important doctrinal issue to the medieval Vaiṣṇava theologians, an issue which ultimately separated the orthodox from the

[74] *Subhāṣitaratnakoṣa* 830 (Ingalls trans.):
patir durvañco'yaṃ vidhur amalino vartma viṣamaṃ
janaś chidrānveṣī praṇayi-vacanaṃ duḥpariharaṃ.

[75] *Nārada Smṛti* xii. 60 cited by Meyer, op. cit., p. 335.

[76] Sir John Woodroffe, *Introduction to Tantra Śāstra*, 5th edn (Madras, 1969), p. 34. The quote is somewhat unusual. Richard Gombrich (personal communication) has commented upon it: 'I expect "*talpāgatāṃ*" refers to his wife. Moreover, I should think this *yogin* is no more than a *vāna-prastha* [a forest-hermit].'

[77] *Daśarūpa* ii. 24; *Sāhityadarpaṇa* 96; Schmidt, op. cit., pp. 256–8.

Sahajiyā Vaiṣṇavas[78] and marked a distinction between the Kṛṣṇa-*bhakti* poets writing in Hindi and those writing in Bengali.[79] If Kṛṣṇa was an adulterer, those holding the *svakīyā* position maintained, he transgressed *dharma* and that he would not, could not, do. But those holding the *parakīyā* position maintained that he was higher than *dharma*, that *bhakti* transcends all rules and precepts, that *bhakti* must be passionate and that the *parakīyā* relationship creates a greater passion than the *svakīyā* one:

If there is no *parakīyā* there can be no birth of *bhāva* [emotion, e.g. Kṛṣṇa-*rati*]. It is in fear of separation that grief [*ārti*] and passionate longing [*anurāga*] grow. . . .[80]

In a *svakīyā* relationship there is no fear of separation; that is why there is no birth of *bhāva* in it. *Anurāga* manifests itself in extra-marital [*aupapatya*] love; that was the cause of the supreme enjoyment of *rasa* in Vṛndāvana.[81]

Adulterous, secret, love was idealized on the same grounds as in the European traditions. In marriage the goal was necessarily domestic accord; devotion was a duty; sex was for procreation; love was possession. In the illicit love-affair the goal was passion; devotion was freely and even dangerously given; sex was for pleasure; love was longing. Because marriage 'is sanctioned by the *Vedas*', according to the *Ratnasāra*, 'one takes the *svakīyā* relationship for granted'.[82]

She who never encounters rapture outside marriage . . . A girl who in unaware of the happiness of love with at least five men is as unholy as an evil spirit. . . . A woman of one husband knows nothing of love.[83]

[78] A formal debate to decide whether Rādhā was *svakīyā* or *parakīyā* was held in 1717 at the Court of Nabāk Jafārā Khāna and those holding the *parakīyā* position won (Dimock, *Place of the Hidden Moon*, pp. 201–15).

[79] 'Jayadeva's influence on Bengali Vaiṣṇavism was great and it also extended to the Hindi poets of the Vallabha sect, but there is an important difference between Bengali Vaiṣṇavism and the Krishna poets of Hindi insofar as Rādhā is concerned. Rādhā is always described by the poets of the Hindi group as the wife of Krishna, but the Bengal Vaiṣṇava poets usually describe her as someone else's wife.' S. M. Pandey and Norman Zide, 'Sūrdās and his Krishna-*bhakti*' in Singer (ed.), *Krishna: Myths, Rites and Attitudes*, p. 183.

[80] *Durlabhasāra*, cited by Dimock, *Place of the Hidden Moon*, p. 211.

[81] Ibid., cited by Dimock, loc. cit.

[82] Cited by Dimock, op. cit., p. 213.

[83] Vidyāpati, cited by Bhattacharya, *Love Songs of Caṇḍīdās*, pp. 162–3.

There are two 'worlds of value'—the external, social, universal values established in the *dharma-śāstra*, in Vedic ordinance and the internal and deeply personal values established spontaneously in love. The illicit relationship between lovers could be considered prototypical of the relationship between the devotee and the Bhagavat—the devotee might find that his relationship with the Lord conflicted with social rules, with family and friends, in which case he was to consider his loving relationship with Kṛṣṇa, his *bhakti*, as the higher value. The feeling of peril, passion, fear, the intensity of love must remain in *bhakti*; the devotee must adore Kṛṣṇa not out of duty, but because he is overwhelmed and enraptured. The cowherd-women provided the model:

> They are oppressed by the idea that they have been owned by males other than Kṛṣṇa, that they have been given in marriage to the Gopas. . . . In their eagerness to serve Kṛṣṇa they transgress the social laws and customs of married life. Although they externally show much regard for the rules and customs of married life and behave as faithful wives of the Gopas, yet in their heart of hearts they pant for Kṛṣṇa.[84]

Caitanya is said to have relished a particular story about two young lovers who joyfully trysted each day and who were deeply in love; one day the girl's father discovered their love-making and forced them to marry—'Their bed of flowers turned to thorns and their love faded away.'[85]

The literary motif of adulterous love, the *abhisārikā*'s surreptitious journey through the darkness, became a religious motif. Similarly in the European tradition Saint John of the Cross identified his soul with the lady of the medieval erotic *alba* who seeks the tryst with her 'Lord' at night:

> Upon a gloomy night,
> With all my cares to loving ardours flushed,
> (O venture of delight!)
> With nobody in sight
> I went abroad when all my house was hushed.
> In safety, in disguise,
> In darkness up the secret stair I crept,

[84] S. C. Chakravarti, *Philosophical Foundation of Bengal Vaiṣṇavism*, p. 371.

[85] The story is related in many texts. Cited by Dimock, *Place of the Hidden Moon*, p. 10. Dimock compares the Indian attitude with that of Marie and the Court of Champagne.

(O happy enterprise!)
Concealed from other eyes
When all my house at length in silence slept.
. . . .[86]

Rādhā is clearly a *parakīyā nāyikā*—her love-making with Kṛṣṇa is in defiance of Nanda, Kṛṣṇa's foster-father, who as a representative of authority exemplifies the social order, the ideal of *dharma.* She disobeys Nanda and Jayadeva cheers the secret-love-play (I. 1). Kṛṣṇa conveys a message of assignation to Rādhā by means of a pun, again, in defiance of Nanda:

'Why do you rest beneath the banyan tree which is the abode of black snakes (*or* of Kṛṣṇa the enjoyer)? O Brother! Why don't you go to the joyful house of Nanda which is within sight's range from here?' Concealing a message for Rādhā, in the presence of Nanda, from the mouth of a traveller, Govinda's words, filled with excellence for evening guests, prevail! [VI. 12]

Their union is the conventional 'secret', hidden, enjoyment of love; Rādhā is the conventional *abhisārikā*:

I went to his hut in the secret thicket; secretly at night he remained hiding; I looked fearfully in all directions; he laughed with an abundance of passion for the pleasure-of-love. . . . [II. 11]

The *Rāmāyaṇa* presents the highest ideal of *svakīyā* love—Sītā's excellence is her fulfilment of her duty to her husband, Rāma. Rādhā's excellence is her passion, the gratuitousness of her love—she is not bound by duty to love Kṛṣṇa—she loves him against Nanda's will and against her own:

[My mind] counts the multitude of his virtues, it does not think of his roaming-about even by mistake, and it possesses delight, it pardons [him] his transgressions from afar; even while fickle Kṛṣṇa delights among the girls without me, yet again my perverse mind loves him! What am I to do? [II. 10]

The misery of love-in-separation experienced by Rāma and Sītā is inflicted upon them by Rāvaṇa; that of Kṛṣṇa and Rādhā is self-inflicted—the *parakīyā* woman may torment her lover, repulse him (VIII), thereby increasing his desire, in a way which would be unpardonable in a dutiful wife. The pain undergone

[86] Saint John of the Cross, 'Canciones del alma que se goza . . .', op. cit., p. 27 (trans. Roy Campbell).

by the *parakīyā* woman in love is pain enjoyed, prized as passion, as a symptom of love; that of a *svakīyā* woman is pain suffered, endured as a necessity, as a symptom of her oppression:

> At first our bodies knew a perfect oneness,
> but then grew two with you as lover
> and I, unhappy I, the loved.
> Now you are husband, I the wife,
> what's left except of this my love [*or*: life],
> too hard to break, to reap the bitter fruit,
> your broken faith?[87]

Jayadeva does not explicitly indicate whether or not Rādhā is actually married to another man (as she is in the poems of both Caṇḍīdāsa and Vidyāpati) but that she is entrusted with seeing Kṛṣṇa home and that she is beyond the age of puberty indicates that she was not the *kanyakā* type of *parakīyā nāyikā*, the woman dependent on her father. Jayadeva avoids clarity; the relationship is ambiguous. Rādhā lives in Nanda's house and is thus both dependent upon him and under his jurisdiction; Kṛṣṇa betrays the authority of his foster-father in his liaison with Rādhā—he makes love to the woman who has been given the maternal role of protecting him at night. This situation, the structure of the relationship between Rādhā and Kṛṣṇa and Nanda, is analogous to the Oedipal situation, the structure of what Freud felt was the crucial and universal infantile experience. And that 'their secret love-games prevail' is in Freudian terms the triumph of the infantile fantasy, the victory of the *id* over the external reality. Celebration and consummation of the victory is a central feature of the cult of the Goddess, the Mother, a cult always particularly influential in north-eastern India, a cult which absorbed and deified Rādhā. Just as the veneration of woman in medieval European poetry correlated with the proliferation of the cult of the Madonna, the veneration and apotheosis of Rādhā suggests the impact of the Indian matriarchal religion. The sacred and profane dimensions converge: Rādhā is a goddess on earth (x. 15).

[87] Attributed to Bhāvakadevī (Ingalls trans.), *Amaruśataka* 81:
purābhūd asmākaṃ prathamam avibhinnā tanur iyaṃ
tato nu tvaṃ preyān vayam api hatāśāḥ priyatamāḥ
idānīṃ nāthas tvaṃ vayam api kalatraṃ kim aparaṃ
hatānāṃ prāṇānāṃ kuliśa-kaṭhinānāṃ phalam idam.

2 *Hail Rādhā full of grace*

A thirteenth-century Archbishop of Amiens prayed to Mary the Madonna:

My joy, my love, my life, my peace, my light . . . I lay my wounded heart at your feet . . . make it whole again. . . . Royal maiden, loyal queen, noble mother, precious vessel, shining crystal full of holiness, temple richly adorned, tabernacle illuminated by the great Light, comfort my soul, gentle lady. . . . Storehouse of spiced wine to sweeten the life of the sober heart, key to the balm which can bring the dead back to life, great is your savour and your sweetness . . . your love delights. . . . I should have all I need, if I had you.[88]

The cult of the Madonna assimilated the adoration of the Lady in the thirteenth century in Europe; the language of profane love which had been used to hymn the beloved became the conventional mode of praising the Blessed Virgin. Likewise in India the Goddess was worshipped with the language which had been conventionalized in *kāvya*.[89] And as the Madonna practically superseded the Father, the Son, and the Holy Ghost, in the popular religious tradition of southern Europe, so Rādhā, aligned and identified with the Goddess, began to ascend as an object of adoration even above Kṛṣṇa. In the *Devī-Bhāgavata Purāṇa* Nārāyaṇa makes obeisance to Rādhā:

Salutation to thee the supreme goddess, who resides at *Rāsamaṇḍala*, who lords over the *Rāsa* and is dearer to Kṛṣṇa than his own life. Salutation to the mother of all the three worlds, whose lotus-like feet are worshipped by gods headed by Brahmā and Viṣṇu. Be propitiated, Oh, Ocean of mercy.[90]

The identification of Rādhā and the Great Goddess could take

[88] Thibaut d'Amiens, 'Prière' in the *Penguin Book of French Verse*, vol. I, p. 135.

[89] Many *Śakti*-cult devotional works consist largely of extolling the beauty of various parts of Devī's body, e.g. *Saundaryalaharī* 42–91.

[90] *Devī-Bhāgavata Purāṇa* IX. 50. 46–47 (cited by B. Majumdar, *Kṛṣṇa in History and Legend* [Calcutta, 1969], p. 187). Majumdar comments on the development of the Rādhā-cult (p. 232) that the Rādhāvallabha sect of the sixteenth century gives Rādhā a higher place than Kṛṣṇa and a nineteenth century member of the sect Vaṃśī Ali, 'to avenge the wrong that had been done to their deity by the author of the *Bhāgavata* in not mentioning her name in his work. . . . composed a poem entitled *Śrī-Rādhikā-mahārasa*, in which the name Kṛṣṇa does not find any place at all. In it we find *Rādhā* playing on the flute and calling her female friends to the forest. . . .'

place easily because of Rādhā's ideality, and because the devotees of the Goddess conceived of Devī as the ultimate female principle behind *all* goddesses. As Kṛṣṇa became identified as the Bhagavat, Rādhā became identified as Bhagavatī:

> Just as Śrī Kṛṣṇa, who is *Brahman* in reality, is above *Prakṛti* [nature] and free from attachment, so Rādhikā, who is also *Brahman* in reality, is above *Prakṛti*, and free from attachment. Just as for the sake of work to be done, He at times assumes forms with attributes, so Mahāprakṛti Rādhikā also, for the same purpose assumes the form of gross *Prakṛti*. . . . Like Bhagavān Hari, Bhagavatī Rādhikā is also eternal and truth in substance. O *Muni*! the Mahāśakti who presides over the life of Bhagavān Śrī Kṛṣṇa appears as Rādhā. . . . *Brahman* is both Rādhā and Kṛṣṇa. She who is Rādhikā is also Kṛṣṇa and he who is Kṛṣṇa is also Rādhikā.[91]

Rādhā and Kṛṣṇa supplanted Lakṣmī and Viṣṇu as the primary forms of the Goddess and the God.

Lakṣmī (or Śrī, Kamalā, Padmā), the goddess of fortune and beauty, is the traditional consort of Viṣṇu and their love-making is a conventional subject in *kāvya* literature:

> May Lakṣmī's body bless you
> as after intercourse she rests
> with one hand leaning on the serpent, a garment in the other,
> her heavy hair fall'n loose upon her shoulder:
> but then once more forced back to the couch, her
> graceful arms protesting not the god's embrace,
> whose lustiness had doubled at the sight
> of beauty so revealed.[92]
>
> may those rites of copulation by which Mādhava
> amuses Lakṣmī be for your protection.[93]

Jayadeva declares the *Gītagovinda* to be 'comprised of tales about the love-play of Śrī and Vāsudeva' (I. 2) and various

[91] *Nārada-pañcarātra* (2nd *rātra*), cited in the *Tantratattva*, pp. 393–4.

[92] *Subhāṣitaratnakoṣa* 125 (Ingalls trans.):
uttiṣṭhantyā ratānte bharam uraga-patau pāṇinaikena kṛtvā
dhṛtvā cānyena vāso vigalita-kabarī-bhāram aṃśaṃ vahantyāḥ;
bhūyas tat-kāla-kānti-dvi-guṇita-surata-prītinā śauriṇā vaḥ
śayyām ālambya nītaṃ vapur alasa-lasad-bāhu lakṣmyāḥ punātu.

[93] Ibid., 132 (Ingalls trans.):
. . . *lakṣmīṃ narmayanto nidhuvana-vidhayaḥ pāntu vo mādhavasya.*

commentators (Mānāṅka, Vanamālibhaṭṭa, *et al.*) take Śrī to refer to the ultimate form of Rādhā—Viṣṇu becomes incarnated as the cowherd Govinda and Śrī-Lakṣmī as the cowherdess Rādhā. That Viṣṇu in his various incarnations unites with Lakṣmī in correlative incarnations is an idea expressed in the *Viṣṇu Purāṇa*:

For when the master of the world, the god of gods, Janārdana, makes a descent then Śrī is his female-companion: and when Hari was Āditya's-son she was again brought forth from a lotus; when he was Rāma of the Bhṛgu-family she was Dharaṇī; when he was [Rāma of the] Raghu-family she was Sītā; he was born as Kṛṣṇa [and then she was] Rukmiṇī; and in the other descents of Viṣṇu she does not leave him—when he has a god's body she is divine, when he is mortal she is mortal. . . .[94]

Śrī was born when the gods, desiring the nectar of immortality, churned the ocean—Viṣṇu incarnated as a tortoise (I. 6) and Mount Mandara, placed on his back, was used as a churning stick; as they churned the ocean of milk the moon appeared, then Śrī, then other treasures and finally the nectar.

You are beautiful like a young rain-cloud, you supported [Mount] Mandara, O *cakora* [bird] to the moon which is the face of Śrī! Victory! Victory! O God! Hari! [I. 23]

In a minimum of syllables there is a range of connotative and denotative contrasts and harmonies; the images are interlinked: Śrī and the moon are born from the nectarous ocean; her face is like the moon; the moon is the store-house of nectar; the *cakora* bird drinks nectar in the form of moon-rays; Hari is beautiful and she is Beauty; she is bright, he is dark; he is in the sky as a cloud and in the sea as a tortoise.

Jayadeva maintains the identity of Kṛṣṇa as the lover of Śrī-Lakṣmī: he is the Lord of Lakṣmī (VIII. title); Kamalā dwells in

[94] *Viṣṇu Purāṇa* I. 9. 142–145:
evaṃ yadā jagat-svāmī deva-devo janārdanaḥ
avatāraṃ karoty eṣā tadā śrīs tat-sahāyinī.
punaś ca padmād utpannā ādityo'bhūd yadā hariḥ;
yadā tu bhārgavo rāmas tadābhud dharaṇī tviyam.
rāghavatve'bhavat sītā rukmiṇī kṛṣṇa-janmani;
anyeṣu cāvatāreṣu viṣṇor eṣānapāyinī.
devatve deva-deheyaṃ manuṣyatve ca mānuṣī.

his heart (vii. 27); he 'lay upon the roundness of Kamalā's breasts' (i. 17); his chest 'is marked with saffron, fixed [there] from embraces on the surface of Padmā's breasts' (i. 26). A verse (possibly a late interpolation) tells of the cosmic love-making of Hari and Śrī-Lakṣmī:

By the mingling of reflected-images joined in the multitude of jewels lining the hoods of the Serpent Lord who was made [Hari's] couch, [Hari] wishes to see with hundreds of eyes the lotus-footed daughter of the ocean [Śrī] who is the production of the sustaining Lord, performing emanations of his body as if in her honour; may Hari protect you very much. [xii. 26]

The relationship between Viṣṇu and Śrī had been described as 'the same and yet different' (*bhedābheda*), the goddess was the 'power' (*śakti*) of the god. In the *Lakṣmī Tantra* Śrī declares:

I am his absolute *Śakti*, (his) I-hood consisting of bliss and consciousness. I am identified with and (at the same time) different from Him, like the moonlight of the moon. This unique existence of ours, though single, appears to be dual.[95]

Viṣṇu and Srī had been identified as the male and female principles in the *Viṣṇu Purāna*; they formed the Vaiṣṇava equivalent of the *Śiva-śakti* duality established in the Śaiva *Purānas* and *Tantras*:

The eternal world-mother Śrī is the constant companion of Viṣṇu; as he pervades all so does she . . . Viṣṇu is meaning, she is speech . . . Viṣṇu is the creator, she is the creation; Śrī is the earth, Hari is the earth's support. . . . Śrī is wish, the Lord is desire. . . . Lakṣmī is the altar [*or*: pyre], Hari is the sacrificial-stake; Śrī is the sacrificial-fuel, the Lord is the *kuśa*-grass. . . . Lord Śauri [Viṣṇu] is Śaṅkara [Śiva], Lakṣmī is Gaurī . . . Keśava is the sun, the lotus-abiding goddess is its light. . . . the supporter of Śrī is the moon, Śrī, even thus his concomitant, is the brightness. . . . Govinda is . . . the ocean, Śrī is its shore. . . . Lakṣmī is the light, Hari, this all, Lord of all, is the lamp; the world-mother is like the creeper, Śrī Viṣṇu is known as the tree. Śrī is the bright night, this god who bears the discus and the mace is the day; Viṣṇu the wish-giver is the husband, the lotus-abiding goddess is the bride. . . . Lakṣmī and Govinda also are love-pleasure and passion. . . . Among men, beasts and gods Lord Hari is all that is

[95] *Lakṣmī Tantra* xv. 9–10 (trans. S. Gupta).

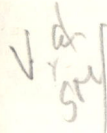

called male and Śrī is all that is known by the name female. There is nothing other than they.[96]

Jayadeva declares himself to be a singer at Padmāvatī's feet (I. 2) whose songs make Padmāvatī happy (XI. 21) suggesting devotion to the Goddess.[97] Such devotion is the basis of a song attributed to Jayadeva, the '*Gaṅgāstavaprabandha*', a work in praise of Gaṅgā, the Goddess and the River Ganges:

(1) O Best Manifestation of the Killer-of-Madhu: O You of lovely ripples, you bear much water, [water that] longs for moonbeams! You appeared from the world-egg when it was broken on Hari's toe-nail! You manifest [or cling to] the vessel of holy water!

Homage to You, O Goddess Gaṅgā!
Homage to You, O Goddess Gaṅgā!
May you remove all sin, O Alakanandā!

(2) You are purifying to men, you flow to the four oceans, O You who have the essence of the merit of the Seven Sages! You are spanned by mountains of gold! You are adored by all the seers! O garland of white flowers upon Hara's head!

Homage to You, O Goddess Gaṅgā! . . .

(3) O messenger for shining, immortal, beautiful women in their meetings with the sons of Sagara! You are inaccessible to evil-doers, you who prevent going to Hell! O producer of *Nirvāṇa*! You take away distress from devotees!

Homage to You, O Goddess Gaṅgā! . . .

[96] *Viṣṇu Purāṇa* I. 8. 16–36:
nityaivaiṣā jagan-mātā viṣṇoḥ śrīr anapāyinī; yathā
sarva-gato viṣṇus tathaiveyaṃ . . . . artho viṣṇur iyaṃ
vāṇī . . . sraṣṭā viṣṇur iyaṃ sṛṣṭiḥ; śrīr bhūmir bhūdharo
hariḥ . . . . icchā śrīr bhagavān kāmo . . . . citir lakṣmīr harir
yūpa; idhmā śrīr bhagavān kuśaḥ . . . . śaṅkaro bhagavāñ-chaurir
gaurī lakṣmīr . . . . keśavaḥ sūryas tat-prabhā kamalālayā . . . .
śaśāṅkaḥ śrī-dharaḥ kāntiḥ śrīs tathaivānapāyinī . . . .
jaladhir . . . . govindas tad-velā śrīr . . . . jyotsnā lakṣmīḥ
pradīpo'sau sarvaḥ sarveśvaro hariḥ; latābhūtā jagan-mātā
śrī-viṣṇur druma-saṃjñitaḥ. vibhāvarī śrīr div asau
devaś cakra-gadā-dharaḥ; vara-prado varo viṣṇur vadhūḥ
padma-vanālayā . . . . ratī rāgaś ca . . . . lakṣmīr govinda eva ca
. . . . deva-tiryaṅ-manuṣyādau punnāmā bhagavān-hariḥ;
strīnāmnī śrīś ca vijñeyā; nānayor vidyate param.

[97] In the popular Vaiṣṇava legends Padmāvatī is said to be the wife of Jayadeva (see Chapter VI, 'The Saint', page 213, below).

(4) You make the best pilgrimage place, O saver of ancestors! You erode the stones of the Himalayas! You flood the city and fill the sea! You are joined in the Vindhya foot-hills!

Homage to You, O Goddess Gaṅgā! . . .

(5) Your sandbanks are washed by your lovely water! When I praise you, look at you, think of you, bathe in you, sink in you, by making the pure speech which is your name you take away my sin!

Homage to You, O Goddess Gaṅgā! . . .

(6) O Exalted One! Here you are drunk by all the groups of sages and here you are sung by all the hosts of gods and here you are celebrated by all the *Vedas*! O daughter of Jahnu-muni! O essence of empirical-existence!

Homage to You, O Goddess Gaṅgā! . . .

(7) Here in your water deep with its clustered rays may a mouthful vanquish a hundred of my sins! You are akin to the other world! You are a beautiful river! O River, you are the means to attain the dwelling of the gods!

Homage to You, O Goddess Gaṅgā! . . .

(8) [Since he has] here spoken respectfully after worshipping the divine river with the permission of Rāma, Lakṣmaṇa and Kauśika, you convey the poet, resolute Jayadeva, across the waters of existence!

Homage to You, O Goddess Gaṅgā! . . .[98]

[98] Attributed to Jayadeva and printed with a commentary in the Nirṇaya-Sāgar Press edn of the *Gītāgovinda*:

(1) madhu-mathana-mūrti-vara indu-kara-kāmukaṃ vahasi bahu vāri sutaraṅge;
hari-carana-nakha-bhidhura-jagad-aṇḍa-nirgatā brahma-jala-patra-kṛta-saṅge.
namo devi gaṅge namo devi gaṅge
hara nikhilam agham alakanande. (dhruvam)

(2) tvam asi jana-pavīnī catur-udadhi-gāminī saptarṣi-kula-sukṛta-sāre;
kanaka-giri-laṅghite sakala-muni-vandite hara-mukuṭa-sita-kusuma-māle.

(3) sagara-suta-saṅgame lasad-amara-sundarī-dūtike duṣ-kṛti-dur-āpe;
niraya-gati-bādhike nirvāṇa-sādhike bhakta-jana-hṛta-saṃtāpe.

(4) tīrtha-vara-kariṇī pūrva-joddhāriṇī hima-śaila-pāṣāṇa-bhede;
nāgaram plāvayasi pūrayasi sāgaraṃ saṃgatā vindhya-giri-pāde.

The opening phrase, the identification of Gaṅgā as a manifestation of Kṛṣṇa is unusual, but the identification of the Goddess as a manifestation of the God is traditional in the *Tantras* and in the *Pañcarātra* literature. The Goddess and the God are *śakti* and *śaktimat*; they are inseparably interpenetrating, interconnected, each inherent in each other—they are 'the same and yet different'. This phrase (*bhedābheda*) expressed a metaphysical concept, a religious structure, a sacred relationship; but, at the same time, it suggests a psychological concept, a secular structure, a profane relationship—lovers are 'the same and yet different'. The sacred and profane, the metaphysical and the carnal, are brought together in the relationship between Rādhā and Kṛṣṇa, at least as understood by the medieval Vaiṣṇavas:

Rādhā-Krishna are one and the same eternally. But for divine sport they separated themselves into two bodies.[99]

Jayadeva's Rādhā says, 'I am the Enemy-of-Madhu' (VI. 5) and the cowherd women in the *Viṣṇu Purāṇa* also imagine that they are Kṛṣṇa:

Crowds of *gopīs* whose bodies were practising the deeds of Kṛṣṇa went into the Vṛndā-forest, when Kṛṣṇa had gone to another place. (Their hearts fixed upon Kṛṣṇa, they said this to each other:) 'I am this Kṛṣṇa, I walk (in his) amorous (manner); look at my gait!' Another says: 'Listen to the songs of mine, Kṛṣṇa's!'—'Wicked Kāliya! Stand still here; I am Kṛṣṇa!' (says) another one, and slapping on her arm she began (to act? recite?) the whole of Kṛṣṇa's *līlā*. Another girl

---

(5) śubha-salila-nirdhauta-pulinavati kurvatas tava nāma-nirmalālāpam;
vanditā vīkṣitā cintitāvagāhitā majjitā mama harasi pāpam.

(6) iha nikhila-muni-gaṇair iha nikhila-sura-gaṇair iha nikhila-vedair udāre;
pīyase gīyase stūyase jahnu-muni-kanyake saṃsāra-sāre.

(7) gaṇḍūṣam iha me'gha-śataṃ adhijayatu tava kara-nikara-gambhīra-toye;
para-loka-bandhur asi sundara-sindhur asi taṭini sura-sadana-sādhanopāye.

(8) iti rāma-lakṣmaṇa-kauśikānujñayā sura-sarit-kṛta-namas kāram;
bhaṇantam iha sādaraṃ dhīra-jayadeva-kavim uttārayasi bhava-jaladhi-pāram.

[99] *Caitanya-Caritāmṛta, Ādi.* I. 5 (Ray trans., p. 2). The notion is repeated throughout the text together with the assertion that the two became one again in the body of Lord Caitanya.

says: 'You cowherds! Be now without any fear; the danger from the rains is over: I am holding up Govardhana!'[100]

Imitation of the lover is described in the rhetorical texts as the sportiveness (*līlā*) which is one of the natural graces of a young woman in love.[101] As Rādhā imagines herself to be Kṛṣṇa the implications are sexual—Kumbha explains that the jingling of her ornaments reminds her of making-love with Kṛṣṇa in the 'inverse posture'; he takes her 'I am the Enemy-of-Madhu' to indicate that she is imagining herself to be on top, playing the part of the man, in coition. But the sexual is the expression of one level of the larger erotic structure—the identification of the lover and the beloved through love is the expression of the very meaning of love; Isolde cries out to Tristan:

> We are one life and flesh. . . . You and I . . . I am yours . . . you are mine . . . one Tristan and Isolde.[102]

And the language of profane love is adaptable to the sacred; a German nun cries out to Christ:

> I You, You I, we two are one
> And so one is made from us two.[103]

And in the Indian traditions the two-in-one-ness of lovers could express the metaphysical two-in-one-ness: the goal of the teaching of the *Upaniṣads* had been the realization of the identity of the Self, the *ātman*, with *brahman*; the Kṛṣṇa-cult provided a new means for that realization—passionate devotion, an erotic *bhakti*. The devotee could identify with Rādhā in order to experience Kṛṣṇa as the lover, 'the same and yet different'.

[100] *Viṣṇu Purāṇa* v. xiii. 24–28 (Hardy trans.):
gopyaś ca vṛndaśaḥ kṛṣṇa-ceṣṭāsv āyatta-mūrtayaḥ;
anya-deśa-gate kṛṣṇe cerur vṛndāvanāntaram.
kṛṣṇe nibaddha-hṛdayā idam ūcuḥ parasparam;
kṛṣṇo'ham eṣa lalitaṃ vrajāmy ālokyatāṃ gatiḥ.
anyā bravīti kṛṣṇasya mama gītir niśamyatām.
duṣṭa kāliya tiṣṭhātra kṛṣṇo'ham iti cāparā;
bāhum āsphoṭya kṛṣṇasya līlā-sarvasvam ādade.
anyā bravīti bho gopā niḥśaṅkaṃ sthīyatām iha;
alaṃ vṛṣṭi-bhayenātra dhṛto govardhano mayā.

[101] *Daśarūpa* II. 60.

[102] See above p. 2.

[103] Cited by Perella, op. cit., p. 97.

The followers of Caitanya, as we have seen, conceived of Rādhā as the *hlādinī-śakti* of Kṛṣṇa; they are eternally one, but they separated in order to taste the sweetness of each other in love. They conceived of her furthermore as 'the true form of the highest development of love':

The essence of *hlādinī* is *prema*; the essence of *prema* is *bhāva*; the highest state of *bhāva* is called *mahābhāva*. The true form [*svarūpa*] of *mahābhāva* is Rādhikā.[104]

As Kṛṣṇa is the embodiment of the erotic sentiment (*śṛṅgāra-rasa*) Rādhā is the embodiment of the emotion which expresses that sentiment.

The concept of the *hlādinī-śakti*, similar to that of the *Shekhinah* in Kabbalistic Judaism, 'God's Divine Presence', an aspect of God, a feminine element in Him, 'the same and yet different', homologizes the sacred and profane dimensions of love.[105] The Sahajiyā Vaiṣṇavas stressed the homology; they merged tantric doctrines with those of the Caitanya school of Vaiṣṇavism, moving away from the ideal of chastity which characterized the orthodox school: sexual union, temporal, fleshy, became the eternal union, the *hieros gamos*, when the human lover and human beloved recognized themselves as Kṛṣṇa and Rādhā:

In the Vaiṣṇava Sahajiyā school the two aspects of Sahaja or the absolute reality are explained as the eternal enjoyer and the enjoyed, as Kṛṣṇa and Rādhā; and it is further held that all men and women are physical manifestations of the ontological principles of Kṛṣṇa and Rādhā. The men and women can, therefore, realise themselves as the manifestations of Kṛṣṇa and Rādhā through a process of attribution (*āropa*), the love of any human couple becomes transformed into the divine love that is eternally flowing on between Kṛṣṇa and Rādhā.[106]

The deification of woman as the exalting, cherishing, pleasing goddess invests her with a power which makes her not only the object of love, but also the object of terror. As she enchants she not only delights but also enslaves. The ambivalent attitude toward woman is characteristic of the Indian traditions. Devī has two aspects—her beautiful and her hideous manifestations:

[104] *Caitanya-Caritāmṛta, Ādi.* IV. 59 (cited and trans. by Dimock, *Place of the Hidden Moon*, p. 134).

[105] See above pp. 11–12.

[106] S. B. Das Gupta, *Obscure Religious Cults*, p. xxxvii..

Heads of thy sons, daily, freshly killed, hang as a garland around thy neck. How is thy waist adorned with human hands! little children are thy ear-rings. Faultless are thy lovely lips; thy teeth are fair as the *kunda* in full bloom. Thy face is bright as the lotus flower and terrible is its constant smiling. Beautiful as the rain-clouds is thy form; all blood-stained are thy Feet.[107]

3 *The ambivalent image of woman*

'You wander in the forest to devour young-women', Rādhā says to Kṛṣṇa, 'Pūtanikā indeed reveals your character as the cruel infant who is a lady killer' (VIII. 8). Pūtanā, the horrible ogress sent by Kaṃsa to destroy Kṛṣṇa, disguised herself as a beautiful woman and went to Nanda's village where she took the infant Kṛṣṇa and put him to her breast:

The horrible one, taking him on her lap, gave the baby her breast, which had been smeared with a virulent poison. But the lord, pressing her breast hard with his hands, angrily drank out her life's breath with the milk.[108]

The cowherds cut up her body and burned it:

The smoke that arose from Pūtanā's body was as sweet-smelling as aloe-wood, for her sins had been destroyed when she fed Kṛṣṇa. Pūtanā, a slayer of people and infants, a female Rākṣasa, a drinker of blood, reached the heaven of good people because she had given her breast to Viṣṇu—even though she did it because she wished to kill him.[109]

There is a series of paradoxes—the beautiful woman is a hideous Rākṣasa, the nurturing breast is poisonous, the attempt to murder Kṛṣṇa causes Pūtanā's own death, her evil action leads to

[107] Rāmprasād Sen, 'Kālī the Battle-Queen' in E. J. Thompson and A. M. Spencer (ed. and trans.), *Bengali Religious Lyrics, Sākta* (Calcutta, 1923), p. 46.

[108] *Bhāgavata Purāṇa* x. vi. 10 (trans. O'Flaherty, *Hindu Myths*, p. 215):
tasmin stanaṃ durjara-vīryam ulbaṇaṃ ghorāṅgam
ādāya śiśor dadāv atha;
gāḍhaṃ karābhyāṃ bhagavān prapīḍya tat-prāṇaiḥ
samaṃ roṣa-samanvito'pibat.

[109] Ibid. x. vi. 34–35 (trans. idem, p. 217):
dahyamānasya dehasya dhūmaś cāguru-saurabhaḥ;
utthitaḥ kṛṣṇa-nirbhukta-sapadyāhata-pāpmanaḥ.
pūtanā loka-bālaghnī rākṣasī rudhirāśanā;
jighāṃsayāpi haraye stanaṃ datvāpa sad-gatim.

her attainment of the 'heaven of good people'. Rādhā's allusion to the story gives the Pūtanā–Kṛṣṇa relationship the status of the erotic man–woman relationship. Pūtanā exemplifies an ambivalent, paradoxical, image of woman which runs throughout Indian literature. 'There is no nectar or poison', Bhartṛhari says, 'except a fair-hipped-woman.'[110] That is, all women are like Pūtanā—their beauty conceals terrors. In the *Ṛg-Veda* Urvaśī tells Purūravas not to expect lasting friendship from women for they have the 'hearts of hyenas'.[111]

Love-making is a 'battle of love' in that the man is resisting or attempting to overcome the terrible aspect of woman, the binder, the devourer, while endeavouring to enjoy the beautiful aspect of woman, the delighter. Both aspects are present in Rādhā—by making Kṛṣṇa desire her she binds him and he is tormented with longing.

> When the beloved [Kṛṣṇa] is tender you [Rādhā] are rough, when he bends-down [in obeisance] you are unbending, when he is passionate, you are hostile, when he has his face raised [in expectation] you have your face-turned-away [in aversion] . . . O perverse-woman. . . . [IX. 10]

Kṛṣṇa addresses her as '*caṇḍī*' (x. 11, etc.) which in the context of the literary tradition qualifies her as the fierce, cruel, violent, passionate woman and in the religious tradition as the Goddess (Caṇḍī being one of the epithets of Devī). It is in the Goddess that the two aspects of woman are reconciled, the Goddess 'beautiful and blood-stained':

> In her 'hideous aspect' (*ghora-rūpa*) the goddess, as Kālī, 'the dark one', raises the skull full of seething blood to her lips; her devotional image shows her dressed in blood red, standing in a boat floating on a sea of blood—in the midst of the life flood, the sacrificial sap which she requires that she may, in her gracious manifestation (*sundara-mūrti*) as the world mother (*jagad-ambā*), bestow existence upon new living forms in a process of unceasing procreation, that as world nurse (*jagad-dhātrī*) she may suckle them at her breasts and give them the food that is 'full of nourishment' (*anna-pūrṇā*).[112]

[110] Bhartṛhari 91:

nāmṛtaṃ na viṣaṃ kiṃcid ekāṃ muktvā nitambinīm. . . .

[111] *Ṛg-Veda* x. 95.15.

[112] Heinrich Zimmer, 'The Indian World Mother' in *The Mystic Vision, Papers from the Eranos Yearbooks*, vol. 6, ed. Joseph Campbell (Princeton, 1970), p. 74.

The ambivalent image, seemingly universal, often expressed in the Christian context in terms of Eve ('through a woman we were sent to destruction . . .') and Mary ('. . . through a woman salvation was sent to us'),[113] is a psychological phenomenon, an extension of the ambivalence which the child instinctively feels for the mother:

One might even believe that this first love relation of the child is doomed to extinction for the very reason that it is the first, for these early object-cathexes are always ambivalent to a very high degree; alongside of the child's intense love there is always a strong aggressive tendency present, and the more passionately a child loves an object, the more sensitive it will be to disappointments and frustrations. In the end, the love is bound to capitulate to the accumulated hostility.[114]

The positive aspect, the 'beautiful form' (*sundara-mūrti*) becomes the negative, by fascinating, luring, by *binding*:

With a smile and with love, with modesty, thought, half-side-long-glances from averted faces, with words, jealous-quarrels, and with love-play; with all emotions indeed *a woman is binding.*[115]

And in response to her power there may be either a depreciatory attitude, an attempt to subdue her, or a devotional attitude, an attempt to elevate her:

The depreciatory attitude which many a man takes towards women is an unconscious attempt to control a situation in which he feels himself at a disadvantage, or he seeks to undercut the dreaded power of the woman by inducing her to act towards him as a mother. In this way he is released in large measure from his fear, for in his relation to his mother nearly every man has experienced the positive aspect of woman's love. Even so, he is not entirely free from apprehension, because in making the woman the mother he . . . once more leaves the woman all-powerful in the situation.[116]

[113] Saint Augustine cited by Morton Hunt, *The Natural History of Love* (New York, 1967), p. 127.

[114] Freud, *New Introductory Lectures* . . . , p. 124.

[115] Bhartṛhari 79:

smitena bhāvena ca lajjayā dhiyā
parāṅ-mukhair ardha-kaṭākṣa-vīkṣitaiḥ;
vacobhir īrṣyākalahena līlayā
samastabhāvaiḥ khalu bandhanaṃ striyaḥ.

[116] M. Esther Harding, *Woman's Mysteries,* Rider edn (New York, 1971), pp. 34–5. The author is a Jungian psychoanalyst.

The two responses are exemplified by Kṛṣṇa in relation to Pūtanā, in killing her, and to Rādhā, in venerating her as the mother, in making obeisance at her feet; the two responses are coalesced in the aggressive sexuality of the battle-of-love—in biting, scratching, hair-pulling, the hostile becomes the amorous.

The ambivalent attitude toward woman reflects an ambivalent attitude toward nature:

There exists a very general association on the one hand between the notion of mind, spirit or soul and the idea of the father or of masculinity; and on the other hand between the notion of the body or of matter (*materia*—that which belongs to the mother) and the idea of the mother or feminine principle. The repression of the emotions and feelings relating to the mother has, in virtue of this associating, produced a tendency to adopt an attitude of distrust, contempt, disgust or hostility towards the human body, the Earth, and the whole material Universe, with a corresponding tendency to exalt and over-emphasize the spiritual elements, whether in man or in the general scheme of things.[117]

In the Indian traditions, the prevalently negative attitude towards the material Universe, *saṃsāra*, *prakṛti*, and the body as the part of the individual which is an aspect of that Universe, is naturally connected with the prevalently negative attitude towards woman as the one who binds man to *saṃsāra* with desire. Liberation is of spirit from matter, *puruṣa* from *prakṛti* in the Sāṃkhya system, the male from the female; and Rādhā, the chain binding Kṛṣṇa to *saṃsāra* (III. 1), is equated with nature:

This lower-lip is akin to the *bandhūka* [flower]; your shiny cheek has the complexion of the *madhūka* [flower] . . . Your eye, which emits the lustre of a blue-lotus, shines; your nose resembles the sesame flower; your teeth are like jasmine. . . . [x. 14]

Jayadeva esteems woman and nature through Rādhā and he does so within the religious domain by virtue of Kṛṣṇa—for Kṛṣṇa has power over the negative aspect of woman, Pūtanā, power to transform her evil into good. Woman, nature and love are redeemed and affirmed against the grain of traditional,

[117] J. C. Flügel, *The Psycho-Analytic Study of the Family*, cited by Joseph Campbell, *Hero With a Thousand Faces*, Meridian edn (New York, 1956), p. 113.

orthodox, Hindu values. The sacred and profane are homologized as they are in Whitman's joyous song:

Sex contains all, bodies, souls,
Meanings, proofs, purities, delicacies, results,
promulgations,
Songs, commands, health, pride, the maternal mystery, the
seminal milk,
All hopes, benefactions, bestowals, all the passions, loves,
beauties, delights of the earth,
All the governments, judges, gods, follow'd persons of the earth,
These are contain'd in sex as parts of itself and justifications
of itself.[118]

THE MESSENGER

Messenger, go run, tell the Loveliest one of the pain and the sorrow I endure for her sake, and the torment. . . .[119]

The messenger or friend or go-between (*dūtī, sakhī*) plays a literary role, fulfils a dramatic function; the role, the function, was conventional, established in the rhetorical and erotic textbooks. Vātsyāyana says that the love-messenger should be a friend of both the man and the woman, but particularly trusted by the woman, that she (or he) ought to be clever, clearheaded, aware, that she should tell the lady of the man's good qualities, his sexual aptitudes, his finesse in the arts and sciences of love, that she should tell of his affairs with other women but point out that now he wants to possess only her.[120] The speech of the go-between became a standard subject for *kāvya*; she tells the man of her lady's misery:

Fearing the moon, she dares not view her own reflection;
frightened of the cuckoo's call, she utters not a word.
How strange, then, that swearing enmity of Kāma
who furnishes such fires,

[118] Walt Whitman, 'A Woman waits for me' in the *Children of Adam*. See above p. 12.

[119] Bernard de Ventadorn, 'Tant ai mo cor . . .' in R. T. Hill and T. G. Bergin, *Anthology of Provençal Troubadours* (New Haven, 1957), p. 46.

[120] *Kāmasūtra* I. v. 39 ff ˙ ıd v. iv. 10, 15, 64, etc.

the artless maiden's love, oh handsome one,
grows, ever more for you.[121]

The friend is a device whereby the suffering of love-in-separation can be described in both the first person and the third person, both with the passion of the sufferer and the detachment of the observer.

The female-friend in the *Gītagovinda* conforms to the conventions—she urges Rādhā to follow Kṛṣṇa by playing upon Rādhā's jealousy and upon her memory of the joys of union, promising Rādhā that those joys will be experienced again; she incites Rādhā's sympathy with descriptions of the pining Kṛṣṇa, points out the fruitlessness of separation, and finally takes on an angry tone as she reprimands Rādhā for delaying. In her attempt to move Kṛṣṇa she simply tells him of Rādhā's desolation.

The love-messenger was considered by the followers of Caitanya to exemplify the *bhakti* of service. She has a gratuitous dedication to prompting the union of Rādhā and Kṛṣṇa above all else, to actualizing within time and space the eternal reality which is Rādhā-Kṛṣṇa. She mediates and reveals, and because it is absolutely without selfish motive, without any desire for joy, it brings infinite joy:

For the love-*līlā* of Lord Krishna and his beloved Rādhā is indeed too deep to be easily realised. . . . For the loving friendliness of Sakhīs alone may reach it. And this love-*līlā* spreads and expands through loving, female friends—Sakhīs alone. It does not grow without such friends. And the female friends enjoy it as they spread it. And none else but such loving, female-friends can realise it. And he or she who would gauge the depths of it must be in the position of such a loving friend. And he or she alone can know the sublime service of Rādhā and her beloved Krishna, which they enjoy in the bower. For there is no other way to obtain it. . . . the love between Rādhā and Krishna though self-revealed, universal and sublime cannot be seen except through the aid of female-friends. And the nature of a loving-female-friend is indeed inexplicable. For she herself does not desire any loving

[121] *Subhāṣitaratnakoṣa* 536 (Ingalls trans.):
sodvega mṛga-lāñchane mukham api svaṃ nekṣate darpane
trastā kokila-kūjitād api giraṃ nonmudrayaty ātmanaḥ;
citram duḥsaha-dāha-dāyini dhṛta-dveṣāpi puṣpāyudhe
bālā sa subhaga tvayi prati-padaṃ premādhikaṃ puṣyati.

association with the Lord Krishna. And she helps Rādhā to enjoy such loving association. And from it she receives a joy that is a hundred times sweeter than any joy of her own association with the Lord Krishna could be.[122]

S. B. Das Gupta has suggested that the female-friend is the key to the devotional attitude of Jayadeva:

The attitude of the Vaiṣṇava poets was that of *Sakhi-bhāva* rather than *Rādhā-bhāva*. Śrī-Caitanya placed himself in the position of Rādhā and longed with all the tormenting pangs of heart for union with his beloved Kṛṣṇa; but the Vaiṣṇava poets, headed by Jayadeva, Caṇḍīdāsa and Vidyāpati, placed themselves rather in the position of the Sakhīs, or female companions of Rādhā and Kṛṣṇa, who did never long for their union with Kṛṣṇa,—but ever longed for the opportunity of witnessing from a distance the eternal love-making of Rādhā and Kṛṣṇa in the supranatural land of Vṛndāvana (*aprākṛta-vṛndāvana*). This eternal *līlā* is the eternal truth, and, therefore, it is this eternal *līlā*—the playful love-making of Rādhā and Kṛṣṇa, which the Vaiṣṇava poets desired to enjoy. If we analyse the *Gītagovinda* of Jayadeva we shall not even find a single statement which shows the poet's desire to have union with Kṛṣṇa as Rādhā had,—he only sings praises of the *līlā* of Rādhā and Kṛṣṇa and hankers after a chance just to have a peep into the divine *līlā*, and this peep into the divine *līlā* is the highest spiritual gain which these poets could think of.[123]

[122] *Caitanya-Caritāmṛta, Madhya.* VIII (Ray trans., p. 169).

[123] S. B. Das Gupta, *Obscure Religious Cults*, pp. 125–6. In a modern interpretation the *Sakhī* is considered to represent the *guru* who shows the devotee the way to God. See below, p. 180.

4

THE DYNAMICS OF LOVE

Two natures single grown,
We must abide them,
Eternal love alone
Has power to divide them.[1]

SEPARATION

The recipe for love is this,
That first we must divide you.[2]

Denis de Rougemont's study of the sacred and profane dimensions of love in European traditions as exemplified in the legend of Tristan and Isolde focuses upon passion, 'that form of love which refuses the immediate . . . and if necessary invents distance in order to exalt itself more completely'.[3] The separation of lovers, both temporarily through various obstructions and permanently through death, the suffering which *is* passion, has been, de Rougemont shows, the central motif, the central obsession, in the history of the literature of love in the western world:

Tristan and Iseult do not love one another. . . . *What they love is love and being in love*. . . . Tristan loves the awareness that he is loving far more than he loves Iseult the Fair. And Iseult does nothing to hold Tristan. All she needs is her passionate dream. Their need of one another is in order to be aflame, and they do not need one another as they are. What they need is not one another's presence, but one another's absence. *Thus the partings of the lovers are dictated by their passion*

[1] Goethe's *Faust, Part II*, trans. Philip Wayne, Penguin edn (Harmondsworth, 1959), p. 283.

[2] Ibid., *Part I*, Oberon speaking in the 'Walpurgis Night dream', Penguin edn (Middlesex, 1949), trans. Philip Wayne, p. 181.

[3] de Rougemont, *Love Declared*, p. 41.

itself, and by the love they bestow on their passion rather than on its satisfaction or on its living object.[4]

Passion was cultivated in literature—the enjoyment of the suffering, the delight of the misery, the pleasure of the pain, was celebrated:

From all other ills doth mine differ. It pleaseth me; I rejoice at it; my ill is what I want and my suffering is my health. So I do not see what I am complaining about; for my ill comes to me by my will; it is my willing that becomes my ill; but I am so pleased to want thus that I suffer agreeably, and have so much joy in my pain that I am sick with delight.[5]

The dynamics are clear: if love is desire, a force which 'seeks to couple together some two things'[6] the actual union must end love; the two can love, can desire to be one, but what can the one desire? And if love as such, as passion, is idealized, the attitude toward actually achieving union must be deprecatory (but this attitude must be well hidden, 'unconscious', so as not to undermine the desire). The hunger is more relished than the food. A *Minnesänger* chanted, 'My heart's deep wound, you must *always* stay open . . .'[7] It is an equation which runs throughout European literature—to love is to suffer and suffering is ultimately affirmed as good, as consciousness, as gnosis. 'Suffering is the ancient law of love,' the Blessed Henry Suso proclaims, 'there is no quest without pain . . . no lover who is not also a martyr.'[8] The erotic equation had both sacred and profane manifestations; in both realms love-in-separation was idealized as the necessary condition of the intense yearning which characterizes the lover and the devotee. The goal was the pilgrimage, not the pilgrimage-place; the goal was not the satisfaction *of* the desire for union with the beloved but the satisfaction *in* the desire for union with the beloved. Gregory of Nyssa explains the Shulamite's despair over her separation from her beloved:

The soul . . . looks for Him but does not find Him. She calls on Him

[4] Idem, *Love in the Western World*, p. 43.

[5] Cherétien de Troyes, cited by idem, p. 39.

[6] See above, p. 1.

[7] Walter von de Vogelweide, cited by Perella, op. cit., p. 117.

[8] Cited by Underhill, op. cit., p. 222.

. . . and she is told by the watchman that she is in love with the unattainable, and that the object of her longing cannot be apprehended. In this way she is . . . wounded and beaten because of the frustration of what she desires, now that she thinks that her yearning for the Other cannot be fulfilled or satisfied. But the veil of her grief is removed when she learns that the true satisfaction of her desire consists in constantly going on with her quest and never ceasing in her ascent, seeing that every fulfilment of her desire continually generates a further desire for the Transcendent.[9]

There was the same idealization of love-in-separation (*viraha*) throughout the Kṛṣṇaite devotional tradition. In the *Viṣṇu Purāṇa* a cowherd-girl whose father prevents her from going to tryst at night with Kṛṣṇa meditates upon him; she becomes absorbed in Kṛṣṇa and thus 'all her sins were removed through the great pain of not meeting him'.[10] And when in the *Bhāgavata Purāṇa* the women of Vraja lament over their separation from Kṛṣṇa, he consoles them by explaining to them that although he was hidden from their eyes, he remained devoted to them and hid from them only to increase their love for him.[11] Through external separation they learn that he is forever within: 'You are never separated from me, for I am the Self of all.'[12] Separation from Kṛṣṇa causes contemplation of him, meditation on him (which in turn leads to liberation as a by-product of secondary significance). The passionate cowherd-woman became the prototype of the passionate devotee. The entire life of the Vaiṣṇava *bhakta* was to consist of a 'holy yearning',[13] the intense desire caused by separation:

And during the last twelve years of his life Lord Caitanya felt incessantly the frenzy of separation from his beloved Krishna. . . . He knew no rest . . . He raved like a madman . . . and he mourned saying, 'Alas! Alas! Where is the Lord of my life?. . . My soul is bursting in grief for the Holy Lord.'. . . His heart was burning because he could

[9] Jean Daniélou (ed.), *From Glory to Glory*, p. 270.

[10] See above, p. 22.

[11] *Bhāgavata Purāṇa* x. xxxii. 21 and xlvii. 35.

[12] Ibid., x. xlvii. 29: '*bhavatīnāṃ viyogo me na hi sarvātmanā kvacit. . . .*'

[13] Compare Saint Augustine: 'The entire life of a good Christian consists of a holy yearning. Now what you yearn for you do not yet see . . . By withholding himself, God extends our yearning; through yearning he extends our soul . . .' (cited by Perella, op. cit., p. 86).

not see his Lord. . . . He showed all the symptoms of divine frenzy and he went mad for the love of Lord Krishna.[14]

According to the *Naradīya-Bhakti-Sūtras* the attachment to Kṛṣṇa felt by the separated devotee is the highest stage of *bhakti*:

Attachment to his attributes and greatness, to his beauty, in worship, remembrance, service, friendship, parental-love, love, self-dedication, consisting of him, and in *separation* [which is] the highest. . . .[15]

The devotee is the 'same as and yet different from' the Lord and so even in the joy of union feels the pain of separation and continues in 'holy yearning'—Govindadāsa sang:

While holding the Dark-one in her hands, Rādhā still weeps, 'where is the Lord of my life?'[16]

The highest form of devotion (*parama-bhakti*) according to Yāmunācārya comes not in union but after union in the 'fear of new separation'.[17] The more intense the fear, the greater the love, the attachment, and the greater the love, the greater the fear, and so on in a perpetual spiral into the 'divine frenzy', the devotional ecstasy which characterizes the Kṛṣṇa-cult.

The love-in-separation motif is prevalent in the folk traditions—

In the profane realm, *viraha* is a favorite theme of Indian poetry, especially of folk-ballads and folk-songs. The *virah-gīts*, sung by village women, mostly describe the painful longings and pitiful laments of a young wife waiting in vain for the return of her beloved. . . .[18]

And the folk milieu is reflected in Hāla's *Sattasaī*:

> In the forest filled with the humming of the bees
> and stroked by the wind of spring,

[14] *Caitanya-Caritāmṛta, Madhya.* II (Ray trans., pp. 26 ff.).

[15] Nārada's *Bhakti-sūtra* 82:
guṇa-māhātmyāsakti-rūpāsakti-pūjāsakti-smaraṇāsakti-
dāsyāsakti-sakhyāsakti-vātsalyāsakti-kāntāsakti-
ātma-nivedanāsakti-tan-mayāsakti-parama-virahāsakti. . . .

[16] Cited and trans. Vaudeville, 'Evolution of Love-Symbolism . . .', p. 39: '*rodati rādhā śyāma kari kora, Hari Hari kaṃha geo prānanātha more* . . .'.

[17] Cited ibid., p. 40.

[18] Charlotte Vaudeville, *Kabīr* (Oxford, 1974), p. 146.

the cowherdess sings tunes about separation,
perplexing the minds of travelers.[19]

The motif had already become conventionalized by the time of the *Rāmāyaṇa.*[20] Rāma laments:

This spring, O Saumitri, resounding [with the songs of] various birds, kindles my sorrow—I who am deprived of Sītā. Love inflames me—I am overrun by grief; the cuckoo, proclaiming his rapture, challenges me. . . . The cruel springtime forest-breeze inflames me, O Saumitri —I am overpowered by sorrow and care. . . .[21]

Grief, it is said, vanishes as time passes, but my sorrow, caused by this separation from my beloved, increases. . . .[22]

Blow wind to where my beloved is—touch her and then touch me as well—then in you our bodies will be in contact and in the moon our glances will be united. . . .[23]

The fire of love has as its fuel the separation from her, as its vast

[19] Hāla, *Sattasaī* 128 (Hardy trans.):
mahumāsamāruāhaa-
mahuarajhaṃkāraṇibbhare raṇṇe;
gāi virahakhharāvad-
dhapahiamaṇamohaṇaṃ govī.

[20] Two stories of 'ill-fated' love, love-in-separation, occur as early as the *Ṛg-Veda*: Yamī longs for union with her brother, Yama, but he repulses her and she grieves (x. 10); the mortal Purūravas is separated from his beloved, the goddess Urvaṣī—he pleads with her to return to him and threatens to kill himself if she does not; she replies, 'Die not, Purūravas, fall not, let not the hideous wolves devour thee. Female friendships do not exist, their hearts are the hearts of jackals' (x. 95.15, Wilson trans.).

[21] *Rāmāyaṇa* IV. 1.12, 13, 16:
ayaṃ vasantaḥ saumitre nānāvihaga-nāditaḥ;
sītayā viprahīnasya śoka-saṃdīpano mama.
māṃ hi śoka-samākrāntaṃ saṃtāpayati manmathaḥ;
hṛṣtaḥ pravadamānaś ca samāhvayati kokilaḥ.
. . .
māṃ . . . cintā-śokabalāt-kṛtaṃ;
saṃtāpayati saumitre krūraś caitra-vanānilaḥ.

[22] Ibid., IV. 5.4:
śokaś ca kila kālena gacchatā hy apagacchati;
mama cāpaśyataḥ kāntām ahanyahani vardhate.

[23] *Rāmāyaṇa* VI. 5.6:
vāhi vāta yataḥ kāntā tāṃ spṛṣtvā māṃ api spṛśa;
tvayi me gātrasaṃsparśaś candre dṛṣṭi-samāgamaḥ.

flame thoughts about her; this fire of love burns my body night and day.[24]

As the motif was developed in erotic *kāvya* the rhetoricians classified love-in-separation into various conditions, categorized the various types of women separated from their lovers, and enumerated the various stages and symptoms of the passionate separation of the lover and the beloved. And the medieval Vaiṣṇava theologians adapted the classifications as phases in the devotee's relationship with Kṛṣṇa.

1 *The conditions of separation*

The rhetoricians indicated four conditions of love-in-separation, *viz*., *pūrva-rāga*, *māna*, *pravāsa*, *karuṇa*.

(a) *Pūrva-rāga* (or *pūrvānurāga* or *ayoga*) is the first awakening of love, affection arising before the lovers are actually united, mutual infatuation through having seen each other, having exchanged meaningful glances:

When lovers, deprived of the sight of each other's comely form, which produces love, suffer pain, *pūrvānurāga* becomes manifest.[25]

It is 'love at first sight'. It is like the condition of love, idealized in the European medieval period and Renaissance, exemplified in Dante's longing for Beatrice:

The moment I saw her I say in all truth that the vital spirit which dwells in the inmost depths of the heart began to tremble so violently that I felt the vibration alarmingly in all my pulses, even the weakest of them.[26]

But there is a fundamental distinction between the first blossoming of love in the Indian tradition and in Dante: for Dante the profane becomes the sacred through distance, through sublimation, through a Platonic ascent from 'the beauties on earth' to 'beauty absolute'. Beatrice must transcend flesh and blood

[24] Ibid., vi. 5.8:
tad-viyogendhana-vatā tac-cintā-vipulārciṣā;
rātriṃ-divaṃ śarīram me dahyate madanāgninā.

[25] *Rasikapriyā* of Keshav Dās cited by M. S. Randhawa, *Kangra Paintings on Love* (New Delhi, 1962), p. 87.

[26] Dante, *La Vita Nuova* (Reynolds trans.), p. 29.

and lead Dante beyond it as well—she could hardly sweat, horripilate, let her hair become dishevelled, riding wildly astride her lover in frenetic copulation as Rādhā, the *nāyikā* of *kāvya*, must do to fulfil the ultimate expectations of *pūrva-rāga.* In erotic *kāvya* the ultimate aim of love, of *pūrva-rāga*, is genital intercourse; there is no distinction between *amor purus* and *amor mixtus.* Chastity had been idealized in India, not in connection with love but as a technique of liberation; to store semen was to store power.

> Chastity was characteristic of Indian asceticism from the very start. The Upaniṣads say that one may realize the Self by practicing *tapas* in the forest, free from passion. . . . Sexual excitement represented a threat against which the ascetic must constantly be on guard. When Brahmā desired his daughter, he lost all the *tapas* which he had amassed in order to create, and a nymph fell from heaven when she destroyed her *tapas* by falling in love with a mortal man.[27]

It was among the followers of Caitanya that chaste love became idealized (and this distinguishes them from the Sahajiyās); the sexual excitement which had been a threat, the desire which had been deemed an evil fetter, became a good provided that it was concentrated on Kṛṣṇa. Caitanya was completely passionate and yet strictly celibate:

> And on another occasion it so happened that the Lord was going to a place called Yamesvara Tota. And he heard at the moment a *Devadassi* singing in a sweet tune called *Gurjari*. She sang a few lines from Joydeva's *Gīta-Govinda.* And so sweetly did she sing that the whole world seemed enchanted by her song. And the Lord heard the song from a distance. And he was overpowered with emotion as he heard it. But he knew not who sang the song, nor whether the singer was a man or a woman. The Lord ran towards her to meet her. And as he ran, thorns of the *Sheja* plant pricked into his feet. And yet the Lord felt it not. The servant Govinda saw all these. And he ran after the Lord in haste. The Lord was still running till the maiden was only a short distance from him. And the servant Govinda overtook the Lord, held him in his arms saying, 'It is a woman that is singing, Oh Lord'. The Lord heard the word 'woman' and he at once recovered his senses. So he came back by the route which he took in going. And the Lord now said to the servant Govinda. 'Thou hast saved my life today. Oh

[27] O'Flaherty, *Asceticism and Eroticism* . . . , p. 40.

Govinda! for if I had touched the woman, I would most assuredly have died. . . .'[28]

The anecdote is cautionary—for the followers of Caitanya the *Gītagovinda* inspired devotion but such inspiration was not without dangers; the line between the sacred and profane dimensions of love was fine.

(b) *Māna*, the second condition of love-in-separation, is 'a curious mixture of joy and sorrow, fear and hope, pride and anger, love and repulsion'.[29] It is the 'pique' described by Stendhal:

> Pique is an impulse of vanity . . . . it either comes from feminine pride ('If I let my lover ill-treat me, he will despise me and will no longer love me') or is jealousy in all its fury.[30]

Similarly the Sanskrit rhetoricians subclassified *māna* into the pique of affection (*praṇaya-māna*) and the pique of jealousy (*īrṣyā-māna*), *māna* either spontaneously, without reason, or because the beloved has caused it through infidelity. Rādhā's *māna* is aroused by Kṛṣṇa's dalliance with the other cowherd-women—she goes away from him 'on account of jealousy because she was no longer his favourite-beloved' (II. 1). When Kṛṣṇa tries to pacify her, first through her friend, then in person, Rādhā rebukes him, displays her *māna*. The friend begs, 'O piqued-woman! don't direct your pique against Mādhava, oh!' (IX. 2–8). It is important for Rādha to display *māna*—it is one of the natural graces of the ideal heroine, one of the characteristics of the *padminī*, and according to Vātsyāyana a man despises a woman who is too easily attained, longs for one who is difficult to attain.[31] *Māna* is part of the formalized ritual, the elaborate social game, of love—Vātsyāyana explains that the *māninī* must cry, inflict pain on herself (and/or her lover), shake her head so that her hair becomes dishevelled, throw herself on the ground; the man then goes to her, pacifies her, falls at her feet; once he has placated her, he withdraws and she then goes

[28] *Caitanya-Caritāmṛta*, *Antya.* XIII (Ray trans., p. 238).

[29] De, *Vaiṣṇava Faith and Movement*, p. 164n.

[30] Stendhal, *On Love* (H.B.V. trans.), Universal Library edn (New York, 1967), p. 129.

[31] *Kāmasūtra* v. i. 16.

to him and gives herself to him.[32] This conventional pattern is exactly the plot-structure of the *Gītagovinda.*

Rādhā's jealousy enhances her charm. But Kṛṣṇa, as the conventional *nāyaka,* must never display jealousy—jealousy expressed by a man immediately changes the mood from the erotic (*śṛṅgāra-rasa*) to the comic (*hāsya-rasa*):

Doubtless the reason for this convention is that in a polygamous society the code of love cannot demand a strict fidelity from the lover. His infidelity may cause his mistress or wife to be jealous but does not necessarily lower the nobility (*udāratā*) of his sentiment. His act of infidelity may have been required by social duty or by common civility. On the other hand, if the mistress were to be unfaithful to her lover, she would cease to be a noble mistress. The lover in turn would be demeaned if he expressed emotional concern over the loss of what had thus already lost its value. Accordingly, a heroine when wronged has recourse in Sanskrit literature only to tears, silence, or bitter words, never to retaliation in kind.[33]

Unlike the other conditions of separation, *māna* is caused by the lovers themselves; it is internal rather than external and as such had special significance in the theological rhetoric which the Gosvāmins of Vṛndāvana based upon the literary rhetoric of love. Māna was the separation from Kṛṣṇa caused by the devotee himself. It was not however negative; it was rather 'the highest manifestation of *sneha* [affection] which can contain in itself even the lack of generous response to Kṛṣṇa'.[34] *Māna* symptomized the devotee's ardorous acceptance of Kṛṣṇa in intensely personal terms, the devotee's transfer of *all* the aspects and manifestations of human love onto the human god, the god whose humanness made him all the more exalted. In both the profane and the sacred dimensions *māna* 'is aroused by the fullness of love, and . . . heightens its glory'.[35]

(c) *Pravāsa,* 'dwelling abroad', is the separation of lovers against their will. Conventionally the man must travel on business, or the lovers are separated by some curse: travellers and their wives or lovers lament (I. 28; 37).

In the mythology of Kṛṣṇa *pravāsa* is exemplified by the tor-

[32] Ibid., II. x. 40–49.

[33] Ingalls, *Anthology of Sanskrit Court Poetry,* p. 15.

[34] A. K. Majumdar, *Caitanya . . .*, p. 318.

[35] *Rasikapriyā* of Keshav Dās cited by Randhawa, op. cit., p. 89.

ment which the cowherd-women feel when Kṛṣṇa leaves Gokula and goes to Mathura. But again the greatest misery results in longing for him, concentration upon him, absorption in him, which in turn results in the greatest delight. The profane erotic condition as it becomes sacralized by virtue of Kṛṣṇa resembles the Catholic 'dark night of the soul', the purgative separation from God:

> Whither hast thou hidden thyself, and hast left me,
> O Beloved, to my sighing?
> Thou didst flee like the hart, having wounded me:
> I went out after thee, calling, and thou wert gone.[36]

(d) *Karuṇa* is the 'pitiful' separation of lovers when one has died and the other mourns. The dead lover must, however, be returned to life, for if there is no hope of union the mood changes from the erotic (*śṛṅgāra-rasa*) to the pathetic or tragic (*karuṇa-rasa*). 'Death', it is explained in the *Sāhityadarpaṇa*, 'must never be depicted [in an erotic *kāvya*] because it destroys the sentiment.'[37]

> ... there must be a danger and interference with the course of true love, but the final result must see concord achieved. Hence it is impossible to expect that any drama shall be a true tragedy; in the long run the hero and heroine must be rewarded by perfect happiness and union....[38]

This aesthetic rule marks a significant distinction between the Indian and European conceptions of love and its literary or dramatic portrayal:

> Love and death, a fatal love—in these phrases is summed up ... whatever is popular, whatever is universally moving in European literature.... Happy love has no history. Romance only comes into existence where love is fatal, frowned upon, doomed by life itself.[39]

In the European tradition the profane link between love and death as exemplified by the romance reflects the sacred link established in the Christian theology of martyrdom. The cross

[36] Saint John of the Cross, *The Spiritual Canticle*, in the *Complete Works* (Peers trans.), op cit., vol. II, p. 25.

[37] *Sāhityadarpaṇa* 215: '*rasa-viccheda-hetuvān maraṇaṃ naiva varṇyate.*'

[38] A. B. Keith, *Sanskrit Drama* (Oxford, 1924), p. 278.

[39] de Rougemont, *Love in the Western World*, p. 15.

is the emblem of the link. The death of Christ, and of the martyrs in emulation of him, is the fulfilment of self-sacrificial love, and it is full of glory. Death for love, in love, by love, is glory, a glory which gives meaning to life by giving meaning to death, by making death sweet:

The death of such souls is ever sweeter and gentler than was their whole life; for they die amid the delectable encounters and impulses of love.[40]

The martyr and the mystic (whose experience of martyrdom is internal, psychological rather than physical) thus longs for the love-death:

Love, love, longed for Jesus; love, love, I yearn to die embracing thee; love, love, Jesus, my sweet bridegroom; love, love, I implore thee for death; love, love, Jesus so beloved, thou givest thyself to me by transforming me into thee.[41]

The love-death, in the European traditions, is an exaltation, a dramatic source of ultimate beauty—'What indeed', Walt Whitman asks, 'is finally beautiful except death and love?'[42]

In the Christian system of redemption 'love is strong as death'[43]—love is a soteriological force; it removes 'death's sting', transforms death into eternal life:

> Love, who wilt wake me out of the grave of mortality,
> Love who wilt adorn me with the garland of glory,
> Lord, I give myself to Thee to remain Thine for ever.[44]

Saint John of the Cross says he dies because he does not die, that only in dying of love, losing himself in union with God, will he

[40] Saint John of the Cross, *The Living Flame of Love*, in *Complete Works*, op. cit., vol. III, p. 122.

[41] Jacopone da Todi, cited by Perella, op. cit., p. 67.

[42] Walt Whitman, *Calamus* ('Scented Herbage of My Breast') in *Complete Works* (ed. Emory Holloway, [London, 1928]), p. 107:

> Death is beautiful from you, (what indeed is
> finally beautiful except death and love?)
> O I think it is not for life that I am chanting here
> my chant of lovers, I think it must be for death . . .

[43] Song of Songs 8: 6.

[44] Angelus Silesius, 'Liebe, die du mich zum Bilde . . .' in the *Penguin Book of German Verse*, p. 148.

be able to cry out, 'Now I live!'[45] The same theme runs through Arab mystical poetry: 'Death through love is life,' Ibn al Faridh wrote, 'whoever does not die of his love is unable to live by it.'[46] And the sacred theme was expressed in the profane, secular, realm—the human lover aspires to overcome death in love:

> What ever dies, was not mixt equally;
> If our two loves be one, or thou and I
> Love so alike that none do slacken, none can die.[47]

The human lover in carnal love aspires to the glory which Christian eschatology established as an ultimate value achieved in the love-death. And in the European romance the glory is realized as tragedy, exalting tragedy.

All this is in contrast to the Indian traditions in which it is fatal love which has no history, in which what is popular and moving is the happy ending. It is not glory, exaltation, that is idealized, but peace, quiescence. The divergent attitudes reflect divergent eschatologies. In the Christian tradition love can exist in death, after death, for God is wholly other than the self; the goal is a resurrection of the self in loving relationship to God. In death love may achieve its highest realization. Love is made sacred in death. But in the Indian traditions, in which the goal in death is dissolution of the self, merger with the Self, the All, freedom from rebirth, love 'beyond the grave' makes little sense. There is no love beyond *saṃsāra*; if there is to be any joy or fulfilment in love it must be in *saṃsāra*. There is no love in liberation. And it was precisely for that reason that the Vaiṣṇava *bhakta*s rejected liberation—the true *bhakta* never desires *mokṣa* 'but only seeks to realize his devotional love'.[48] 'At the utterance of the word "*mukti*" (liberation), hatred and fear arise in the mind; at the utterance of the word "*bhakti*", the mind is filled with joy.'[49]

In the European love story the union of lovers is depicted while the audience knows that in the end there will be separa-

[45] Saint John of the Cross, op. cit., vol. II, pp. 427–8.

[46] Cited by de Rougemont, *Love in the Western World*, p. 111.

[47] John Donne, 'The Good-morrow' in *Complete Poems*, Everyman edn (London, 1931), p. 1.

[48] De, *Vaiṣṇava Faith and Movement*, p. 296.

[49] *Caitanya-Caritāmṛta, Madhya.* VI, cited by Kennedy, op. cit., p. 98.

tion (and then the love-death). In the Indian love story the separation of lovers is depicted while the audience knows that in the end there will be union (and then love-in-enjoyment [*sambhoga*]). Throughout the depiction death threatens as a result of separation—Vātsyāyana explains that if union does not take place ultimately death will occur.[50] Rādhā seems 'ordained to die of separation' (IV. 17). The friend tells Kṛṣṇa that he alone can save her from that death (IV. 19–21; VI. 11; etc.); 'How can she live through long separation, having seen the mango branch, its tip in flower?' (IV. 22). The sandal-scented breeze, the 'life-breath of the world', is killing her, taking away her life-breath (VII. 39). And Kṛṣṇa also, through the separation from Rādhā, 'seems to die' (V. 3; cf. X. 11). As a result of the pain of separation the lover calls out for relief in death:

> Make torment O Malayan breeze! O Love! Take my
> life! I shall not seek refuge at home again!
> O Sister-of-Death! Moisten my limbs with your waves!
> Let the fire in my body cease! [VII. 41]

Rādhā cries out that her death would be preferable to enduring the fires of separation (VII. 5).

Like Rādhā's love for Kṛṣṇa, the medieval Vaiṣṇava *bhakta*'s love was characterized by unbearable pain.

> For the nature of the holy love for Lord is wonderful; it is full of joy within. And yet it produced the worst poisonous effects on the body. . . . For in the love for Lord both poison and nectar are wonderfully commingled. . . . Love for the Lord is indeed as full of pain as of pleasure.[51]

Rādhā as the lady-separated-from-her-lover was the prototype of the devotee, always longing for the beloved Kṛṣṇa.

2 *The lady-separated-from-her-lover*

The rhetoricians and writers on the erotic sciences classified the lady, in terms of her relationship with her lover, as either united with him or separated from him. As opposed to the 'playful, beaming, joyous' (*krīḍojjvala-praharṣitā*) heroine, the *svādhīna-*

[50] It is the last of the 'ten stages of love'; *Kāmasūtra* v. i. 4–5. See below, pp. 153 ff.
[51] *Caitanya-Caritāmṛta, Madhya.* II (Ray trans., p. 32).

bhartṛkā whose lover sits at her side and serves her, is the *virahiṇī*, or *viyoginī*, who suffers separation from her lover. Conventionally she is anxious, depressed, exhausted, she weeps, grows thin and pale, sighs, and feels that her ornaments are useless. The *virahiṇī* was subclassified by the rhetoricians into seven types and in the Bengal Vaiṣṇava devotional rhetoric each of these types represented a phase of relationship with Kṛṣṇa; as Kṛṣṇa encompassed all the possible types of heroes, Rādhā (and through her, the empathetic devotee) encompassed all the types of heroines: (a) the *proṣita-priyā* is the loving woman whose lover is in some distant land (I. 28); (b) the *vāsaka-sajjā* awaits her lover's arrival, preparing their meeting place and adorning herself for love-making (VI. 8, 11); (c) the *vipralabdhā* is insulted and offended because her lover does not come to the tryst as agreed (VII. 3); (d) the *virahotkaṇṭhitā* is distressed when her lover does not come through no fault of his own (VII. 11); (e) the *khaṇḍitā* is enraged because of the evidence of infidelity marking her lover's body (VIII); (f) the *kalahāntaritā*, separated from her lover on account of anger and *māna*, is repentant (XI. 24 ff.); (g) the *abhisārikā* goes to meet her lover (XI. 23; etc.).[52]

In the devotional songs of the medieval period the *virahiṇī* became a symbol for the devotee, for his soul, longing for the Lord. The exterior space which separated the lover and beloved then becomes a metaphor for the interior psychological obstacle between the individual self and the 'Self of all'. The removal of that obstacle was the goal of various systems of *yoga* —*yogic* meditation and the lover's longing are in a sense parallel activities, the one inward, the other outward. For the poets of the courtly tradition the parallel was ironic and they played upon it:

> Averse from eating, turned from every object of the senses
> and this too, that your eye is fixed in trance;
> again your mind is single-pointed,
> and then this silence, and the fact that all the world to
> you seems empty:—
> tell me gentle friend, are you a *yoginī* [a female
> practitioner of *yoga*]

[52] For a collation of the Sanskrit passages (with German translations) from all the major erotic and rhetorical texts dealing with the *virahiṇī*, see Schmidt, op. cit., pp. 287–310.

or a *viyoginī* [a woman separated from her lover]?[53]

This debility of body and lack of all desire,
this fixing of your eye in trance, and perfect silence:
this state bespeaks a heart fixed on one single object.
What is that one, fair lady; brahma or your lover?[54]

And in the *Gītagovinda* travellers, separated from their women, meditate upon them and obtain union only in that way (I. 37); Kṛṣṇa meditates on Rādhā, chanting *mantras* as an invocation to her, his mind fixed in *samādhi*, his mind fixed on her alone (III. 15; V. 7; XII. 6). The single-mindedness of the longing lover is like the single-mindedness of the *yogin*; the lover, like the ascetic, is averse to sensual delights; the lover burns with *kāma*, the *yogin* burns with *tapas*. The *virahiṇī* or *viyoginī* meditating upon her beloved in an attempt to experience union with him, absorption in him, as a literary convention mocks the ascetic tradition. But when the beloved is Kṛṣṇa the implications are broader—the ironic analogy becomes the crucial homology of the sacred and profane dimensions of love. The *Gītagovinda* is a meditation consecrated to Viṣṇu (XII. 28) for those whose minds are fixed on Hari (XI. 9), that is, a devotional work for devotees, and it is about the erotic contemplation of Kṛṣṇa—a young girl contemplates his lotus-face (I. 41); Rādhā's mind is fixed upon him, she is absorbed in meditation (IV. 7; VI. 10, 11); and (following Mānāṅka's gloss) Rādhā's mind does not roam from concentration on him (II. 10). Rādhā's condition and the devotee's condition are the same—separation from, and longing for union with, Kṛṣṇa the lover, Kṛṣṇa the Lord.

O Mādhava! . . . she is merged with (devoted to, absorbed in) you in her imagination—she is distressed in your absence. [IV. 2–8]

Through separation the lover becomes absorbed (*līna*) in the

[53] *Subhāṣitaratnakoṣa* 703 (Ingalls trans.):
āhare viratiḥ samasta-viṣaya-grāme nivṛttiḥ parā
nāsāgre nayanaṃ yad etad aparaṃ yac caikatānaṃ manaḥ;
maunaṃ cedam idaṃ ca śūnyam akhilam yad-viśvam ābhāti te
tad brūyāḥ sakhi yoginī kim asi bhoḥ kiṃ vā viyoginy asi.

[54] Ibid., 715 (Ingalls trans.):
yad daurbalyaṃ vapuṣi mahatī sarvataś cāspṛhā yan
nāsālakṣyam yad api nayanaṃ maunam ekāntato yat;
ekādhīnaṃ kathayati manas tāvad eṣā daśā te
ko'sāv ekaḥ kathaya sumukhi brahma vā vallabho vā.

beloved, and absorption was the goal of *yoga*. A verse attributed to Jayadeva in the *Ādi Granth* describes such *yogic* absorption—the *yogin*, like the lover, becomes weak, and loses the idea 'that thou and I are distinct':

[I] pierced (*bhediyā*) with breath (*sata* = *sattva* = *prāṇa*) the moon (*canda* = *candra* = *Iḍā*, the left nostril: i.e., I performed the *pūraka* movement in breath-control in Yoga); [I] filled (*pūriyā*) with breath the *nāda* (the *suṣumnā*, the space between the two nostrils at the top of the nose: i.e., I performed the *kumbhaka*); [I] gave up (*dattu kīyā*) the breath by the sun (*sūra* = *piṅgalā*, the right nostril: i.e., I performed the *recaka* movement)—sixteen times (*khoḍasā* = *ṣoḍaśa*: i.e., in repeating the *praṇava* or *Oṃ-kāra* sixteen times in each of the processes of taking in, holding and ejecting the breath in performing *prāṇāyāma*).

Without strength (*abala*) [its] strength broken (*toḍiyā*: i.e., the strength of the earthly frame has been broken, and it has been made weak physically); in the unmoving or fixed (*acala*), [my] unfixed or moving or unstable (*cala*) [mind or breath] has been established (*thappiyā*); the unfashioned [mind] (*aghada*) has been fashioned (*ghaḍiyā*); then or there (*tahā*) nectar (*āpiu, apiu* = *amṛta*, according to traditional explanation: *amṛta* = **ambrita* = **ambia*, **ambiu*, **abbiu*, **appiu* = *āpiu, apiu*?) has been drunk [by me] (*pīyā*).

[I have] described [Him Who is] the beginning of the mind (or soul) and of the [three] qualities (*guṇa—sattva, rajas, tamas*). Thy twofold sight (i.e., the idea that thou and I are distinct) has been lost (*sammāniyā*: Panjabi explanation—*samā jāndī hai* = 'enters'). With reference to the adorable one(s) (*ardha* = *ārādhya*), adoration has been made (*aradhiyā* = *ārādhita-*); with reference to that (those) which is (are) to be trusted or believed in (*sardhi* = *śraddhin*) trust has been given (*saradhiyā* = *śraddhita-*); as for the water (*salala* = *salila*), it has become blended (*sammāniyā*) in the water.

Jayadeva says (*badati* = *vadati*): [I] have taken joy (*rammiyā*) in the God who triumphs (*Jaya-deva*); receiving (*liwa*) absorption (*nirbāṇa* = *nirvāṇa*) in Brahman, [I] have received (*pāyā*) final absorption (*līṇa* = *līna*).[55]

[55] Trans. Chattarji, op. cit., pp. 193–4.

Bāṇī Jaidewjīū-kī (*Rāga Mārū*)
Canda sata bhediyā nāda sata pūriyā
Sūra sata khoḍasā dattu kīyā;
abala bala toḍiyā, acala cala thappiyā,
aghaḍa ghaḍiya tahā āpiu pīyā.

This absorption in *brahman* is called in the *Upaniṣads tan-maya*, 'consisting of It'—through *yogic* concentration 'an undistracted man' is absorbed in *brahman*, shares in the nature of *brahman*, becomes *brahman*.[56] The same term was used in the rhetorical texts to describe a symptom of love-in-separation—the pining lover is absorbed in the beloved, 'seeing [the beloved] everywhere, both within and without'.[57] And so Rādhā imagines that she is Kṛṣṇa (vi. 5); she sees him everywhere (vi. 2); she embraces the darkness itself thinking it is him (vi. 7). It is by virtue of this symptom of longing that love-in-separation is idealized. An epigram attributed to Bhartṛhari explains that in conjecturing about separation and union, separation from the beloved is better than union with her, because in union she is just one, in one place, but in separation all things consist of her, she is everywhere (both within and without).[58] *Tan-maya* was considered, in the *Bhakti-sūtras*, a symptom and stage of devotion to Kṛṣṇa[59]—by virtue of his nature both as lover and as lord, the scriptural technical term and the aesthetic technical term became the same term; the sacred and profane were reconciled —to consist of the beloved was to consist of the Lord. The *viyoginī* and *yoginī* were one, the ecstatically longing devotee attained liberation through his passion just as the ecstatically concentrating *yogin* attained it through his passionlessness.

3 *The stages and symptoms of separation*

Vātsyāyana describes ten stages of love: first the lovers see each other and experience love by sight (*cakṣuḥ-prīti*) and then mental attachment *(manaḥ-saṅga)* arises as do the in-

mana ādi guṇa ādi wakhāṇiyā
teri dubidhā dṛiṣṭi sammāniyā. [Rahau]
ardha-kau aradhiyā sardhi-kau saradhiyā
śalala-kau salali sammāni āyā;
badati Jaidewa Jaidewa-kau rammiyā,
Brahma-nirbāṇa liwa līṇa pāyā.

[56] *Muṇḍaka Upaniṣad* ii. ii. 4; cf. *Śvetāśvatara Upaniṣad* v. 6, vi. 17.

[57] *Sāhityadarpaṇa* 222: '*tan-mayaṃ tat-prakāśo hi bāhyābhyntaratas tathā.*'

[58] Bhartṛhari 770:
saṃgama-viraha-vitarke varaṃ hi viraho na saṃgamas tasyāḥ;
saṃgama ekā bhavatī virahe jaganti tvan-mayāni syaḥ.

[59] *Nāradīya-bhakti-sūtra* 70, 82.

tentions of love (*saṃkalpotpatti*); then there is loss of sleep (*nidrācchеda*) and loss of weight (*tanutā*) and soon the lover, if still unable to attain union with the beloved, develops an aversion to all sense objects except the perception of the beloved (*viṣayebhyo vyāvṛtti*); then there is a loss of shame or modesty (*lajjā-praṇāśa*), intoxicated-madness (*unmāda*) and fainting or stupefaction (*mūrcchā*), and finally, if union does not take place, there is death (*maraṇa*).[60] Other texts, erotic and rhetorical, elaborate the list in classifications of the further symptoms of love-in-separation: longing (*abhilāṣa*), anxiety (*cintā*), remembrance (*smaraṇa*), telling the qualities of the beloved (*guṇa-kīrtana*), agitation and fear (*udvega*), delirium and senseless chatter (*pralāpa*), seeing all things as consisting of the beloved (*tan-maya*), sickness and fever (*vyādhi*, *jvara*), stupor or stiffness (*jaḍatā*), languor and displeasure (*arati*), and so forth.[61] The texts establish conventions for the poetry of the sentiment of love-in-separation, conventions to which Jayadeva adhered in his descriptions of Rādhā's love for Kṛṣṇa and Kṛṣṇa's love for Rādhā: Kṛṣṇa is despondent (III. 1), and fearful (III. 2):

> . . . he seems to die . . . he moans all the more distraught . . . . he covers his ears . . . he falls ill . . . . he tosses on his bed which is the ground; he often moans your name; he . . . sadly waits. . . . [V. 3–5]

> At one moment he scatters sighs, then looks towards the heavens, then resorts to the grove humming, then gasps-for-breath, then prepares the bed, then looks around bewilderedly—O lovely-woman, your beloved is wearied by the suffering of love. [V. 16]

And Rādhā suffers:

> . . . she moans, laughs, grieves, weeps, wanders-about, lets-go of her grief; O Mādhava! as if from a fear of love's arrows she is merged with you in her imagination—she is distressed in your absence. [IV. 7]

> O Keśava! In your absence Rādhikā, the slender-bodied one, considers even the exalted necklace placed upon her breasts to be like a burden . . . . she fearfully regards the sandal unguent, truly passionate and smooth, to be like a poison on her body . . . . she bears her sighing breath, incomparably long, burning like the scorching of love . . . . she casts the lotus of her eye, which has a multitude of watery drops, in

[60] *Kāmasūtra* V. i. 4–5.

[61] See Schmidt, op. cit., pp. 124–32, for a collation of all the stages and symptoms of love from all the major Sanskrit erotic and rhetorical texts.

every direction . . . . she does not let her cheek go from the palm-of-her-hand . . . . she considers the bed of sprouts . . . . to be made of fire . . . . she mutters 'Hari, Hari' passionately . . . as if ordained to die of separation. [IV. 11–17]

She bristles, makes love-cries, laments, trembles, gasps-for-breath, ponders, jumps-up, closes-her-eyes, falls, rises, and even faints. . . . [IV. 19 *passim*]

The same symptoms have, of course, typified western descriptions of love since the Shulamite who was 'sick with love',[62] since Sappho 'in the love-trance':

> Yea, my tongue is broken, and through and through me
> 'Neath the flesh, impalpable fire runs tingling:
> Nothing see mine eyes, and a voice of roaring
> Waves in my ear sounds;
> Sweat runs down in rivers, a tremor seizes
> All my limbs, and paler than grass in autumn,
> Caught by pains of menacing death, I falter,
> Lost in the love-trance.[63]

And the symptoms of profane love became the symptoms of sacred love:

The soul, then, has these yarnings and tears and sighs. . . . They all seem to arise from our love . . . they are like a smouldering fire. . . . The understanding is keenly alert to discover why this soul feels absent from God . . . it causes the distress to grow until the sufferer cries out loud. . . . I saw a person in this state who I really believed was dying. . . .[64]

And similarly, as we have seen, Caitanya felt the symptoms of love-in-separation, the symptoms which were identical for both the sacred and profane dimensions of love.

In the Kṛṣṇa-*bhakti* tradition various stages, symptoms, manifestations, of devotion to the Lord, closely related to the stages, symptoms and manifestations of secular love-in-separation as classified in the rhetorical and erotic texts, were enumerated,

[62] Song of Songs 2: 5.

[63] Sappho, 'Ode to Atthis', cited by Hunt, *Natural History of Love*, p. 45.

[64] Saint Teresa, *Interior Castle* VI. xi, in the *Complete Works* (Peers trans.), vol. II (London, 1946), pp. 324–5.

including the proclaiming of his qualities (*guṇa-kīrtana*) and listening to stories about him (*śravaṇa*), remembrance of him (*smaraṇa*) and having him as one's refuge (*śaraṇa*):[65]

(a) Proclaiming Kṛṣṇa's qualities (*guṇa-kīrtana*) is both an erotic activity as a symptom of estrangement from Kṛṣṇa the lover and a religious activity as a manifestation of devotion to Kṛṣṇa the Bhagavat. Rādhā's enumeration of his qualities (ii. 2–8; vii. 31–37; *passim*) is erotic; Jayadeva's enumeration of them (in the work as a whole) is devotional. The sacredness of the love then depends not only on the object of love, but also on the subject, the source. The erotic activity of singing the qualities of the beloved, as an expression of attachment and desire, represents ensnarement in *saṃsāra*, and yet as a devotional activity singing the qualities of the Lord is liberation itself—to have told the Kṛṣṇa stories, to have sung his praises and recited his name, was considered in the *bhakti* tradition to be better than having poured oblations on sacrificial fires, better than having bathed in the holy rivers, better than having recited the *Vedas*.[66]

In the *Bhaktavijaya* of Mahīpati, a Vaiṣṇava hagiography, Viṣṇu instructs Vyāsa, 'the author of the *Purāṇas*', to descend to earth as an *avatāra* in the form of Jayadeva the poet in order to 'bring about the salvation of mankind' by singing the *Gītagovinda*. Viṣṇu explains that the *Vedas* and *Purāṇas*, the orthodox sacred texts, cannot really be understood in the Kali Era—'in the *Kali Yuga* the praising of God is the chief means of salvation.'[67] As a participant in this age of degeneration and above all as a poet, Jayadeva most appropriately expresses his devotion with *guṇa-kīrtana*—the songs spread happiness and prosperity, and dispel the fevers of the age, for they are sung 'with devotion to Hari' (iii. 10; etc.). Jayadeva says that others should recite the songs as well (iv. 9; xi. 9) and they are, in fact, used in

[65] The *Bhaktiratnāvalī* of Viṣṇupuri lists ten stages or expressions of *bhakti*, *viz.*, *kīrtana*, *śravaṇa*, *smaraṇa*, *pāda-sevā* (honouring Kṛṣṇa's feet), *arcanā* (ritual worship), *vadana* (voicing Kṛṣṇa's name), *dāsya* (servitude), *sakhya* (friendship), *ātma-nivedana* (dedicating one's self to him), and *śaraṇāpattiḥ*. Each mode of devotion is described in a separate chapter or 'string' comprised of verses from the *Bhāgavata Purāṇa* extolling the particular mode. (Cf. *Nāradīya-bhakti-sūtra* 82, cited above, p. 140.)

[66] *Bhaktiratnāvalī* v, e. g., *Bhāgavata Purāṇa* iii. xxxiii. 7.

[67] *Bhaktavijaya* of Mahīpati ii. 11, trans. from Marathi by J. E. Abbot and Pandit N. R. Godbole (Poona, 1933), p. 12.

contemporary congregational devotional worship, in the *bhajana*.[68]

(b) Listening to songs which praise Kṛṣṇa (*śravaṇa*) no less than reciting them drives away the evil of the age and brings joy, prosperity and devotion which in turn is liberation.[69] Repeatedly Jayadeva uses the imperative 'listen . . .' or 'let it be heard' (I. 15; VII. 9; etc.); he who listens to the songs has Kṛṣṇa enter his heart, and that is the ultimate goal of *bhakti*—'While the poet Jayadeva sings, may Hari appear . . . in your mind . . .' (V. 6), 'Through this speech as sung by Śrī Jayadeva, may Hari too enter [your] heart!' (VII. 38). As the love-messenger sings the deeds of Kṛṣṇa and recites his qualities to Rādhā, the lover is established in her heart; as it is Jayadeva singing, the Lord is established in the heart of the devotee, the *rasika*. *Śravaṇa* then as an activity *in* the *Gītagovinda* is erotic while as an activity *of* the *Gītagovinda* it is devotional. Listening to his praises causes Rādhā to remember Kṛṣṇa and remembrance is also an expression and manifestation of *bhakti*.

(c) Remembrance (*smaraṇa*) had erotic connotations as early as the *Atharva-Veda*:

Of the Apsarases, chariot conquering, belonging to the chariot-conquering, [is] this the love (*smara*): Ye gods, send forth love; let you [man] burn for me. Let you love (*smṛ*) me. . . .[70]

The chant would be used by a woman to make a man love her —if she could capture his 'memory', he would 'burn' for her. And 'remembrance', as we have seen, became an epithet of the love-god.[71]

Remembrance of the beloved, a symptom of love-in-separa-

[68] Milton Singer notes the use of songs from the *Gītagovinda* in this congregational worship in Madras: 'One *aṣṭāpadī* . . . is sung during each weekly *bhajana* until the entire work of twenty-four *aṣṭāpadīs* is finished; then the first *aṣṭāpadī* is repeated and so on.' 'The Rādhā-Krishna *Bhajanas* of Madras City' in Singer's (ed.), *Krishna: Myths, Rites and Attitudes*, p. 93.

[69] *Bhaktiratnāvalī* IV.

[70] *Atharva-Veda* VI. 103.1 (Whitney trans.):

ratha-jitāṃ rātha-jiteyīnām apsarasām ayaṃ smaraḥ
devāḥ pra hiṇuta smaram asau mām anu socatu
asau me smaratāditi priyo me smaratāditi.

[71] See above, p. 76. N.B. *Gītagovinda* I. 31; IV. 20; VII. 13; VIII. 1, 4; X. 8; XI. 8, 10; XII. 1.

tion, is a conventional motif in Sanskrit erotic *kāvya* enabling the poet to combine the moods of separation and union, a combination which Abhinavagupta explained heightens the charm of the verse.[72] There is pain due to the present absence and pleasure due to remembering the past presence. Aristotle delineated the psychology of loving recollection:

> The lovesick always take pleasure in talking, writing, or composing verse about the beloved; for it seems to them that in all this recollection makes the object of their affection perceptible. Love always begins in this manner, when men are happy not only in the presence of the beloved, but also in his absence when they call him to mind. This is why, even when his absence is painful, there is a certain amount of pleasure even in mourning and lamentation; for the pain is due to his absence, but there is pleasure in remembering and, as it were, seeing him and recalling his actions and personality.[73]

And so Rādhā recalls Kṛṣṇa's charms: 'my mind remembers Hari—he joked and played love-games here during the *rāsa* [dance]' (II. 2–8). Her remembrance is erotic but the song ends with Jayadeva's remembrance which is devotional: 'The song of Śrī Jayadeva . . . is exactly suitable for the virtuous for remembrance of the feet of Hari' (II. 9). The *Gītagovinda* is for those whose minds are 'passionate in remembrance of Hari' (I. 3); Jayadeva's description of the forest in the amorous springtime is 'the essence of remembrance of the feet of Hari' (I. 35); the song in which Rādhā asks Kṛṣṇa to adorn her after love-making 'dispels the fever and impurity of the Kali [Era] for it is formed with the nectar which is remembrance of the feet of Hari' (XII. 24). In the *Bhagavad-Gītā* Kṛṣṇa says that he is easily won by the one who remembers him;[74] in the *Viṣṇu Purāṇa* remembrance of Viṣṇu is said to dispel all evil;[75] remembrance of Kṛṣṇa yields liberation according to the *Bhāgavata Purāṇa*.[76]

The term *smaraṇa* 'reflects through its two connotations the dialectic of the *Gītagovinda*: erotic poem as religious text'.[77]

[72] Cited by Ingalls, *Anthology* . . . , p. 216.

[73] Aristotle, *Rhetoric*, I. xi. 11–12, trans. J. H. Freese, Loeb Library (London, 1926), p. 121.

[74] *Bhagavad-Gītā* VIII. 14.

[75] *Viṣṇu Purāṇa* II. vi. 40, etc.

[76] Passages dealing with *smaraṇa* collected in *Bhaktiratnāvalī* VI.

[77] Friedhelm Hardy, personal communication.

The ambiguity of the term reflects the ambiguity of the work.

(d) Taking refuge (*śaraṇa*) in Kṛṣṇa is listed in the devotional texts as a symptom and expression of *bhakti*[78]—the impulse to seek the refuge comes out of the devotee's despair at separation from the Lord or generally out of the despair which typifies the Kali Age. Jayadeva links this despair with the pain of love-in-separation in the thirteenth song of the *Gītagovinda*—when Kṛṣṇa does not come to the tryst, Rādhā cries out 'to whom shall I . . . go for refuge . . .?' (vii. 3–9). Jayadeva then juxtaposes his devotion with her erotic longing by declaring, in the last verse of the song, that he takes 'refuge at the feet of Hari' (vii. 9). Separation is valued because it makes lover-devotee realize the need for refuge in, union with, Kṛṣṇa:

> In Him alone seek refuge with all your being, all your love; and by his grace you will attain an eternal state, the highest peace.[79]

UNION

> Grant me one hour on love's most sacred shores
> To clasp the bosom that my soul adores,
> Lie heart to heart and merge my soul in yours.[80]

The *Sāhityadarpaṇa* catalogues four kinds of union, four conditions of love-in-enjoyment, one for each of the four conditions of separation.[81] Union is made meaningful by separation; the pain of separation yields the joy of union. 'Love-in-enjoyment without love-in-separation cannot prosper',[82] cannot reach fullness. And so Rādhā and Kṛṣṇa, 'though one, become two', though eternally united, according to the medieval Vaiṣṇavas, become separated in order to taste the ravishing sweetness of reunion. The *Gītagovinda* begins with their union (i. 1) and ends with their union (xi. 23 ff.) and between union and reunion there is separation filled with remembrance and anticipation. This pat-

[78] *Bhaktiratnāvalī* xiii.

[79] *Bhagavad-Gītā* xviii. 62 (Zaehner trans.):
tam eva śaraṇaṃ gaccha sarva-bhāvena . . . ;
tat-prasādāt parāṃ śāntiṃ sthānaṃ prāpsyasi śāśvatam.

[80] *Faust, Pt. I* (Wayne trans.), op. cit., pp. 154–5.

[81] *Sāhityadarpaṇa* 227.

[82] Ibid., loc. cit., commentary: '*na vinā vipralambhena sambhogaḥ puṣṭim aśnute.*'

tern, union–separation–reunion, is the conventional pattern in Indian erotic literature. But it is also the archetypal structure in Indian ontology: in the beginning was the All, the One, *Brahman*, *Ātman*, *Puruṣa*, the sacred power; creation meant separation, duality, multiplicity; and then at the end of each cosmic era, there is reunion, reabsorption into the One; and then it starts again and again and again—reunion–re-separation–reunion–re-separation–reunion. . . . For the individual liberation is reunion, a return to the primordial unity, a recovery of the unborn Self:

> Myself immortal, wise, I know Him
> The Self—Immortal—Brahman!
> Breath of breath, eye of eye,
> Ear of ear, and mind of mind:
> Who knows him thus, has understood
> Primeval Brahman who from the beginning is.
> Descry This with your mind:
> Herein there's no diversity at all.
> Death beyond death is all the lot
> Of him who sees in This what seems to be diverse.
> Descry It in its Oneness,
> Immeasurable, firm,
> Transcending space, immaculate,
> Unborn, abiding, great,—
> [This is] the Self![83]

In the Vedānta system the 'separation' is *māyā*, the world-illusion, and the highest goal is to overcome that illusion, to experience the All, the One, the only Reality, 'The Self—Immortal—Brahman!' In the Sāṃkhya system the 'separation' is of the self, the *jīva*, from the Primal *Puruṣa*, it is the entanglement of the *jīva* in *prakṛti*, nature; again the highest goal is reunion, the isolation of the self from nature. The structure of the relationship between Rādhā and Kṛṣṇa reflects, echoes, the structure of

[83] *Bṛhadāraṇyaka Upaniṣad* IV. vi. 17–20 (Zaehner trans.):
tam eva manya ātmānam vidvān brahmā'mṛto'mṛtam.
prāṇasya prāṇam uta cakṣuṣaś cakṣuḥ uta śrotrasya śrotram;
manaso ye mano viduḥ te nicikyur brahma purāṇam agryam.
manasaivānudraṣṭavyam na iha nānāsti kiṃ cana;
mṛtyoḥ sa mṛtyum āpnoti ya iha nāneva paśyati.
ekadhaivānudraṣṭavyam etad aprameyaṃ dhruvam;
virajaḥ para ākāśād aja ātmā mahān dhruvaḥ.

the relationship between the *jīva* and the *Puruṣa* and the temporal enactment of the relationship, the movement from union to separation to reunion, reflects the process, individual or cosmic, of entry into *saṃsāra* and liberation from it. The structure of the profane experience is identical to the structure of the sacred and by virtue of that structure the profane could be interpreted as, or *be*, the sacred.

Each phase of the pattern as it applies to the sexual life was scrutinized by the eroticians and for each phase they delineated a system of rules. There were correct and proper ways for lovers to meet, to make love, to depart. And following the eroticians the rhetoricians delineated a system of rules for the depiction of that behaviour in art. Love-in-enjoyment was as conventionalized in *kāvya* as love-in-separation. The behaviour pattern of lovers and the representation of that pattern in art was highly formalized, 'elevated to the level of a rite' as it was in medieval Europe:

> To formalize love is the supreme realization of the aspiration to the life beautiful. . . . To formalize love is a social necessity, a need that is the more imperious as life is more ferocious. Love has to be elevated to the level of a rite. The overflowing violence of passion demands it. Only by constructing a system of forms and rules for the vehement emotions can barbarity be escaped.[84]

Forms and rules typify the enjoyment of united lovers in *kāvya*. Dhanaṃjaya says there are ten actions of women which occur during love-in-union: she plays, shows delight, is bashful, pretends anger, manifests her affection and so forth and 'her lover, flattering her, should delight her with love-play and the amatory-arts and so forth. . . .'[85] Love-in-union was 'love-play', a game with rules, and it was an 'amatory-art', a science to be learned and practised with refinement.

1 *The art, the science*

The second part of the *Kāmasūtra* (the *Saṃprayogika*) is an annotated catalogue of the 'sixty-four' arts of love-making, eight

[84] Johan Huizinga, *The Waning of the Middle Ages*, trans. F. Hopman (London, 1924), p. 96.

[85] *Daśarūpa* IV. 76–78: '. . . *ramyec cāṭu-kṛt kāntaḥ kalā-krīḍābhiś ca tām* . . .'.

basic procedures (each in turn subclassified into eight types), *viz.*, embracing, kissing, scratching, biting, coital postures, striking and love-cries, the woman assuming the man's role, and oral copulation: Vātsyāyana declares that mastery of these arts enables one to fulfil the 'three aims of life' (*kāma*, *artha*, and *dharma*), to win a lover and to command a leading position in cultured society.[86] The descriptions of sexual union in Sanskrit *kāvya* draw upon Vātsyāyana's exposition of the arts and sciences of love:

There is no question of the importance of knowledge of this topic for the writers of erotic poetry and there is abundant proof that the *Kāmasūtra* was studied as eagerly by would-be poets as were grammar, poetics, and lexicography.[87]

The technical terms from the erotic textbooks could be used for puns and suggestive allusions. For example, Jayadeva describes a woman's breasts as being like the heavens—as the firmament is adorned with clouds, stars and the moon, her breasts are adorned with musk, jewels and a nail-mark known in the erotic texts as 'the moon' (VII. 24).[88]

In that the poet apparently used the language and conventions of the erotic *śāstra* (rather than his own experience) to depict love-in-union, the lovers in *kāvya* often have the manners and sexual sensibilities of those lovers for whom the *Kāmasūtra* was compiled: the educated and accomplished courtesan, the 'loving-woman' of the royal harem, and the *nāgaraka*, the polished 'man-about-town', the urbane, cosmopolitan, cultured gentleman whose daily life consisted in adorning himself, bathing, eating, resting, teaching parrots to speak, gaming, listening to and playing music, discussing art and literature, indulging in witty repartee, going to social gatherings, and above all love-making, love-making with style and refinement.[89] It is a textbook on sensual enjoyment for those for whom sensual enjoyment would constitute a fulfilment of *dharma*.[90]

[86] *Kāmasūtra* II. x. 49 ff.

[87] Keith, *History of Sanskrit Literature*, p. 51.

[88] Śaṅkara Miśra glosses the verse with a citation from the *Pañcasāyaka* describing the 'crescent-moon' (*ardhendu*) scratch; cf. *Kāmasūtra* II. iv. 14.

[89] *Kāmasūtra* I. iv.

[90] It is also a 'marriage manual', but the wife was expected to fulfil her '*dharma*' in a service to her husband extending well beyond sexual service.

Kṛṣṇa, although he is raised among the cowherds, and Rādhā, although she is a milkmaid, become the refined lovers of the erotic tradition as they are united in the *Gītagovinda*, which is about the 'amatory arts' (I. 3). Jayadeva's muse is 'like a young girl who is skilled in the charming arts [of love] (VII. 10). Rādhā is accomplished in those arts (X. 15); she prepares the bed to practise them with Kṛṣṇa (IV. 4). Kṛṣṇa is *Nāgara-Nārāyaṇa*—the Lord is a *nāgaraka* (VII. title); he is proficient in the arts of love and Rādhā longs to experience that proficiency (VI. 4); a cowherdess pulls him into a grove with 'curiosity in the arts of love-play' (I. 43). Kṛṣṇa has 'mastered the procedures of the science (*tantra*) of love' (II. 5), the *kāma-śāstra*, but also 'Tantra' might be taken as a proper noun—he is proficient in the erotic practices which make up Tantric *sādhanā*, the sexual union in which the human couple becomes the divine couple, uniting *all* opposites, cosmic and biological, external and internal:

It is the coincidence of time and eternity, of *bhāva* and *nirvāṇa*; on the purely 'human' plane, it is the reintegration of the primordial androgyne, the conjunction, in one's own being, of male and female—in a word, the reconquest of the completeness that precedes all creation.[91]

The art of love is a battle (XII. 10).[92] The purpose of the arts and sciences of love-in-union is, of course, to produce pleasure, but the pleasure is often achieved only through inflicting and receiving pain. The poet Bhāravi declared that love, although renowned for tenderness, is 'truly cruel (or paradoxical [*vāma*]) during sexual-union'.[93] The arts and sciences of sexual union are explicitly sado-masochistic:

If you are *truly* angry, O lovely-toothed-woman, give me a wound with the arrows that are your sharp nails, bind me with fetters which are your arms or bite me, by which all-the-conditions of pleasure come to be. . . . [X. 3]

O artless-woman! give me squeezes against your ample breasts and

[91] Mircea Eliade, *Yoga, Immortality and Freedom*, trans. Willard Trask, 2nd edn (Princeton, 1969), p. 271.

[92] See above, pp. 100–1.

[93] Bhāravi's *Kirātārjunīya* (IX. 49), cited by Yaśodhara in his commentary on *Kāmasūtra* II. vii. 1–2: '*saukumārya-guṇa-saṃbhṛta-kīrtir vāma eva surateṣv api kāmaḥ*.'

bind me with the creepers of your arms and bite me with cruel teeth! . . . [x. 11]

Held-captive by her arms, pressed by the weight of her breasts, pierced by her finger-nails, the cup of his lower-lip bitten by her teeth, crushed by the slope of her hips, bent-down by her hand on his hair, crazed by the trickling-flow of honey from her lower-lip, the lovely-beloved (Kṛṣṇa) somehow obtained delight—so, oh! the way of love is paradoxical (or cruel [*vāma*])! [xii. 11]

The sado-masochistic trend underlines the arts of love-making as delineated by Vātsyāyana:

(a) Embraces (*upagūhana*)[94]—Vātsyāyana stresses the pleasure derived from the application of great pressure; there is violence inherent in the technical vocabulary—among the embraces catalogued are the 'piercing' embrace (the *viddhaka* from √*vyadh-*, 'to pierce, transfix, strike, hit, wound . . .'), the 'rubbing' embrace (the *udghṛṣṭaka* from √*ghṛṣ-*, 'to rub, grind, crush, pulverize . . .'), and the 'squeezing' embrace (*pīḍitaka* from √*pīḍ-*, 'to press, squeeze, hurt, injure, pain . . .').[95] Jayadeva uses the terminology unequivocally—Kṛṣṇa asks Rādhā to give him the squeezing-embrace (*pīḍana*) (x. 11) and in union with him she does so (xii. 11); squeezing (*pīḍayan*) Rādhā he delightedly 'fears' that her breasts may break his back (xi. 34) and in the subsequent verse the same term (*pīḍa*) is used to describe the way in which he kills the war-elephant of Kaṃsa (xi. 35). This close juxtaposition of the term in an erotic and in a martial usage emphasizes the affinities of the sexual and aggressive instincts, of love and war—the goal is to overwhelm the partner whether that partner is the beloved or the enemy. Similar to the *pīḍitaka* embrace is the act of *mardana* (from √*mṛd-*, 'to press, squeeze, crush, destroy, kill . . .') which is deemed appropriate to love-making in various erotic texts.[96] Kṛṣṇa's chest is 'cruel in crushing (*mardana*) the swollen breasts' of the cowherd-women (ii. 6); 'his hands are ever-moving in squeezing (*mardana*) the cowherdesses' swollen breasts' (v. 8 ff.).

[94] On embraces see *Kāmasūtra* ii. ii; Schmidt, op. cit., pp. 431–5, has collated all the descriptions of the various embraces from the major Sanskrit erotic texts.

[95] These definitions (and below) are from Monier-Williams, op. cit., *s.v.*

[96] On *mardana* see the *Nāgarasarvasa* and the *Smaradīpikā*, cited by Schmidt, op. cit., p. 527.

As pleasurable things are painful in separation (II. 20; *passim*) so painful things are pleasurable in union.

(b) Kissing (*cumbana*)[97]—'He gave me kisses', Rādhā remembers, 'and pulled my hair' (II. 16); hair-pulling should, according to the eroticians, accompany kissing:

The glossy, thick, curling, black hair of young women is praised! In order to increase passion it should be pulled slowly by men at the time of kissing, according to the precepts [of the erotic texts].[98]

There is a blend of pain and pleasure. Kissing is usually described as drinking the honey, mead or nectar from the beloved's lower-lip. The metaphor suggests the life-giving nature of the kiss—'Bring the elixir or nectar from your lower-lip', Kṛṣṇa exclaims, '. . . vivify me . . .' (XII. 6). It is the ambrosial kiss which, conventionally, keeps the lovers alive in the battle which is love-making:

> Struck on all sides in the amorous battle,
> her body scarred from stroke of nail and tooth,
> she would perish surely in an instant
> did she not quaff ambrosia from her lover's lip.[99]

(c) Scratching and biting (*nakhacchedya* and *daśanacchedya*)[100]—according to Vātsyāyana 'there is no sharper erotic stimulant than the effects of nails and teeth'.[101] The 'scars from stroke of nail and tooth' which the lovers inflict upon each other were felt to have sentimental and ornamental value, value as a love-stimulant and indication of intense passion. Lovers scratch each

[97] See *Kāmasūtra* II. iii.

[98] The *Anaṅgaraṅga* of Kalyānamalla IX. 37:
snigdhā ghanāḥ kuñcita-nīla-varṇāḥ keśāḥ praśastās
taruṇī-janānām;
rāga-pravṛddhyai vidhinaiva mandaṃ grāhyā narais
cumbanadānakāle.

[99] *Subhāṣitaratnakoṣa* 586 (Ingalls trans.):
nakha-daśana-nipātaja-jarjarāṅgī rati-kalahe
paripīḍitā prahāraiḥ;
sapadi maraṇam eva sā tu yāyād yadi na pibed
adharāmṛtaṃ priyasya.

[100] On nail-marks see *Kāmasūtra* II. iv and Schmidt, op. cit., pp. 478–96, and on love-bites *Kāmasūtra* II. v and Schmidt, pp. 496–508.

[101] *Kāmasūtra* II. iv. 31:
nānyatpaṭutaraṃ kiñcid asti rāgavivardhanam;
nakha-danta-samutthānāṃ karmaṇāṃ gatayo yathā.

other, Vātsyāyana says, as their passion increases;[102] and this tender aggression is said to be appropriate particularly:

. . . at the time of first union, when returning from or embarking on a journey, when placating a woman after a quarrel with her, when she is drunk.[103]

Kṛṣṇa's body is scratched (VIII. 4) and his lip is bitten (VIII. 6) by a cowherdess; Rādhā scratches him at their first union (II. 15) and he scratches her at their re-union (XII. 13). The loving assault is reciprocal, as it should be according to the *Kāmasūtra*, 'blow for blow, kiss for kiss'.[104]

The commentator Vanamālibhaṭṭa (author of the *Sañjīvanī*) sees in Jayadeva's eulogistic verses of Kṛṣṇa in the forms of the Man-lion and Boar suggestions of the erotic activities of scratching and biting.

The nail on your fair lotus-hand has a marvellous point which tore open the bee-like body of Hiraṇyakaśipu; O Keśava who bore the form of the Man-lion! . . . [I. 8]

The bee and the lotus typify the erotic mood—the masculine bee drinks honey from the feminine flower. Jayadeva establishes the metaphorical erotic relationship behind the relationship between Kṛṣṇa and Hiraṇyakaśipu, the hero and the demon; the juxtaposition provides a poetic tension. The poet reverses the conventional image of the bee entering the flower—he endows the flower with the sting, the 'marvellous point'. It is a subtle presentation of the same juxtaposition of the erotic and heroic which Māgha makes more directly in the *Śiśupālavadha*:

O man-lion, when thou didst assume that mighty lion form and cleft with thy tawny mane the clouds, thou didst tear him to pieces, rending asunder his breast with those nails which bend so gently on a loving maiden's bosom.[105]

[102] Ibid., II. iv. 1: '*rāga-vṛddhau saṃgharṣātmakaṃ nakha-vilekhanam.*'

[103] Ibid., II. iv. 2:

> tasya prathama-samāgame pravāsa-pratyāgamane pravāsagamane
> kruddha-prasannāyāṃ mattāyāṃ ca prayogo na nityam
> acaṇḍavegayoḥ.

[104] *Kāmasūtra* II. iii. 34:

> kṛte pratikṛtaṃ kuryāt tāḍite pratitāḍitam;
> karaṇena ca tenaiva cumbite praticumbitam.

[105] Cited and trans. Keith, op. cit., p. 128:

Similarly the earth on the tusk of the Boar *avatāra* 'fixed like the dark-spot upon the moon' (I. 7) suggests to Vanamālibhaṭṭa the amatory 'art' of biting—the moon typifies the erotic sentiment and is often compared to the face of the beloved; the 'boar-mark' is a particular kind of love-bite.[106]

In addition to these explicitly aggressive 'arts' of love-making the eroticians prescribe thrashing, thumping, and slapping, and describe the appropriate moans, cries and shrieks.[107] Vātsyāyana attributes roughness (*pāruṣya*) and violence or impetuosity (*rabhasatva*) to the man and weakness (*aśakti*) and gentleness (*abalatva*) to the woman, but he explains that in the full rage of passion these attributes can be reversed[108]—the woman does not take the assault lying down; in the battle of love she rides astride her lover; 'she falls upon his chest' (VII. 19).

(d) The woman assuming the man's role in copulation (*viparīta-rata*)[109]—Rādhā's friend urges her to make love to Kṛṣṇa in this 'inverse' mode (V. 12); when Rādhā does finally go to him she mounts Kṛṣṇa 'for victory over her lover' in the battle of love (XII. 12) and in this manner tries to demonstrate the 'heroic sentiment'; the woman is active and vigorous in the sexual siege.

These scenes [conventional in *kāvya*] are used to furnish an impression of intimacy between lovers, born of long affection, and of the heroine's desire to please her lover rather than herself. . . . Much of the charm of the *viparītarata* verse to the Indian reader was the masculine one of finding the woman all the more feminine by her attempt to imitate the man.[110]

In this coital posture the woman identifies with the man—'I am the Enemy-of-Madhu', Rādhā thinks, as she imagines making love to Kṛṣṇa in the inverse position (following a consensus of commentators) (VI. 5). The reversal of roles between Rādhā and Kṛṣṇa became a conventional motif in Kṛṣṇaite literature:

---

saṭācchaṭā-bhinna-ghanena bibhratā nṛ-siṃha saiṃhīm
atanuṃ tanuṃ tvayā;
sa mugdha-kānta-stana-saṅga-bhaṅgurair urovidāraṃ prati-
caskare nakhaiḥ.

[106] The '*varāha-carvitaka*', *Kāmasūtra* II. v. 17.

[107] Ibid., II. vii.

[108] Ibid., II. vii. 20–21.

[109] Ibid., II. viii.

[110] Ingalls, *Anthology* . . ., p. 200.

O Lord, I would play on your flute! Oh Beloved, those same notes which I heard you playing I would also play! I will put on your jewelry, and I will put my own jewelry upon you. You will sit as if you were angry with me, and I will assuage your anger by touching your feet. . . . You become Rādhā and I will become Mādhava, truly Mādhava; this is the reversal which I shall produce. I would braid your hair and will put (your) crown upon my head. Sūr Dās says: Thus the Lord becomes Rādhā and Rādhā the son of Nanda.[111]

The motif most likely represents a popular religious adaptation of a standard literary convention whereby the lovers, dressing in the dark after love-making, accidentally put on each other's clothes (VII. 42). When the lovers become Rādhā and Kṛṣṇa the reversal of roles, the exchange of garments, the *viparīta-rata*, could suggest identification of devotee and the Lord—the sexual utterance, 'I am the Enemy-of-Madhu', could have sacred, devotional, significance. The identification of one's self as the Self with *Brahman*, the All, saying, '*Aham Brahmāsmi*—I am *brahman*', had been the ultimate realization of the *Upaniṣads*:

In the beginning this [universe] was brahman alone, and he truly knew [him]self (*ātman*), saying: 'I am Brahman.' And so he became the All. Whichever of the gods became aware of this, all became that [All]: so too with seers (*ṛṣi*) and men. Seeing this, Vāmadeva, the seer, realized [this, saying]: 'I became Manu and the sun!' This is true even now. Whoso thus knows that he is Brahman, becomes this whole [universe]. Even the gods have not the power to cause him to un-Be, for he becomes their own self. So whoever reveres another deity, thinking: 'He is one and I am another,' does not [rightly] understand. . . .[112]

This proposition became the crucial truth of the Vedānta

[111] Sūr Dās, cited by Walter Spink, *Krishnamandala* (Ann Arbor, 1971), p. 88.

[112] *Bṛhadāraṇyaka Upaniṣad* I. iv. 10 (Zaehner trans.):

brahma vā idam agra āsīt tad ātmānam evāvet; aham
brahmāsmīti; tasmāt tat sarvam abhavat; tad yo yo
devānām pratyabudhyata; sa eva tad abhavat; tathā
ṛṣīṇāṃ tathā manuṣyāṇām; taddhaitat paśyan ṛṣir
vāma-devaḥ pratipede; aham manur abhavam sūryaś ceti;
tad idam api etarhi ya evaṃ veda; aham brahmāsmīti; sa
idaṃ sarvam bhavati; tasya ha na devāś ca nābhūtyā īśate;
ātmā hy eṣāṃ sa bhavati; atha yo anyāṃ devatām upāste;
anyo'sau anyo'ham asmīti; na sa veda . . . .

system; but the identification in this context excludes loving, excludes even the notion of union:

> The difficulty is that in this case it is not strictly proper to speak of union at all; for according to the proposition 'I am Brahman', which means that I am the sole unqualified Absolute, One without a second, I cannot logically speak of being united to Brahman, since I am already He (or It).[113]

But Kṛṣṇa, according to the Bengal Vaiṣṇavas, encompassed *brahman*, was higher than *brahman*; in the context of their teaching the goal remained identification with God, but it was identification (through separation) based on love and including the notion of separation-and-union:

> As the mirror to my hand, . . . so you to me. But tell me, Mādhava, beloved, who are you? Who are you really? Vidyāpati says, 'they are one another.'[114]

Rādhā's profane sexual fantasy of being Kṛṣṇa in the *viparīta* sexual posture fulfils the sacred ideal of loving identity.

The inverse copulatory position had already established religious ramifications in Hindu Tantric inconography—the female is frequently depicted astride the male in the '*maithuna viparīta*', the sexual union of the two fundamental principles of the universe, the passive male, *puruṣa*, Śiva, and the active female, *prakṛti*, Śakti:

> The quiescent Shiva-aspect is by its definition inert. It is because of this that the Devī is in the Tantras symbolically represented as being above the body of Shiva. . . . Activity is in the nature of Prakriti. . . . For the same reason the female form is represented in sexual union as being above (*viparīta*) the male.[115]

Vātsyāyana explains that the reversal of roles can only be temporary;[116] the woman must finally surrender to the man, the human being to the god; activity must finally cease. In the erotic battle Rādhā becomes weak and weary, she is wounded and her cosmetics are worn off, her hair is dishevelled, her

[113] Zaehner, *Mysticism Sacred and Profane*, p. 32.

[114] Trans. Edward Dimock and Denise Levertov, *In Praise of Krishna* (New York, 1967), p. 15.

[115] Sir John Woodroffe, *The Serpent Power*, 5th edn (Madras, 1953), p. 27.

[116] *Kāmc ūtra* II. vii. 22.

garments thrown off (xii. 12–15). It is as if she has been slain in battle: 'the surface of her hips was motionless, the creepers of her arms were loosened, her breast was shaken, her eyes were closed . . .' (xii. 12). Sexual satisfaction is a death:

> The ejection of sexual substances in the sexual act corresponds in a sense to the separation of soma and germ plasm. This accounts for the likeness of the condition that follows complete sexual satisfaction to dying and for the fact that death coincides with the act of copulation in some lower animals. These creatures die in the act of reproduction because, after Eros has been eliminated through a process of satisfaction, the death instinct has a free hand for accomplishing its purposes. . . .[117]

'We die', John Donne wrote, 'and rise the same, and prove mysterious by this love.'[118] Death has been in European erotic literature generally a metaphor for the orgasm, the ecstatic loss of tension. The satirical Śanskrit poet Dāmodaragupta played upon the 'likeness of the condition that follows complete sexual satisfaction to dying':

> Let me tell you, friend, of a singular thing that a boorish fellow of a lover did to me today; I had closed my eyes in the ecstasy of the moment, when thinking me dead he took fright and let go of me.[119]

The correlation of orgasm and death was made in some Tantric texts:

> . . . *bodhicittam notsṛjet,* 'the semen must not be emitted,' the texts repeat. Otherwise the yogin falls under the law of time and death, like any common libertine.[120]

The *yogī* retains his seed, to store power, to transcend the world of becoming. But the court poet, the love poet, celebrates the world of becoming, the world evolved 'under the law of time

[117] Sigmund Freud, Standard Edition xix. 47, cited by David Bakan, *The Duality of Human Existence* (Boston, 1966), p. 170.

[118] John Donne, 'The Canonization' in *Complete Poems,* p. 7.

[119] Cited and trans. by Keith, op. cit., p. 237:
śṛṇu sakhi kautukam ekaṃ grāmyeṇa kukāminā yad adya kṛtam
surata-sukha-mīlitākṣī-mṛteti bhītena muktāsmi.

[120] Eliade, *Yoga, Immortality and Freedom,* pp. 267–8. Perhaps Eliade overstates the case—in many Tantric texts ejaculation is condoned and the semen even used in ritual; orgasm is the *mahāsukha,* the 'great delight'. But this is not to say that the correlation of ejaculation and death was not made in those texts as well. In the orgasmic death time is also obliviated, there is a 'suspension of ego activity'.

and death'. As a court poet Jayadeva implicitly sings acceptance of time and death for the sake of human love; and as a Vaiṣṇava poet he implicitly sings acceptance of empirical existence—Kṛṣṇa after all dances and plays in the world 'like any common libertine'.

An art, a science, a battle to the death, a ritual, love-in-union was also conceived as play, as a delightful game. The battle of love-pleasure is mixed with play (XII. 12).

2 *The game*

Words for play (*līlā*, *krīḍā*, *vilāsa*, etc.) are used generally in Sanskrit *kāvya* to indicate sexual activity.

> . . . in Sanskrit . . . *krīḍati* (play) is frequently used in the erotic sense: e.g. *krīḍaratnam* ('the jewel of games') means copulation . . . love-play [is] the most perfect example of all play, exhibiting the essential features of play in the clearest form. . . . It is not the [sexual] act as such that the spirit of language tends to conceive as play; rather the road thereto, the preparation for and introduction to 'love', which is often made enticing by all sorts of playing. . . . The dynamic elements of play . . . such as the deliberate creation of obstacles, adornment, surprise, pretence, tension, etc., all belong to the process of flirting and wooing.[121]

Dhanaṃjaya says that the lover should delight his beloved with play (*krīḍā*)[122] and Vātsyāyana, in his list of arts auxiliary to the principal arts of love, enumerates a multitude of games in which the lover should be proficient; he advises the lover to court his beloved by playing various games with her and he asserts that the hearts of women may be won over by such playfulness.[123] Play is an 'erotic mode of activity':

> Play is the essential character of activity governed by the pleasure-principle rather than the reality principle. Play is 'purposeless and yet in some sense meaningful'. It is the same thing if we say that play is the erotic mode of activity. Play is that activity which, in the delight of life, unites man with the objects of his love. . . .[124]

[121] Johan Huizinga, *Homo Ludens*, Paladin edn (London, 1970), p. 63.

[122] *Daśarūpa* IV. 78 (cited above, p. 161).

[123] *Kāmasūtra* I. iii, iv; III. iii. 7; etc.

[124] Norman O. Brown discussing Freud's notion of infantile sexuality in *Life Against Death*, op. cit., pp. 39–40.

The playful attitude pervades the sexuality of the *Gītagovinda*—it is light-hearted, delighted sexuality, never austere; the sexuality is not for procreation, not for anything but itself; it is play precisely because it is purposeless:

> Play . . . is an activity which proceeds . . . outside the sphere of necessity or material utility. The play-mood is one of rapture and enthusiasm, and is sacred or festive in accordance with the occasion. A feeling of exaltation and tension accompanies the action, mirth and relaxation follow.[125]

In the Kṛṣṇa-*bhakti* cult the pleasure-principle triumphed, became the very basis of reality; the festive became the sacred; worship was full of rapture and enthusiasm, full of play.

The vocabulary of play implies 'beauty'[126] as well as sexuality, beauty as the erotically playful; the playfully erotic. *Lalita*, the past participle of √*lal*- ('to play, sport, dally, frolic, behave loosely or freely')[127] means 'sported, played, wanton, amorous, voluptuous . . . desired . . . innocent, soft, gentle, charming, lovely. . .'.[128] Sexuality, beauty, playfulness are intermingled in the word—Rādhā remembers Kṛṣṇa's cheeks as *lalita* at their first tryst (II. 14); his mouth (VII. 32) and garland (I. 17) are *lalita*; his forest-hut is *lalita*—a lovely hut and a hut for love-play (V. 5). Rādhā's hair is *lalita*, lovely on account of love-play (XII. 22).

Playfulness enhances the beauty of the woman—the ideal lady as described by the rhetoricians has among her natural graces: *lalita*, 'a gentle movement of her limbs',[129] *vilāsa*, 'a change of appearance, actions and so forth at the sight of her lover'[130] and *līlā*, 'an imitation of her lover',[131] all words for play.

Beautiful objects are those which play with the senses, heigh-

[125] Huizinga, *Homo Ludens*, p. 154.

[126] Daniel Ingalls, in his 'Words for Beauty in Classical Sanskrit Poetry', op. cit., includes a number of words which denote play as well as 'beauty'—*līlā, lasati* (*ullasati, parilasati, vilasati*), *laḍita*, etc.

[127] Monier-Williams, op. cit., *s.v.*

[128] Ibid.

[129] *Daśarūpa* II. 68: '*sukumārāṅgavinyāso masṛṇo lalitaṃ bhavet.*' *Lalita* is also a quality of the hero—it is lightheartedness, a natural and sweet demeanour (*Daśarūpa* II. 22).

[130] *Daśarūpa* II. 61: '*tātkāliko viśeṣas tu vilāso'ṅgakriyādiṣu.*' As a quality of the hero *vilāsa* indicates a firm step and glance and a laughing voice (*Daśarūpa* II. 17).

[131] *Daśarūpa* II. 60: '*priyānukaraṇaṃ līlā* . . .' (N.B. *Gītagovinda* VI. 5).

tening desire. There is a play of light; sounds are played upon the flute. Kṛṣṇa's ear-rings shake in play (*keli*) (I. 38), arousing desire in Rādhā. The playful (*vilāsa*) movements of his eyes inspire love (I. 41). The friend tells Rādhā to go to Kṛṣṇa playfully (*salīlam*) (XI. 8). Playfulness inspires love—the arrows of love fall upon victims in play (*līlā*) (VII. 8); the love-god conquers all the world for sport (*krīḍā*) (III. 12).

Playfulness, as the sexual, the beautiful, the mirthful, the joyous, is the quintessential aspect of Kṛṣṇa's personality—he takes on a human form for play;[132] he is the 'primary-cause of games in the community of the gods' (I. 20); he plays like the erotic sentiment incarnate (I. 47); Kṛṣṇa plays (*vilasati*) with women who are *vilāsinīs*, who are *kelipara*, women who are playful in the sense of wanton, aroused, lovely, coquettish (I. 39 ff.). And as he plays in springtime, the mirthful sexuality pervades the world—the *karuṇa* trees laugh (I. 32) and the cuckoos frolic (√*krīḍ*-) (I. 37)—all of nature is at play; the bank of the Yamunā is a 'playground'; play is the spring rite, the festive rite, the erotic rite.

The *Gītagovinda* is 'comprised of tales about the love-play (*rati-keli*) of Śrī and Vāsudeva' (I. 2), about the secret love-play of Rādhā and Mādhava (I. 1), for those who are curious about the arts of love-play (*vilāsa*) (I. 3). To ease the pain of separation Rādhā indulges in imagining games (*līlā*) with Kṛṣṇa (VI. 11). In separation the pleasure of play (*krīḍā*) is but torture (IX. 10). Union is play; love-making is love-play upon a bed of play (*keli*) (XI. 2) in a hut of play (*keli*) (XI. 14) and after union the playfulness persists—Rādhā asks Kṛṣṇa to fix her ornaments after love-making while he, 'the joy of her heart, was playing (*krīḍati*)' (XII. 17 ff.).

The playfulness of Kṛṣṇa is an expression of the child-aspect of his nature, the divine child who stole curds from the kitchens of the cowherds, the jubilant adolescent who sported in the forest with Balarāma. And the sexuality is 'infantile' and 'polymorphous':

Freud . . . distinguishes fore-pleasure and end-pleasure in sexual inter-

[132] Kṛṣṇa is called '*līlā-mānuṣa-vigraha*' (Monier-Williams, *s.v.*). In the *Viṣṇu-Purāṇa* Viṣṇu is described as the one who sports like a child (*krīḍato bālakasyeva*) (*V.P.* I. ii. 18).

course. The fore-pleasure is the preliminary play with all parts of the body, and represents a perpetuation of the pure polymorphous perverse play of infantile sexuality. The end-pleasure in the orgasm is purely genital and post-pubertal.[133]

Kṛṣṇa is embraced by the cowherd-women 'all-over-his-body' (I. 47); the emphasis is upon fore-pleasure, the joy of play, of sportive dance in a flock of women.

Kṛṣṇa's play is heroic as well as erotic; his conflicts as well as his love-making are playful ('. . . both conflict and love imply rivalry or competition, and competition implies play').[134] Kṛṣṇa kills the war-elephant of Kaṃsa for play (*krīḍā*) [XI. 35) and the various incarnations of Kṛṣṇa were considered 'descents in play' (*līlāvatāra*), parts which the Lord 'plays' on earth.[135]

This *līlā* or sport consists in the descent of the supra-cosmic into the cosmic, in the coming down of the sport of the spiritual region into the material plane. Although the supra-cosmic and the cosmic, the spiritual and the material, [the sacred and the profane,] are wide apart, yet under exceptional circumstances the supreme Spirit manifests Himself in the material world with a material body and constructs a bridge, as it were between the two regions. . . . In different epochs of the history of the world Kṛṣṇa, the supreme Spirit, creates different types of rebellions against Himself and playfully makes his appearance on earth . . . in order to bring them down. This peculiar game of creating rebels against himself and fighting them down is called the *Avatāra-līlā*.[136]

Kṛṣṇa's play is an expression of his nature not only as child, hero and lover but also as the Lord (the all-encompassing aspect). Play was considered *the* creative and 're-creative' activity of the Lord by the teachers of the Vedānta school and by the Bengal Vaiṣṇavas following them. The creation of the world is play:[137]

[133] Brown, *Life Against Death*, p. 37.

[134] Huizinga, *Homo Ludens*, p. 155.

[135] In the *Subhāṣitaratnakoṣa* the word *krīḍā* is frequently used in relation to the *avatāras* of Viṣṇu—the Boar is the *krīḍākroḍatanur* (104) or the *krīḍāvarāho* (1201); the cowherd is the *krīḍāgopālamurtiḥ* (129) and the Fish is the *krīḍājhaṣaḥ* (135). Ingalls comments: 'The qualifier *krīḍā* indicates that the form was assumed temporarily and with no personal object in view; that his form in no way limits the god's essential nature.' (*Anthology* . . . , p. 477).

[136] S. C. Chakravarti, op. cit., pp. 166–7.

[137] *Brahma-Sūtra* II. i. 33: '*lokavat tu līlā kaivalyam*' ('But [creation] is mere sport [of Brahman] such as we see in ordinary life').

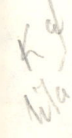

This is the One who, by his own free will and for the sake of his own sport (*līlā*), constitutes with a portion of Himself the peculiar structure of the world with all its infinite variety of animate and inanimate beings in which it consists.[138]

When the Bengal Vaiṣṇavas adopted this conception of creation, *līlā* acquired erotic implications by virtue of Kṛṣṇa's essential nature (implications which it did not have in Vedānta). In the Vaiṣṇava context *līlā* 'refers specifically to the revelation of Krishna's eternal self and his eternal surroundings in the earthly Vṛndāvan'.[139] The revelation is full of delight, of love. The eternal activity is expressed in time and space as love-play, human, carnal and jubilant:

The Lord's intrinsic self consists of nothing but a spontaneous sport of his own infinite bliss. This sport must be understood to be non-phenomenal, but similar in form to that of phenomenal beings. In the phenomenal world pleasure derived from conjugal love is reckoned as the highest fruition of sensuous pleasure; it is only natural that the Bhagavat should also display in his own Śaktis supersensuous pleasure of a similar character. The sex instinct is thus acknowledged in this theology as one of the highest human instincts which finds a transfigured counterpart or ideal in the highest sportive instinct of the divine being.[140]

Rādhā and Kṛṣṇa are one, eternally united but they become two 'to taste the sweetness of their *līlā*',[141] to experience the *rasa* of human love-play. Devotion, worship, became playful activities in medieval Vaiṣṇavism—'worship and salvation are regarded as . . . a blissful enjoyment of the divine sports'.[142] The

[138] *Vedārthasaṃgraha* 14 of Rāmānuja (ed. with an English trans. by J. A. B. van Buitenen):

. . . sva-līlāyai sva-saṃkalpenānanta-vicitra-sthira-trasa-
sva-rūpa-jagat-saṃsthānaṃ svāṃśenāvasthitam . . .

Van Buitenen comments on the concept of *līlā* (p. 192): 'The important conception of God's sport is best understood by its opposite *karman*. It contains free action (an action not resulting from a preceding action in an endless retrogressive succession) performed to no purpose at all: no purpose that of necessity would result in new *phalas* for the agent to enjoy or to suffer. Hence it is compared to the literally inconsequential playfulness of a child. . . . In creating, sustaining and resorbing the world God has no cause to effectuate and no end to achieve.'

[139] Dimock, 'Doctrine and Practice . . .', p. 220.

[140] De, *Vaiṣṇava Faith* . . . , pp. 287–8.

[141] *Caitanya-Caritāmṛta, Ādi.* IV (cited above, p. 57).

[142] De, *Vaiṣṇava Faith* . . . p. 168; cf. p. 419: 'The Bengal School of Caitanya . . .

devotee could express his *bhakti* by play, singing and dancing. The saint Svarūpa is said to have sung the fifth song of the *Gītagovinda* (II. 2 ff.) to Caitanya and as he sang the refrain ('my mind remembers Hari—he joked and *played love-games* in the *rāsa* [dance]') Caitanya was transported:

> And in the ecstasy of love he stood and began to dance. . . . The feelings of joy and wantonness all swelled. And these feelings rose and combined; and they became supremely high. . . . And the Lord [Caitanya] asked the saint Svarūpa time and again to sing a *pada*. And time and again he tasted the sweetness of the song. And as he sang, the dance of the Lord became all the more intense. And the Lord danced thus for a long time. . . . Thus I have narrated the Lord's *līlās* in the garden. . . .[143]

Play as the creative activity, as the erotic activity of Kṛṣṇa, as the expression of Vaiṣṇava devotion, was sacred. Similarly Plato had seen play as the activity whereby one could gain access to the domain of the sacred:

> What, then, is the right way of living? Life must be lived as play, playing certain games, making sacrifices, singing and dancing and then a man will be able to propitiate the gods. . . .[144]

Likewise the German mystic Jacob Boehme saw 'man's perfection . . . in the transformation of the bodily life into joyful play'.[145]

> As God plays with the time of this outward world so also should the inward divine man play with the outward in the revealed wonders of God in this world, and open the Divine Wisdom in all creatures, each according to its property.[146]

Jayadeva expresses his devotion through aesthetic activity—his play is with words and sounds—the *Gītagovinda* is a 'playful-creation' (*līlāyita*) (XII. 28). '*Poiesis* . . . is a play-function':[147]

---

emphasizes the inward realization of the divine sports in all their erotic implications. . . .' In the Sahajiyā school the realization could be outward—in love-play a man and woman could experience the love-play of Kṛṣṇa and Rādhā.

[143] *Caitanya-Caritāmṛta, Antya.* XV (Ray trans., p. 271).

[144] Plato, *The Laws* VIII. 803, cited by Huizinga, *Homo Ludens*, p. 38.

[145] Brown, *Life Against Death*, pp. 40–1.

[146] Cited by Brown, loc. cit.

[147] Huizinga, *Homo Ludens*, p. 141.

The rhythmical or symmetrical arrangement of language, the hitting of the mark by rhyme or assonance, the deliberate disguising of the sense, the artificial or artful construction of phrases—all might be so many utterances of the play spirit.[148]

In play the erotic, the religious, and the literary converge, and that convergence is the very meaning of the *Gītagovinda.*

[148] Ibid., p. 155.

5

THE MEANINGS OF LOVE

Lovers, indeed, if only they could might utter
strange things in the midnight air. For it seems that
 everything's
trying to hide us. . . .

Lovers, to you, each satisfied in the other,
I turn with my question about us. . . .

. . . I ask you about us. I know
why you so blissfully touch: because the caress withholds,
because it does not vanish, the place that you
so tenderly cover; because you perceive thereunder
pure duration. Until your embraces almost
promise eternity. . . .
Lovers, are you the same? When you lift yourselves
up to each other's lips—drink unto drink:
oh, how strangely the drinker eludes his part![1]

ALLEGORY

In history you have the deeds of God to wonder at,
in allegory his mysteries to believe. . . .[2]

GG / intro

The *Gītagovinda* is not so much an allegorical work as an allegorically interpreted work. In the context of the Sanskrit literary tradition a *kāvya* could be erotic, literally sexual, without being in conflict with religious ideals; the *Gītagovinda* is literally about carnal love but it is also literally devotional. There was no contradiction—just as Kṛṣṇa was at once a playful lover and *the* transcendental God, a poem could celebrate

[1] Rainer Maria Rilke, *Duino Elegies*, the second elegy, trans. J. B. Leishman and Stephen Spender, 4th edn (London, 1963), pp. 35–7.

[2] Hugh of Saint-Victor, *Didascalicon*, cited by Aelfred Squire, *Selected Writings of Hugh of Saint-Victor* (London, 1962), p. 23.

both the sensual delights of the world and the ultimate mysteries transcending the world. The need to read the poem allegorically, to interpret the sexuality as a mere analogy for the spiritual relationship, wholly differentiated from it, arises only when celibacy is idealized, when sexuality becomes a transgression against religious ideals.

> . . . dubious are the attempts which have been made to interpret the eroticism of the poem [the *Gītagovinda*] not in a literal but in an allegorical sense, exactly as in the case of the Biblical *Song of Songs*. . . . Whatever may be the case with its Biblical parallel . . . it seems to me that the eroticism of the poem is genuine eroticism. There is no more allegory in the story of Rādhā and Krishna than in many other legendary adventures of Krishna.[3]

It was perhaps the obvious thematic similarities between the *Gītagovinda* and the *Song of Songs* which inspired the first Europeans who read Jayadeva's poem to consider it an allegory. Sir William Jones, the eighteenth-century pioneer of Indological studies, in his essay 'On the Mystical Poetry of the Persians and Hindoos' described the 'little Pastoral Drama, entitled *Gītagovinda*' as a piece of 'emblematic theology' the subject of which was the 'reciprocal attraction between divine goodness and the human soul'.[4] Christian Lassen, who translated the *Gītagovinda* into Latin in 1836, in his allegorical interpretation made Kṛṣṇa the 'divinely-given soul':

> To speak my opinion in one word, Krishna is here the divinely given soul manifested in humanity . . . The recollection of this celestial origin abides deep in the mind, and even when it seems to slumber—drugged as it were by the fair shows of the world, the pleasures of visible things, and the intoxication of the senses—it now and again awakes, . . . full of yearning to recover the sweet serenity of its pristine condition. Then the soul begins to discriminate and to perceive that the love, which was its inmost principle, has been lavished on empty and futile objects; it grows a-wearied of things sensual, false, and unenduring; it longs to fix its affection on that which shall be stable, and the source of true and eternal delight.[5]

[3] Franklin Edgerton, 'The Hindu Song of Songs' in Wilfred Schoff (ed.), *The Song of Songs: a Symposium* (Philadelphia, 1924), pp. 45–6.

[4] Cited by Duncan Greenlees, *The Song of Divine Love* (Madras, 1962), p. v. (Greenlees is described on the dustjacket as 'an Oxonian of repute'!)

[5] Cited by Lal, op. cit., p. 13.

Following suit Edwin Arnold entitled his Victorian translation of the *Gītagovinda*, *The Indian Song of Songs*, thereby imposing analogical implications upon it from the start. Whatever the *Song of Songs* was originally, whether a marriage song or a liturgy of the Adonis–Tammuz fertility cult, or *whatever*,[6] its canonization demanded allegorical interpretation, demanded theological defence; allegory was the means of demonstrating how the love-song was the revealed word of God. The *Gītagovinda* needs no such defence and yet it has been given it, largely out of preconception, out of an assumption that the allegory of love must be a universal phenomenon. Duncan Greenlees, a disciple of Annie Besant and an exponent of the *philosophia perennis*, insists on the pure 'spirituality' of the *Gītagovinda*:

Some in their blind folly have dared to spit filth at this holy thing, naming this wonderful poem erotic . . . in their false self-righteousness. They cannot approach the sweet darkness of the dim arbour or shrine wherein the soul of Man is at-oned with his eager all-loving God, and in their madness try to besmirch the devotee with the slime of their own defilement. Alas for an age wherein man denies his purest treasure![7]

Having once tasted the sweetness of God's love, Rādhā, the Soul, is ceaselessly seeking to know it again and more fully. To her comes the dear friend, the Guru, who points out to her that God belongs to all and she can have no monopoly on his Love.[8]

The hero of our story is no man dallying with a young girl, it is the supreme Lord who from time to time appears in the world to restore the rule of righteousness and overthrow all evil.[9]

[There is] nothing carnal in His love, nothing to cause a blush in the

[6] A. Robert, in *Cantique des Cantiques* (Paris, 1963), gives a complete history of the interpretation of the *Song of Songs* as well as an exhaustive bibliography, French translation and commentary.

[7] Greenlees, op. cit., p. vi.

[8] Ibid., p. xv.

[9] Ibid., p. 2. I think it is here that Greenlees misses the whole point of the *Gītagovinda* which is, I believe, that 'supreme Lord who appears in the world to overthrow all evil' *is* a 'man dallying with a young girl'. Kṛṣṇa embodies the homology of sacred and profane love, the homology of divinity and humanity, of *nirvāṇa* and *saṃsāra*.

tenderest virgin, [it is] but the eternal urge of Spirit to unite with Spirit.[10]

Greenlees interprets the grove in which Rādhā and Kṛṣṇa make love as the 'Interior Castle', the inner self,[11] and he interprets the removal of Rādha's girdle as the falling away of all that veils us from the 'all-seeing-eye' of God.[12] He describes Rādhā and the Gopīs as 'little souls whom He has made'[13]—a clearly Semitic, a very non-Indian, idea. Modern Indian writers have been influenced in their approaches to the *Gītagovinda* much more by these European notions than by Indian concepts; Indian concepts are westernized:

The erotic elements in his poem may be explained away as purely allegorical or symbolical representation of highly spiritual ideas.[14]

Radha is not a woman but a thing representing the materialism, and the whole is a gradual story of the pilgrimage of the soul up to the path of glory. . . . Mystically viewed Jumna is that portion between the two eye brows.[15]

In his *Geeta Govinda*, Jayadeva weaves these bare strands [of the story of Rādhā and Kṛṣṇa] into what is one of the finest religious-romantic poems of the world . . . the love-play of Krishna and Rādhā is an allegory depicting the trials and temptations of the human soul before it finds, in its union with God, bliss eternal and joy ineffable.[16]

Krishna is the human soul engaged in amorous sports with the *gopīs*, who represent the delights of the illusory world, ignoring Rādhā, the personification of intellectual and moral beauty. He is 'reminded' by the messenger, and returns to Rādhā, who comes to his rescue and weans him from the pleasures of the world of the senses. Ultimately Krishna is freed from sensuous distractions, and his love with Rādhā, the personification of divine love and beauty, takes place.[17]

This Lover-Beloved relation is not the allegorical dualism of Greek *agape* or *eros*, or Spenserian Earthly and Heavenly Love, but rather a

[10] Ibid., pp. 18–19. [11] Ibid., p. 40. [12] Ibid., p. 43.

[13] Ibid., p. 42.

[14] S. C. Mukherji, *A Study of Vaiṣṇavism in Ancient and Medieval Bengal* (Calcutta, 1966), p. 101.

[15] C. R. Iyengar, cited by M. Krishnamachariar, *History of Sanskrit Literature* (Delhi, 1970), p. 341n.

[16] Lal, op. cit., p. 68.

[17] Randhawa, *Kangra Paintings of the Gita Govinda*, p. 52.

union of the two and a simultaneous recognition of the divine in the human in which *ātman* consciousness transcends to *Brahman* superconsciousness.[18]

The subject matter of Gita Govinda is the dalliance (Ras) scene in the spring season in lovely Vṛndaban. That season and that spot is most suited to the Divine lovers to offer spiritual treat to the adepts who after a brief sojourn in this uneternal earthly abode are to return to the Divine realm Gauloka.[19]

The Gītagovinda, we may conclude, is certainly an erotic poem, but it is much more: it is symbolic, it opens up a world of mystical perception. One may only enjoy its music, another may enjoy the erotic sentiment (*vilāsa-kalā*) but a real *rasika* will find also the joy of the mystical marriage of the individual soul and the Deity.[20]

The allegorical nuptial song of the mystical marriage of the soul and God is based upon a conception of the sacred and profane dimensions of love as analogous—the attraction of the soul to God is *like* the attraction of the lover to the beloved. The *Song of Songs* could be interpreted pneumatically, could be made an allegory, because both the soul and God, like the lover and beloved, in the western traditions generally, have attributes, are personal, and distinct from one another. The soul includes the personality; it can desire, it can love, it can be 'clothed . . . in the garment of a bride to prepare for a pure and spiritual marriage with God'.[21]

The soul, then, we define to be sprung from the breath of God, immortal, possessing body, having form, simple in its substance, intelligent in its own nature, developing its powers in various ways, free in its determinations, subject to growth by opportunity, in its faculties mutable, rational, supreme, endued with an instinct of presentiment, evolved out of one (original).[22]

[18] Alphonso-Karkala, op. cit., pp. 502–3. He insists that the *Gītagovinda* is based on 'the experience of metaphysical love in the union of *ātman* and *Brahman*'.

[19] Bankey Behari, *The Geet Govind* (Jodhpur, 1964), p. 1.

[20] Sarkar, op. cit., p. 42; see also Monika Varma, *The Gīta Govinda of Jayadeva* (Calcutta, 1968), 'Introduction'; M. Mukherji, 'Jayadeva, Poet and Mystic', *Journal of the Department of Letters*, University of Calcutta, 1935. Practically all secondary material dealing with the *Gītagovinda* perpetuates the notion that it is an allegory.

[21] Gregory of Nyssa, op. cit., p. 153.

[22] Tertullian, *De Anima* 22, cited by H. Wheeler Robinson, 'The Soul (Chris-

This entity, the self as inclusive of the personality, the essential principle of an individual's nature, would in the Indian traditions generally be considered an aggregate of intellect (*buddhi*), mind (*manas*), and I-ness (*aham-kāra*), material qualities, part of *prakṛti*, not the ultimate ground and essence of the individual's being. In other words, what is generally considered the soul in western traditions, is in Indian traditions a phase of the temporal, mundane, bodily existence. The eternal, 'spiritual' principle in man, the 'soul' beyond empirical existence is the *ātman* or the *jīva* or the *puruṣa*. The *ātman* as the essential human 'soul' was considered in the *Upaniṣads* and in the Vedānta texts to be absolutely one and the same as *brahman*, spaceless, eternal, without qualities, impalpable, devoid of duality. It is impossible to conceive of the relationship between the *ātman* and brahman as like the relationship between lover and beloved; there is absolutely no basis for the analogy. The Sāṃkhya system developed the *Upaniṣadic* concept of *puruṣa*s, individual souls, infinite in number, indestructible, eternal entities, pure spirit which in the human being finds itself conjoined with *prakṛti*, nature. All *puruṣa*s or *jīva*s are the same; they have no volition, no attributes. So again the analogy of love does not work: the 'soul', or 'spiritual monad' does not seek union with God—what is sought is simply its release from the world of coming-to-be and passing-away.

> It is somewhat strange that a poem [the *Gītagovinda*] which describes the transports of sensual love with all the exuberance of an Oriental fancy should . . . have received an allegorical explanation in a mystical religious sense.[23]

It is strange, not because of the exuberant sensuality, but because, in Indian traditions, there is little basis for an analogy between the 'soul' and the lover. The 'transports of sensual love' could however, in the Kṛṣṇaite tradition, have a 'mystical religious sense', but not in the same way as the *Song of Songs* has in the Judeo-Christian exegetical tradition. Kṛṣṇa, as the personal God, was considered by the medieval Vaiṣṇavas higher than Brahman precisely because Kṛṣṇa, unlike Brahman, could be

---

tian)' in James Hastings (ed.), *Encyclopedia of Religion and Ethics*, vol. XI (London, 1920), p. 735.

[23] A. A. Macdonell, *A History of Sanskrit Literature* (London, 1900), p. 345.

loved. Kṛṣṇa does not 'stand for' the Absolute, as the bridegroom 'stands for' Jahweh, God, or Christ; Kṛṣṇa is not an allegorical *potentia animae*—the cowherd lover *is* God. Kṛṣṇa was loved by Rādhā, is loved by her in the *Gītagovinda*, and he should be loved by the *bhakta*—it is loving that makes the person a *bhakta*. Rādhā does not 'stand for' or personify the *bhakta* or his 'soul' in the Bengal Vaiṣṇava teaching but their attitudes are the same; Rādhā's sexual longing for Kṛṣṇa runs parallel to the devotee's religious longing. The *bhakta* places Kṛṣṇa in his heart as Rādhā did. It is *rasa* theory rather than allegory (although there is an allegorical tradition in Sanskrit literature)[24] which invests the *Gītagovinda* with its sacred dimension—the *bhakta*, the devotional *rasika*, tasting the flavour of the poem, experiences the great joy of love, the loving relationship with Kṛṣṇa, in its various phases. The *Gītagovinda* was a favourite text of Caitanya, not because it allegorically described the 'human soul's pilgrimage up the path of glory', not because of any abstract concept, but because it possessed and conveyed the 'sweet sentiment' of love of Kṛṣṇa.

1 *Caitanya*

'And our Lord [Caitanya]', Kṛṣṇadāsa Kavirāja relates, 'tasted night and day the songs of Vidyāpati, Caṇḍīdāsa and Jayadeva with Śrī Rāmānanda and Śrī Svarūpa Dāmodara.'[25] As he listened he would fall into rapture, be transported in ecstatic delirium. The *Gītagovinda* was full of power and meaning for Caitanya and his followers. In part the power and meaning had their source simply in the subject—'Which is the sweetest melody?', Caitanya asked Śrī Rāmānanda, and the man

[24] The allegorical drama existed as a genre at least from the eleventh century—the *Prabodha-candrodaya* of Kṛṣṇamiśra is an allegory in the strictest sense in which characters are simply and explicitly personifications of abstract concepts. The play is about Mind (Manas)—he has two sons Discrimination (Viveka) and Confusion (Moha) and these sons are at odds; ultimately through uniting with Sacred-knowledge (Upaniṣad) Discrimination has two offspring, True-knowledge (Prabodha) and Wisdom (Vidyā); Confusion is overcome. The genre gives dramatic form to the personifying tendency in Hindu mythology (i.e., Śrī is Beauty, Kāma is Love and his wives are Pleasure and Delight, etc.).

[25] *Caitanya-Caritāmṛta, Ādi.* XII (Ray trans., p. 158); cf. *Madhya.* II (Ray, p. 37) and *passim.*

answered: 'Those which are turned to the love-songs of Rādhā and Kṛṣṇa.'[26] *Any* song about Kṛṣṇa's *līlā*, about his love-making with Rādhā, would be sacred. Nothing connected with Kṛṣṇa could be profane. But the sacredness, the power and the meaning are enhanced by Jayadeva's skill as a poet; the enchanting, enrapturing rhythms of the songs invest the story with magic—devotees are drawn to Kṛṣṇa by the sweet sounds of the *Gītagovinda*, just as the cowherd-women were drawn from their homes by the mellifluous notes from Kṛṣṇa's flute. Popular legends tell of a Mughal who read the *Gītagovinda*:

One day while riding he was singing its verses, when he fell into an ecstasy of pleasure, and thought that, though a Moslem, he felt communion with Krishna.[27]

The *Gītagovinda* provided the followers of Caitanya with songs for their *kīrtanas*; it is a musical expression of devotion.

O Mind! devotion associated with the ambrosia of the notes and melodies of music is verily paradise and salvation. . . . Through philosophical knowledge one attains salvation only gradually after several births; but he who has knowledge of melodies along with natural devotion to God becomes a liberated soul here and now.[28]

But in addition to the literal and musical dimensions of the poem there is, Śrī Rāmānanda tells Caitanya, an 'inner sense'; he recites two verses of the *Gītagovinda* (III. 1–2) and comments:

And these two verses, O Lord, are full of meaning. And the more we look at the inner sense, the sweeter will be the meaning that they will reveal.[29]

The verses quoted present the crucial mystery of the Kṛṣṇa-cult —the highest Lord becomes truly and completely human; he who is 'liberation from phenomenal-existence' (I. 21) enters into

[26] Ibid., *Madhya*, cited by A. K. Majumdar, op. cit., p. 184.

When Caitanya met Rāmānanda, the governor of Vidyānagara province, he asked him, 'Are you Rāya-Rāmānanda?' and the governor answered, 'I am indeed that vile slave of a *Śūdra*', at which Caitanya fell on the ground in a devotional rapture (see Majumdar, p. 178).

[27] Macauliffe, op. cit., p. 9.

[28] The south Indian composer Tyāgarāja (1767–1847) cited in de Bary, op. cit., pp. 360–1.

[29] *Caitanya-Caritāmṛta*, *Madhya*. VIII (Ray trans., p. 156).

phenomenal-existence, becomes bound and subject to pain like a man, as a man:

Moreover, the Enemy-of-Kaṃsa, having placed Rādhā in his heart as the chain binding him with desire for the world, abandoned the beauties of Vraja. Having pursued Rādhikā here and there, his mind suffering from the wounds of love's arrows, repentant in the grove on the bank of the Kalinda-Nandinī, Mādhava was despondent. [III. 1–2]

Kṛṣṇa is the Bhagavat, the Lord, and Rādhā is the *mahābhāva*, the incarnation of the highest loving sentiment; Kṛṣṇa is the *saktimat* and Rādhā is the *hlādinī-sakti*; but while Rādhā and Kṛṣṇa are understood as more than themselves (that is, more than lover and beloved), they are not envisioned as other than themselves (that is, they are actual, not allegorical, figures). The followers of Caitanya do not interpret the *Gītagovinda* allegorically but historically, literally:

It is important to note that the Vṛndāvana-*līlā* is not a mere symbol or divine allegory, but a literal fact of religious history. The Rādhā-Kṛṣṇa myth . . . is taken as a vivid historical, as well as super-historical, reality. . . .[30]

The *Gītagovinda* is quoted throughout the *Caitanya-Caritāmṛta* and in the theological texts of the followers of Caitanya to illustrate or prove various points; it is cited as authoritative as if it were a historical document rather than a literary text. Rādhā, who historically sport*ed* with Kṛṣṇa in the earthly Vṛndāvana, who eternally sport*s* with Kṛṣṇa in the heavenly Vṛndāvana, is not an allegorical representation of the devotee or the devotee's soul (*jīva*), but her relationship with Kṛṣṇa is paradigmatic of what the devotee's relationship with Kṛṣṇa ought to be. The devotee must realize that the *jīva* and the Bhagavat are the 'same and yet different', like lovers, like Rādhā and Kṛṣṇa. The *Gītagovinda* had the power, for the

[30] De, *Vaiṣṇava Faith and Movement*, p. 169. There does not seem to have been any allegorical commentary on the *Gītagovinda* before the modern period. In the commentary *Bālabodhinī* by Pujārī Gosvāmin (or Caitanyadāsa), who was a follower of Caitanya, I expected and hoped to find an allegorical exegesis, but found instead the traditional linguistic and rhetorical gloss. But that in itself is revealing of the tradition.

medieval Vaiṣṇavas of Bengal, to prompt that realization and the power to manifest the Lord.

On one occasion [in spring] . . . the Lord [Caitanya] went to a garden [in Purī] at night . . . with his devotees. Beautiful indeed was the garden. It was almost like a second Vṛndāvana. It bloomed with the beauty of trees and creepers. Through it the wind from the mount of Malaya blew with the fragrance of flowers. . . . And as the Lord saw this garden his heart was filled with joy. So the Lord sang in the garden a song with his devotees. And the song began with the words: *lalita-lavanga-latā* [i.e., the third song of the *Gītagovinda*]. The Lord sang this song with his devotees and he went round the garden as he sang. And the Lord went to every tree and every creeper in the garden. And he went at last near an *aśoka* tree. Under the tree the Lord suddenly saw his beloved Lord Kṛṣṇa. And the Lord ran towards his beloved one as he saw Him. As the Lord Kṛṣṇa saw our Lord before Him He laughed. And at a moment He vanished from sight. So the Lord got his beloved Kṛṣṇa but he lost Him immediately. He fell down at once in a trance on the ground there in that beautiful garden.[31]

By singing, in spring, Jayadeva's song about Kṛṣṇa's vernal play, Caitanya receives a vision of the Lord. The song has the power to pierce time, to reactualize the past 'amorous spring when Hari dances' in the present springtime. There is no attempt to allegorize the song, to do with it what Catholic exegetes have done with spring imagery in the *Song of Songs*, to make of it a 'springtime of the spirit'—'Do you see chastity, shining like a fragrant lily?', Gregory of Nyssa asks, 'Do you see the rose of modesty, and the violet, the *good odor of Christ*?'[32] The followers of Caitanya did not make Rādhā an allegorical figure; she does not *represent*, she *is*, and her being encompasses, rather than points to, the transcendental.

In Tantric literature however, which is often centrally concerned with indicating correspondences between various levels of existence, Rādhā became a representative on the human plane of both macrocosmic and microcosmic entity. The Tantric conception of reality is not unlike the metaphysical view implicit in the European allegorical tradition:

Part of the function of an allegory is to make you feel that two [or more] levels of being correspond to each other in detail and indeed

[31] *Caitanya-Caritāmṛta, Antya.* XIX (Ray trans., pp. 331–2).

[32] Gregory of Nyssa, op. cit., p. 189.

that there is some underlying reality, something in the nature of things, which makes this happen.[33]

2 *Tantra*

A passage in the *Tantras* instructs the initiate to indulge in forbidden, sexual, incestuous, blasphemous activity in order to secure liberation:

> . . . inserting his organ into the mother's womb, pressing his sister's breasts, placing his foot upon his *guru*'s head, he will be reborn no more.[34]

Perhaps this means simply that he who is liberated is beyond all morality, all good and evil, beyond even *dharma;* but the verse is also taken by commentators to be written in a code language, the 'twilight language' or 'intentional language', *sandhyā-bhāṣā*:

> In the Hindu tantric parlance, the 'organ' is the contemplating mind, the 'mother's womb' is the *mūlādhāra* or base centre of the *yoga* body, the 'sister's breasts' are the heart and the throat centre (*anāhata* and *ājñā*) respectively, and the '*guru*'s head' is the brain centre (i.e. the thousand-petalled lotus *sahasrāra-cakra*, also called *śūnya-cakra*), and the implied instruction is thus translatable: he practices mental penetration through the successive centres, and when he reaches the uppermost centre, he will not be reborn, as he has thereby attained *nirvikalpa samādhi*.[35]

Songs, *caryā-padas*, bearing close formal resemblance to the *Gītagovinda*, were written by Tantric Buddhist *siddhācāryas* from the eighth century on in Bengal in this secret language and one such verse is attributed to Jayadeva.[36] This is not to suggest that the *Gītagovinda* is written in *sandhyā-bhāṣā*, but Jayadeva was most likely well aware of Tantric doctrine, practice and

[33] William Empson's *Structure of Complex Words*, cited by Angus Fletcher, *Allegory* (New York, 1964), p. 70.

[34] Cited by Bharati, op. cit., p. 171:

> mātṛyonau liṅgaṃ kṣiptvā bhaginī-stana-mardanam [kṛtvā?]
> guror mūrdhni pādaṃ dattvā punar janma na vidyate.

See *Tantric Texts* IX, p. 10, ed. Arthur Avalon (Sir John Woodroffe).

[35] Bharati, loc. cit.

[36] See above, p. 40 ff. and pp. 152–3.

language, a language which has an allegorical quality, a kind of transparency, a polydimensional, multi-referential, vitality.

> We find ourselves in a universe of analogies, homologies, and double meanings. In this 'intentional language' any erotic phenomenon can express a Hatha-yogic exercise or a stage of meditation, just as any symbol, any 'state of holiness', can be given an erotic meaning. We arrive at the result that a tantric text can be read with a number of keys: liturgical, yogic, tantric, etc.[37]

This kind of ambiguity between the sexual and the ascetic is typical of the *Gītagovinda*.

The basic structure of the *Gītagovinda*—the union, separation and reunion of male and female—lends itself to Tantric interpretation,[38] for in Tantrism absolute reality is conceived as a unity which in the process of phenomenalization becomes separated into two, the male and the female; the goal is liberation from phenomenal existence by uniting the male and female, both within and without, both physiologically (the union of the *kuṇḍalinī*[39] and the brain) and cosmically (the union of Śakti and Śiva). Rādhā goes to Kṛṣṇa for union with him, just as the *kuṇḍalinī* goes to the thousand-petalled lotus in the brain; Rādhā is active, as is the Śakti in Hindu Tantrism. Paṇḍit Sadashiv Rath Sharma[40] of the Jagannātha temple in Purī explained to me that the *Gītagovinda* is understood by the *yogins* of Orissa to be a Tantric text:

> Rādhā is the realization of the fourth *cakra* which is shaped like a pillar and situated in the heart. When the *yogī* realizes the light in this centre he realizes the great pleasure of Supreme Cheerfulness. Rādhā is the manifestation of light. Rādhā is the *sudarśana-cakra*. The *Gīta-*

[37] Eliade, *Yoga, Immortality and Freedom*, p. 252.

[38] For example: 'There is every possibility that Rādhā represents the consecrated woman of the Tantras chastened by the divine love of Kṛṣṇa', Moti Chandra, 'notes' to the portfolio of paintings from the *Gītagovinda*, Lalit Kalā Akademi (New Delhi, 1965).

[39] The *Kuṇḍalinī* is 'a potent occult energy symbolized by a serpent having three and a half coils, and sleeping with its tail in its mouth. It is often referred to as a goddess and has its home in the subtle body of man, occupying a point near the base of the spine at the *mūlādhāra* plexus.' Benjamin Walker, *Hindu World*, vol. I (London, 1968), p. 547.

[40] For several hours each afternoon in August of 1974 I met with Paṇḍit Rath Sharma in his home in Purī, Orissa; he dictated stories about the life of Jayadeva. (See Chapter VI, 'The saint', page 213 below.)

govinda describes the *maithuna* of Tantra. *Maithuna* is not just the union of man and woman; when the *kuṇḍalinī* unites with the centre of self-realization that is *maithuna*. And the *Uddiṣa Tantra* says that when the tip of the tongue touches the tonsil that is also *maithuna*.

Paṇḍit Rath Sharma had explained also that the *Gītagovinda* is a great devotional work, 'pure *bhakti*'; I asked him how Tantrism and *bhakti* could be reconciled (as the two systems seem to me fundamentally so different); the question did not make a great deal of sense to him (as the two systems seemed to him fundamentally so similar):

In *bhakti* everything is *śuddha-prema* [pure love]. Without *śuddha-prema* there is nothing. No relationship to God. There are two types: *śārīrika-prema* [carnal love] and *ātmika-prema* [spiritual love]. *Śārīrika-prema* can lead to supreme realization when the devotee imagines the *maithuna* of Rādhā and Kṛṣṇa. When the *yogī* imagines this combination in his mind it is the *maithuna* of human and superhuman thought. It is the *maithuna* of Tantra. That is also *rasa*. Such *rasa* is in the *Gītagovinda*. Jayadeva related to Kṛṣṇa as Rādhā to Kṛṣṇa. Jayadeva Gosvāmin was a great *yogī* and he was full of devotion to Lord Jagannātha.

Historically the reconciliation and coalescence of Kṛṣṇa-*bhakti* and Tantrism seems to have occurred in the Sahajiyā-cult.

3 *Sahajiyā*

From those works of Jayadeva which we have, we can understand that he was a Vaiṣṇava-sahajiyā. He worshipped a joint image of Rādhā and Kṛṣṇa. . . . There are other indications of Sahajiyā feeling in his poetry. . . . Vanamāli-dāsa, who wrote [Jayadeva's] biography, leaves no doubt that he was a pure Sahajiyā.[41]

Perhaps the primary 'indication of Sahajiyā feeling in his poetry' is Jayadeva's positive, idealizing, attitude toward human love. The Sahajiyās did not make the strict distinction between human and divine, between profane and sacred love. The followers of Caitanya insisted that the only sacred love was that which had Kṛṣṇa as its object; the goal, for a man or a woman, was to long like Rādhā for union with

[41] Haraprasād Śāstrī, an article on Caṇḍīdāsa, cited by Dimock, *Place of the Hidden Moon*, pp. 56–7; cf. S. C. Mukherji, op. cit., p. 100; S. B. Das Gupta, op. cit., p. 115.

Kṛṣṇa. But in the Sahajiyā school all men were considered embodiments of Kṛṣṇa and all women embodiments of Rādhā; the goal then for a man was realization of himself as Kṛṣṇa together, in love, with a woman whose goal was realization of herself as Rādhā. Together, in a ritual based on the sexual *vāmācāra* Tantric rite, the *maithuna* in which the man becomes Śiva uniting with Śakti, the man and woman experienced the union of Rādhā and Kṛṣṇa. The esoteric Tantric practice was infused with a notion of human love by the Vaiṣṇava-sahajiyās: 'Using one's body as a medium of prayer and loving spontaneously', Caṇḍīdāsa sang, 'is the *sahaja* love.'[42] *Sahaja* means 'inborn, natural, spontaneous, easy'—spontaneity, naturalness, was the means and the end of the Sahajiyā celebration of love. In the ascetic tradition, the 'natural', the sensual, was to be overcome, restrained, struck down. But for the Sahajiyās the senses could lead the devotee to the great delight, the eternal Vṛndāvana, spontaneous human love would guide the heart to God, the deity which was understood as the ultimate fulfilment of one's own, innate, self. Of all the various Indian traditions the Sahajiyā school seems most perfectly to conceive of 'the sacred and profane dimensions of love as homologous and and not differentiated'.

The Sahajiyā Vaiṣṇavas, having a Tantric legacy, did believe in the identity of human and divine. They used the Vaiṣṇava poetic paraphernalia of human love. . . . To both orthodox and Sahajiyā Vaiṣṇavas, the love of Rādhā and Krishna is the guiding principle of the universe. To the Sahajiyās, however, men and women are microcosms, and have within them the ultimate reality of the divine pair in both phases of their love: in separation and in union. Thus love between man and woman duplicates, not symbolically but actually, the love between Rādhā and Krishna, the nature of which is transcendent joy.[43]

The Sahajiyās differed from the orthodox Vaiṣṇavas and from Indian religious traditions generally in their attitude toward women. Women were esteemed as in the literary tradition, as in the Śākta tradition, as Rādhā is esteemed in the *Gītagovinda*:

She is of the greatest beauty, and she has a husband at home [i.e., she

[42] Cited above, p. 26.

[43] Dimock, 'Doctrine and practice . . .', p. 62.

is *parakīyā*]. She is most wonderful, in beauty and in qualities. Suddenly, by *bhāva*, she will come to unite with him. Her beauty will pass through his eyes into his heart, and when it enters his heart it will draw his mind. There will be *sādhana* with her.[44]

The vessel of sahaja is fresh and young, and wounds with the arrows of her glance. She possesses all the marks of beauty, and the clothes and jewels upon her body are bright and colored. Her lips are full of nectar and her body such that a golden creeper cannot compare with it. Her *alaka* and *tilaka* are flattering to her body. Such a nāyikā is a sahaja-nāyikā; serve such a one and know her excellence and greatness.[45]

The *nāyaka* serving the *nāyikā*, the man serving the woman, the god serving the human being and 'knowing her excellence and greatness'—it is a motif running throughout Sahajiyā literature; Caṇḍīdāsa addressed the washerwoman, Rāmī:

I knew thy feet to be a cool retreat and so I took shelter there. Thou art to me the revealer of the Vedas, thou art to me as the consort of the Savior Lord Śiva,—thou art the iris of my eyes;—my worship of love toward thee is my morning, noon-tide and evening services,—thou art the necklace of my neck. The body of the washerwoman is of the nature of the eternal maid Rādhā. . . .[46]

The song elaborates the motif as established in the *Gītagovinda*; Kṛṣṇa the Lord bows down to a milkmaid and sings to her: 'You are my adornment, you are my life, you are my jewel in the ocean of existence' (x. 4).

For the Sahajiyās, the *sahaja*, the 'natural' state, was the blissful state, love as a state of being, ultimate and wholly sacred, the state in which Rādhā and Kṛṣṇa are eternally united (and yet longing for each other as if separated). The goal was the full realization of that state, experience of the *sahaja* nature of the self.

The Sahajiyā-sect 'regards Jayadeva as its *Ādi-guru* and one of its nine *Rasikas*'[47]—he had ardently sung of the carnal love of Rādha and Kṛṣṇa with religious, devotional, fervour. His influence upon the sect is marked; the medieval Sahajiyā songs bear close resemblance to those of Jayadeva both formally and

[44] *Nāyikā-sādhana-ṭīkā*, cited by Dimock, *Place of the Hidden Moon*, p. 218.
[45] *Nigūḍhārtha-prakāśāvalī*, cited ibid., loc. cit.
[46] Cited by S. B. Das Gupta, op. cit., p. 144.
[47] De, *Vaiṣṇava Faith and Movement*, p. 436n.

thematically. In a sense the *Gītagovinda* is, in Sahajiyā terms, allegorical, for to the Sahajiyās Rādhā and Kṛṣṇa were personifications of principles within the human being. But in another sense it is existence itself which is a kind of lived, enacted, allegory—the man and the woman, in turn, are but the ephemeral personifications of the eternal reality of Kṛṣṇa and Rādhā.

SYMBOL

> Though the reflection in the pond may dissolve
> before us: *know the symbol!*
> Only in the double realm will the voices be lasting
> and gentle.[48]

In the symbol various and diverse realms of thought, feeling, experience can merge—the symbol can establish a relationship between those realms. The image becomes invested with power, becomes a symbol, becomes aesthetically, religiously, psychologically potent, when it is used or interpreted to suggest phenomena on several levels of experience; personal, cultural and universal experiences may be linked, held together, encapsulated in the symbol, expressed as one experience.

> In symbol there is concealment and yet revelation: here therefore, by Silence and Speech acting together comes a double significance.[49]

The symbolic dimension of an image extends its implications and ramifications, enables it to have various meanings at once. The abstract, transcendental, conceptual, universal, through symbolization may be expressed concretely, specifically sensually.

> The symbol is characterized by a translucence of the special [the species] in the individual, or of the general [the genus] in the special, or of the universal in the general; above all by the translucence of the eternal through and in the temporal.[50]

[48] Rainer Maria Rilke, *Sonnets to Orpheus* I. 9, in *The Penguin Book of German Verse*, p. 404.

[49] Carlyle, *Sartor Resartus* III. 3, cited by Norman O. Brown, *Love's Body* (New York, 1966), p. 190.

[50] Coleridge, *The Statesman's Manual*, cited by William Wimsatt and Cleanth Brooks, *Literary Criticism: A Short History* (New York, 1957), p. 400.

Translucent, but not transparent, revealing and yet concealing, the symbol, whether used or interpreted consciously or unconsciously, tends to integrate within itself a multitude of elements, situations and ideas, and each element suggests in turn other elements. 'As the mind explores the symbol it is led beyond the grasp of reason . . . beyond the range of human understanding.'[51] The immensity, diffusion, ambiguity, openness of symbols makes their interpretation difficult:

If we are to interpret the 'sense' of the symbol we must expand it, and this must be in terms of literal sentences. If, on the other hand, we thus expand it we lose the 'sense' or value of the symbol as *symbol*. The solution of this paradox seems to me to lie in an adequate theory of interpretation of the symbol. It does not consist in substituting *literal* for symbol sentences, in other words substituting 'blunt' truth for symbolic truth, but rather in deepening and enriching the meaning of the symbol.[52]

Symbols occur in Vedic literature as magical devices; objects were invested with symbolic value, identified with a god or a cosmic force in the elaborate system of identifications and correspondences which yielded symbols as interpenetrations of various levels of reality.

A very striking feature of these works [the *Vedas* and *Brāhmaṇas*] is their passion for identification of one thing with another. . . . The purpose was strictly practical; more specifically magical. It was to get results by setting cosmic forces in motion. To this end a cosmic force was said to 'be' this or that other thing, which other thing we can control. . . .[53]

In the *Upaniṣads* symbols became gnostic devices—certain images were used to point or draw the mind to ultimately unknowable truths—knowledge of the unknowable was symbolic knowledge. Symbols led beyond 'reason and the range of human understanding'.

The Indian word for symbol, '*pratīkam*', . . . denotes originally (from *prati-añc-*) the side 'turned towards' us, and therefore visible, of an object in other respects invisible. In this sense the teachers of the Vedānta often speak of symbols (*pratīkāni*) of Brahman. They under-

[51] C. G. Jung, *Man and His Symbols*, Laurel edn (New York, 1968), p. 4.
[52] Wilbur Urban, *Language and Reality* (New York, 1939), pp. 434–5.
[53] Franklin Edgerton, *The Beginnings of Indian Philosophy* (London, 1965), p. 21.

stand by the term definite representations of Brahman under some form perceptible by the senses . . . which for the purpose of worship . . . are regarded as Brahman and are related to the latter as the images of the gods (*pratimā, arcā*) to the gods that they represent. . . . By symbol . . . we understand all the representations conceived with a view to the worship of Brahman, himself incapable of representations, under some one of his phenomenal forms.[54]

Symbols, in this sense, sacred images, were used in meditation —the mind was concentrated upon the sensuous emblem of the god in order to be drawn to the supra-sensible abstract realization.

Thus the Yogi . . . says to himself, 'I meditate upon the jewel of Vishnu's brow, as the soul of the world; upon the gem on his breast, as the first principle of things;' and so on: and thus through a perceptible substance proceeds to an imperceptible idea.[55]

In the context of the *Gītagovinda* Viṣṇu's brow and the gem of his breast would be emblems not only of his divinity but also of his sexuality, features which infatuate Rādhā. Various images in the poem suggest[56] simultaneously religious and erotic motifs and as such they may be read as symbols, transactions between realms of experience, interconnections between contexts. Symbolization is the 'transformation of things into *something other* than what they appear to profane experience to be'.[57] The profane object can embody the sacred truth—the symbol is capable of spanning and reconciling the sacred and profane dimensions.

1 *The lotus* (for example)

> The lotus that grows from Viṣṇu's [Hari's] navel
> and holds within its greatness all the wonders of the world

[54] Paul Deussen, *The Philosophy of the Upanishads*, Dover edn (London, 1966), pp. 99–101.

[55] H. H. Wilson, *The Vishnu Purana*, 3rd edn (Calcutta, 1961), p. 130n.

[56] The suggestive power of words was the concern of the theorists of the Dhvani school; *dhvani*, suggestion or overtone, was considered the power of words, figures of speech or sentences, often compared to the persisting resonance of a bell after it has been rung, by which one understands more than has explicitly been stated.

[57] Mircea Eliade, *Patterns in Comparative Religion*, Meridian edn (New York, 1963), p. 452.

Symbols
Lotus

may not be blamed—for in all things
fixed laws apply—that at one place
Brahmā should drink of immortality
and bees but take the honey,
tasty with ambrosia and with sweat.[58]

The lotus has symbolic value here—the cosmic process and the natural process of bees gathering honey (with all its erotic associations) occur 'at one place', in the lotus—the cosmic process is made natural and the natural process is given cosmic significance through the lotus as a symbol. 'At one place' the specific and the universal, the erotic and the religious, converge.

The lotus is a significant image in Indian erotics—the ideal woman for love is the lotus-lady, the *padminī*,[59] who is delicate as a lotus, whose hands, feet, and eyes are like lotuses, whose perspiration and whose vulva have the fragrance of the lotus. Lotuses are used in the preparation of aphrodisiacs, love-potions, concoctions to ensure potency and fertility, cosmetics and perfumes to attract a lover.[60] The lotus-seat (*padmāsana*) is a prescribed sexual posture (as well as a meditational posture).[61]

The lotus also has extensive mythological and religious ramifications. In Tantrism the bodily centres where the elements are thought to reside are imagined and described as lotuses; in the 'twilight language' of the *caryāpada*s the 'lotus' refers both to the vulva and the universe.[62] The lotus as a symbol, as an image which helps to centre the *yogin*'s concentration, occurs repeatedly in *yantra*s, the diagrams which aid in meditation, and in their verbal equivalent, *mantra*s. It is a symbol of the sacred.

. . . The unfoldment of spiritual vision in meditation is symbolized by fully-opened lotus-blossoms. . . . Just as the lotus grows up from the darkness of the mud to the surface of the water, opening its blossoms

[58] *Subhāṣitaratnakoṣa* 1067 (Ingalls trans.):
upālabhyo nāyaṃ sakala-bhuvanāścarya-mahimā
harer nābhī-padmaḥ prabhavati hi sarvatra niyatiḥ;
yad atraiva brahmā pibati nijam āyur madhu punar
vilumpanti svedādhikam amṛta-hṛdyaṃ madhu-lihaḥ.

[59] See above pp. 112–13.

[60] For example see *Kāmasūtra* VII. i. 6, 8, etc.; cf. Schmidt, op. cit., pp. 825–914.

[61] See Schmidt, pp. 534–601.

[62] See M. Shahidullah, op. cit., p. 9.

only after it has raised itself beyond the surface . . . in the same way the mind . . . unfolds its true qualities . . . after it has raised itself beyond the turbid floods of passions and ignorance. . . .[63]

The lotus is particularly associated with Viṣṇu: in the *Harivaṃśa* the lotus is designated as his first *avatāra*;[64] among his epithets are: Padmanābhi (the lotus-navelled-one), Padmabhāsa (he-who-is-brilliant-like-a-lotus), Padmahāsa (he-whose-smile-is-like-a-lotus), etc. The lotus is his emblem: he holds it in one of his four hands (along with the conch, mace and discus). And his beloved Śrī, Lakṣmī, is called Lotus (Kamala, Padmā). A lotus grows from his navel bearing Brahmā, giving birth to the universe:

> Its seed is the god Brahmā
> its nectar are the oceans and its pericarp Mount Meru,
> its bulb the king of serpents
> and the space within its leaf-bud is the spreading sky;
> its petals are the continents, its bees the clouds,
> its pollen are the stars of heaven:
> I pray that he, the lotus of whose navel forms thus
> our universe,
> May grant you his defence.[65]

Viṣṇu is the source of the lotus as a symbol of the creative principle:

> The lotus is, indeed, a representative of the force and energy inherent in the waters and of the humidity of the soil. . . . Being the first product of the creative principle . . . it was conceived as a sort of generative organ of these waters [the cosmic ocean in which Viṣṇu sleeps]. Water being regarded as a female substance or 'concept', the lotus identified or associated with similar creative entities of a female character . . . could . . . act as the womb of creation, the womb of the universe.[66]

The lotus-image has a dynamic ambiguity, a tensive quality,

[63] Lama Anagarika Govinda, *Foundations of Tibetan Mysticism*, U.S. edn (New York, 1969), p. 89.

[64] Cited by Gonda, op. cit., p. 104.

[65] *Subhāṣitaratnakoṣa* 146 (Ingalls trans.):

> bījaṃ brahmaiva devo madhu jala-nidhayaḥ karṇikā svarṇa-śailaḥ
> kandaṃ nāgādhirājo viyad ativipulaḥ patra-kośāvakāśaḥ;
> dvīpāḥ patrāṇi meghā madhupa-kulam amūs tārakā garbha-dhūlir
> yasyaitan nābhi-padmaṃ bhuvanam iti sa vaḥ śarma devo dadāut.

[66] Gonda, loc. cit.

as a symbol—it suggests the erotic and the religious and as such draws the two dimensions closely together, holds them sensuously together. When Rādhā suffers with thoughts of lotuses (IV. 21) the suggestion is conventionally erotic—lotuses suggest and enhance the erotic mood and hence cause the separated lover to grieve; but as the lotus, with its religious overtones, is also an emblem of Kṛṣṇa as the Lord the suggestion is also devotional. Likewise Rādhā wearing a lotus-tendril bracelet (VI. 4) suggests both the lover trying to allay the fever of passionate longing and the devotee adorned with an emblem of the Bhagavat. The friend urges Rādhā to feast her eyes upon Kṛṣṇa 'upon a cool bed of wet lotus sprouts' (IX. 6)—the image juxtaposes Kṛṣṇa as lover suffering from love-in-separation (lying upon the cool bed for relief from the burning passion) and Kṛṣṇa as Nalineśaya or Padmeśaya, He-who-abides-on-a-lotus, Kṛṣṇa-Viṣṇu, sleeping on a lotus in the cosmic sea before creation. Human loving, through the symbol, is identified with the vast energy at work in the creation of the universe.

The lotus becomes a symbol through a coalescence of metaphors—the sexual metaphor (the beautiful woman as a lotus), the cosmological metaphor (the universe as a lotus), the psychological metaphor (consciousness as lotus) and so forth, form the symbol as they work together, simultaneously.

> All poetic symbols are metaphors and arise out of metaphor. But a symbol is more than a metaphor. The metaphor becomes a symbol when by means of it we embody an ideal content not otherwise expressible.[67]

Sanskrit rhetoricians define a metaphor (*rūpaka*) as the identification of images, the superimposition of the qualities of one object onto another.[68] It is through metaphor (and *kāvya* is highly and predominantly metaphorical) that poetry expresses what Shelley called 'the before unapprehended relations of things'.[69] Metaphor is an apprehension of the similarity of dissimilars, of the unity of plurality—heterogeneous images are brought together, heterogeneous realms of experience are uni-

[67] Urban, op. cit., p. 470.

[68] See George Gerow, *A Glossary of Indian Figures of Speech* (The Hague, 1971), pp. 239–59.

[69] 'Defence of Poetry', cited by Wheelwright, op. cit., p. 72.

fied. The metaphorical vision of the world, a vision at the very heart of Sanskrit poetry, is a vision of the interconnectedness of things. And in the Vaiṣṇava context, Viṣṇu-Kṛṣṇa, the 'Pervader', *is* this interconnectedness, *is* the similarity behind dissimilars, the unity beyond diversity:

The world was produced from the presence of Viṣṇu; it abides in him; he is the cause of its continuation and cessation; he is the world.[70]

Viṣṇu-Kṛṣṇa is the ultimate reality behind *all* symbolic appearances.

2 *The night*

Nocturnal symbolism universally suggests the pre-natal and precosmic darkness, the darkness of unknowing, the darkness of death, death itself. Darkness and light, as an archetypal, primally experienced, fundamental, duality tend to symbolize further dualities—flesh and spirit, female and male, to represent concretely more abstract notions of evil and good, non-being and being, unconscious and conscious, and so forth. Within or beyond such universal associations the specific context in which an image occurs limits or extends its significance—the context defines the symbolic value of an image. And yet, as William Empson has pointed out, *all* the values and meanings of a word come into play when the word is used in a poem.[71] The lotus, the night, and other images have divergent implications in various traditions, but the divergences converge, the dissimilar associations are assimilated, as the image becomes a poetic symbol. The symbol is capable of resolving antagonisms between the traditions—this is the case in the *Gītagovinda* with the images of nocturnal darkness.

In the Indian religious traditions darkness is *tamas*, one of the 'three strands' of empirical existence, the constituents of nature. *Tamas* binds and blinds, stupefies and stultifies:

But from ignorance is Darkness born: mark [this] well. All embodied

[70] *Viṣṇu Purāṇa* I. i. 31:
viṣṇoḥ sakāśād udbhūtaṃ jagad atraiva ca sthitaṃ;
sthiti-saṃyama-kartāsau jagato'sya jagac ca saḥ.

[71] William Empson, *Seven Types of Ambiguity*, New Directions edn (New York, n.d.), *passim*.

> [selves] it leads astray. With fecklessness and sloth and sleepiness it binds.[72]

> Darkness is the fearful gloom, hell, ignorance, pitch dark dread. It suggests evil, sin, degeneration; the *Kali Yuga* is the age of darkness; darkness typifies the fearful sentiment (*bhayānaka-rasa*).

In the Indian literary traditions darkness is erotic. The night enhances the erotic mood (*śṛṅgāra rasa*); the darkness of bees, *tamāla* clusters, blue lotuses, the darkness of cascading black hair and collyrium, is the beautiful darkness of love (XI. 11, 12, etc.).

> It is generally said that in the evening, at night, in the dark, women have little timidity and are more resolute on sexual intercourse and more impassioned—and (at night) they are not inclined to repulse a man; therefore they are to be made use of at that time.[73]

Kṛṣṇa resolves the basic contrariety between the two traditions, the sacred and the profane. He tells a traveller that the banyan tree where he rests is the abode of '*kṛṣṇa-bhogin*', either 'Kṛṣṇa-the-lover' or the 'black-snake', the deadliest of cobras (VI. 12). The pun is indicative of the whole essence of the *Gītagovinda*—that which is deadly (sexual longing, deadly by binding one to *sāṃsāra*) is also liberation in joy by grace of Kṛṣṇa, the 'dark-one'. As the darkness of night is made joyous by the beloved in the tryst, the darkness of the Kali Age is made joyous by Kṛṣṇa in devotion. Although he is dark he is a 'light in the deep darkness'.[74] During the nocturnal union, 'the darkness

[72] *Bhagavad-Gītā* XIV. 8 (Zaehner trans.):
tamas tv ajñāna-jaṃ viddhi mohanaṃ sarva-dehinām
pramādālasya-nidrābhis tan nibadhnāti (bhārata).

[73] *Kāmasūtra* III. iv. 31:
pradoṣe niśi tamasi ca yoṣito manda-sādhvasāḥ
surata-vyavasāyinyo rāgavatyaś ca bhavanti na ca
puruṣaṃ pratyācakṣate tasmāt tat-kālaṃ prayojayitavyā
iti prāyovādaḥ.

[74] Līlāśuka Bilvamaṅgala cited above, p. 98. A modern interpreter, Karapatri, explains the dark colour of Viṣṇu-Kṛṣṇa: '. . . all outward manifestations of Viṣṇu, whose nature is pure white, the cohesive tendency, must be black and all outward manifestations of Śiva, whose inner self is the abysmal darkness, appear white.' (Cited by A. Daniélou, *Hindu Polytheism* [New York, 1963], p. 159.)

Rādhā also implies that Kṛṣṇa's dark complexion is a reversal of his inner self, as she chides him for his infidelity: 'Like your outer [self], O Kṛṣṇa, your inner self will become very dark indeed . . .' (VIII. 7).

was dispelled by the rays of his pearl ornaments' (II. 5). Kṛṣṇa is dark like a rain-cloud (I. 23; VII. 35; etc.); the go-between tells Rādhā that in union with him she'll be like lightning in a rain-cloud (v. 12). That he is dark and she is light (*pīta*, v. 12) suggests a parallel between the carnal relationship, man and woman, and the astronomical relationship of dark and light; their union then suggests the reconciliation of further antipodean elements in the universe, the harmonization of all opposites which in the cosmos is the end of the Kali Era and which in the individual is liberation.

The contraries are harmonized within Kṛṣṇa as well as by him. Though he is dark in accordance with the Kali Age, he is, as he transcends time and space, infinitely luminous—Kṛṣṇa revealed his true nature to Arjuna:

If in [bright] heaven together should arise the shining brilliance of a thousand suns, then would that perhaps resemble the brilliance of that [God] so great of Self.[75]

3 *The spring*

The poem is set at night and in spring, the voluptuous springtime. Leaves, flowers, animals, insects, wind and water, everything takes part in the biological rite of the amorous springtime. Spring, that phase of the immense turning of time, the revolution of the wheel of *saṃsāra*, when life emerges from death, may be *māyā*, may be illusion, but illusion is celebrated by the poet for all its fragrance, brilliance and delight. The goal for the poet seems to be not so much liberation from the world through renunciation of the world, but rather a sort of beatitude achieved through a lyrical participation in the world, achieved through a discovery of harmony.

Spring to him [the court poet] was beautiful not for the beauty of its birds and flowers so much as for the harmony with which human nature accompanied physical nature's changes. . . . every motion of the world of nature meets an exact response in the human heart. . . . and if humans are subject to the laws of nature, at the same time

[75] *Bhagavad-Gītā* XI. 12:
divi sūrya-sahasrasya bhaved yugapad utthitā
yadi bhāḥ sadṛśī sā syād bhāsas tasya mahātmanaḥ.

nature is viewed in wholly human terms. . . . In this atmosphere *everything is symbolic.*[76]

In the vernal symbol there is a translucence of primary principles—the 'description of the forest during the amorous spring season' (I. 28 ff.) is, through symbolic action, a description of the male and female elements which pervade everything, elements which in the context of Vaiṣṇavism must be taken as representative of Viṣṇu and Lakṣmī, for he is 'all that is called male and she is all that is called female'.[77] In spring the breeze (masculine grammatically and mythologically) tenderly touches the creeper (feminine grammatically and through convention, through association with the curving bodies, arms, brows, of loving women) (I. 28); the burgeoning blossoms swarmed and entered by bees[78] suggest the sexual union of men and women (I. 29); the river caresses the bank, the vine embraces the tree (I. 34). The primary tension in the *Gītagovinda* (and generally in *kāvyas* which depict spring) is between the vernal union of male and female on the natural level and their vernal separation on the human level.[79] In this perspective the theme of the poem becomes the reconciliation of human experience with natural rhythm, cosmic process. Jayadeva's pastoral vision of love is similar to that of the poets of medieval Europe who conventionally opened their love poems with a spring song:

> The beautiful time of rebirth
> brings us leaves
> with fresh greenness
> and flowers of varied hues,
> and therefore all lovers
> are gay and singing—
> except for me; I wail and weep
> without a taste of joy.[80]

[76] Ingalls, *Anthology* . . . , p. 112. [77] *Viṣṇu Purāṇa*, cited above, pp. 124–5.

[78] 'The bee', Ingalls points out, 'always considered masculine, is the sensualist of nature. The word bee (*bhramara*) also means "lover, paramour". . .' (*Anthology* . . . , p. 300). The bee also suggests Kṛṣṇa (the dark colour, the fickleness); he is a 'bee on the lotus which is Rādhā's artless face' (v. 20).

[79] Separation is most poignant in spring and during the rains. The monsoon season 'was a season for lovemaking unequalled by any other except early spring' (Ingalls, *Anthology*. . . , p. 126). The *Viṣṇu* and *Bhāgavata Purāṇas* depict Kṛṣṇa's love-play with the *Gopīs* as occurring in autumn.

[80] Bernard de Ventadorn, cited by Wilhelm, op. cit., p. 180.

The troubadour spring songs attained an ambivalence between the sacred and profane, giving religious overtones to erotic poems by way of reference to the *Song of Songs*. The pastoral setting evoked images of 'the Great God Pan' *and* the God of the Twenty-third Psalm at once, images of the Lady who was at once Venus with her son, Cupid, and Mary with her son, the Christ. Jayadeva attains the ambivalence by asserting that his erotic song is 'the essence of remembrance of the feet of Hari' (I. 35)—the opening flowers, the curling tendrils, 'the birds and the bees', all the images of spring, serve to remind the devotee of Kṛṣṇa whose name, Mādhava, can mean 'relating to spring'.

Kṛṣṇa dances and plays in love in an ideal landscape, an ideal spring, described from convention rather than from experience —the poet does not attempt to depict the flora and fauna of the external world realistically, but rather to give poetic life to an interior landscape, an aesthetic reality.

The forest setting suggests sensual activity as the traditional setting for love-meetings, gambols, picnics; and it suggests religious activity as the traditional setting for sages' hermitages. In Jayadeva's spring, because the cowherd women dance and play with Kṛṣṇa, the sensual activity is religious activity, expression of love for Hari. In spring even the sages are infatuated (I. 33).

For the medieval Vaiṣṇavas the forest, Vṛndāvana, became a symbol of Paradise—the earthly manifested love-play of Rādhā and Kṛṣṇa was regarded as the temporal expression of the eternal play between the Bhagavat and his Śakti in the eternal Vṛndāvana. The profane realm is but the temporal and spatial manifestation of the sacred dimension. The symbolic axiom is of correspondence—symbols establish what Baudelaire called the '*correspondances*' between sensuous data and ideas, feelings; in the symbolic vision, Baudelaire explains, 'the earth and its visibilia are a correspondence of Heaven'.[81] For the poet every object in nature attains a sacred reality—the world becomes a temple with living pillars, '*une forêt de symboles*'.[82] The medieval Vaiṣṇavas assumed the correspondence and the world became the living temple for adoration of Kṛṣṇa.

[81] An article by Baudelaire on Gautier, cited by Guy Michaud, *Message Poëtique du Symbolisme* (Paris, 1947), vol. I, p. 70.

[82] Baudelaire, 'Correspondances' in *Les Fleurs du Mal.*

Of all the symbols in the forest, the creeper (*latā*) is perhaps the most explicit symbol of love. The winding, curling, clinging, entwining tendril, in its attachment, its tenacity, displays the erotic tropism. The creeper wrapped around the tree is a metaphor for the sexual embrace as early as the *Ṛg-Veda*;[83] the 'creeper' is a technical term for a particular kind of embrace in the erotic textbooks;[84] the conceit occurs throughout *kāvya* and in European literature as well;[85] in the *Viṣṇu Purāṇa* Lakṣmī is described as the creeper wrapped around the tree which is Viṣṇu.[86] 'Bind me', Kṛṣṇa says to Rādhā, 'with the creepers which are your arms . . .' (x. 11)—to love and be loved are both to bind and be bound. The *Gītagovinda* presents a paradox, the central paradox of Vaiṣṇavism; there is freedom in attachment, liberation in bondage, because of the nature of Kṛṣṇa, because of the nature of love. Kṛṣṇa is the source of liberation (I. 21) and yet he is bound (III. 1). The paradox is symbolically expressed in the image of the *atimukta* creeper (I. 34)—as it embraces the mango tree, the filaments on the tree bristle and its buds close just as the beloved's body hairs bristle and eyes close in embraces; the creeper is named '*atimukta*' because its blossoms are 'beyond the pearl' in whiteness; but '*atimukta*' can also mean 'completely liberated, free from desire, passion'. The creeper is both attached and beyond attachment, bound and liberated. Like Kṛṣṇa, like Rādhā, like the devotee.

The *yogin*, in meditation, moves from the sensuous symbol to the transcendent reality, from the image of the immanent god to absorption in the absolute Godhead, from the domain of the self to the domainlessness of the Self. But the poet stays with the symbol. Jayadeva, as a devotional Vaiṣṇava poet of the erotic sentiment, invests the sensuous symbol with the transcendent reality, revealing the absolute Godhead as the immanent, dancing god, disclosing the presence of the metaphysical in the physical and discovering sacred meanings of the profane. The

[83] *Ṛg-Veda* x. 10, 13: Yamī says to Yama, 'Some other female embraces thee as a girth a horse, or as a creeper a tree' (Wilson trans.).

[84] *Kāmasūtra* II. ii. 9. The '*latāveṣṭhitaka*' embrace.

[85] For example in *Délie* by Maurice Scève, a sixteenth-century work cited by Perella (op. cit., p. 214): 'If while I held him in my arms as a tree is encircled by ivy. . . .'

[86] See above, p. 124.

Gītagovinda is unequivocally about the joy and sorrow of carnal love, about a god who in love is purely human; but it is also a work consecrated to that god as God. The holiest of holies is encountered in the human heart; the most sacred mystery is experienced in the moments, however fleeting, of human love.

6

JAYADEVA

Poet and saint! to thee alone are given
The two most sacred names of earth and heaven,
The hard and rarest union which be
Next that of godhead with humanity. . . .[1]

THE POET

We know nothing of the personal lives of Sanskrit poets, just as they tell us nothing of the personal lives of their patrons. The persons here have melted into the types of poet and king.[2]

Jayadeva refers to himself as a *kavi*, a professional poet (I. 2, 25; V. 6; VI. 9; VII. 10, 29; XI. 21; XII. 28). That profession, perhaps like all professions in traditional Indian society, represented a strictly defined and conditioned social role with very particular demands and rewards. The poet, most often a *brāhman*, most often living under the patronage of a king or wealthy merchant, was usually the son of a poet and was trained for the profession from an early age.[3] The peculiarities of Sanskrit *kāvya*, the complexities of the rhetorical devices, the strict conventions, the particular aesthetic standards, made it necessary for the aspiring poet to study the *Veda*s, *Purāṇa*s, the Epics, the ancient poets, the various *Śāstra*s, to be proficient in all the arts and sciences, lexicography, grammar, poetics, erotics, vernacular languages and so forth.[4] He would, furthermore, be encouraged to travel

[1] Abraham Cowley, 'On the Death of Mr. Crashaw' in the *Norton Anthology of Poetry*, ed. A. M. Eastman *et al.* (New York, 1970), p. 343.

[2] Ingalls, *Anthology* . . . , p. 24.

[3] Jayadeva gives the name of his father as Bhojadeva (XII. 30) and on the basis of this reference the commentator Gopala constructed a genealogy (Nirṇaya Sāgar Press edn, appendix).

[4] See Rājaśekhara, *Kāvyamīmāṃsā* x.

—Bāṇa, in the *Harṣacarita*, describes how he left home to wander throughout India, visiting courts and attending literary meetings (*goṣṭhis*).[5] On such a journey the poet would attempt to display his wit and cleverness, his poetic ability, in order to make a name for himself. He would also attempt to learn, to gain experience and 'sage attitudes':

> . . . by observation of great courts charming the mind with their noble routine, by paying his respects to the schools of the wise brilliant with blameless knowledge, by attendance at the assemblies of able men deep in priceless discussions, by plunging into the circles of clever men dowered with profound natural wisdom, he [Bāṇa writes of himself] regained the sage attitude of mind customary among his race.[6]

The poet's social function was to amuse, entertain, flatter and inspire an elite, and in return he would earn fame and material gain. The poet was above all a man of taste and culture associating with other men of taste and culture, connoisseurs who gathered together to listen to poetry, to discuss art, philosophy, love, to loiter together sublimely. Rājaśekhara describes the ideal life of the poet: he studies and works on his compositions but leaves ample time for leisurely meals and baths, for meeting with friends and enjoying the company of women; he lives in a spacious house, cared for by servants, with gardens, ponds, streams, fountains; he dresses in fine clothes; he passes his time in cultivation of his social (as well as his literary) style.[7] In order to enjoy such a life the poet would have to find a patron, to establish a place for himself in a royal *samāja*:

> It was held to be the duty of a king to maintain a regular assembly *samāja* of scholars and *kavis*. This would meet in a hall *sabhā* built for the purpose, under the chairmanship of the monarch. It was the custom for *kavis* to submit their work to the criticism of the *samāja* by reading or reciting it before the assembled scholars and writers, and also members of the court who might attend. On these occasions the king made suitable awards to successful *kavis*. Sometimes they were given titles of poetic rank. In addition to these practical exercises in

[5] The *goṣṭhi* is described in the *Kāmasūtra* (I. iv. 34–36) as a gathering of people of similar age, wealth, sophistication and character who meet together to discuss poetry, drama, love and the arts.

[6] Bāṇa, *Harṣacarita*, end of first *sarga* (trans. Cowell and Thomas).

[7] Rājaśekhara, loc. cit.

criticism, theoretical discussions concerning *kāvya* were prominent at the meeting of the *samāja.*[8]

The poet of the *samāja* would have been expected to be able to write verses on any of the conventional subjects, verses about any god. Form, the poetic, technical, manipulation of content, took precedence over the content itself. One of the poems attributed to Jayadeva in the *Saduktikarṇāmṛta* invokes Śiva:

Bearing *earth* under the semblance of the ashes [smeared on his body], *water* in the semblance of the Heavenly River [Ganges which flows from his hair], *fire* under the semblance of the eye on his forehead [which burns with the fire of his austerities], *air* in the semblance of the Snake-Lord's breath, *ether* [the atmosphere] in the semblance of the cavity inside great Aghora's [Śiva's] mouth—[thus] perpetually manifesting the universe through the five elements, may the Moon-crested-one [Śiva] grant you success![9]

The mythology of Kṛṣṇa, however, was a particularly suitable subject for a courtly literature because it combined the three favourite themes of that literature—love, war, religion. And Jayadeva's ardent Vaiṣṇavism is made explicit in the signature lines of the songs of the *Gītagovinda* and in a signature verse:

. . . may wise-people joyfully purely-understand all that according to the *Śrī-Gītagovinda* of the poet (*kavi*) and scholar (*paṇḍit*) Jayadeva whose soul (*ātman*) is solely directed to Kṛṣṇa. [XII. 28]

In addition to his veneration of Kṛṣṇa Jayadeva acknowledges his tutelary deity, Sarasvatī—he describes himself as 'an abode of thoughts elaborated by the deeds of the Word Goddess (Vāc)' (I. 2) and he invokes his muse, Bhāratī (VII. 10). The poet's devotion to her is perfectly reconcilable with his devotion

[8] Warder, op. cit., pp. 202–3.

[9] *Saduktikarṇāmṛta* I. iv. 4 (Śarma edn, 19 in Banerji edn):

bhūti-vyājena bhūmīm amara-pura-sarit-kaitavād ambu bibhral
lālāṭākṣi-cchalena jvalanam ahi-pati-śvāsa-lakṣāt samīram;
vistīrṇāghora-vaktrodara-kuhara-nibhenāmbaraṃ pañca-bhūtair
viśvaṃ śaśvad vitanvan vitaratu bhavataḥ sampadaṃ candra-mauliḥ.

Presumably in reaction to this verse Bimanbehari Majumdar (op. cit., pp. 192–3) has challenged the idea that Jayadeva was a Vaiṣṇava at all: 'he was really a worshipper of Śiva and at the same time took delight in depicting the amours of Rādhā and Kṛṣṇa as an erotic mystic.' And S. K. Chattarji (cited by S. C. Mukherji, op. cit., p. 101) comes to the conclusion that Jayadeva was a '*pañcopāsaka-smārta* who was later raised to the status of Vaiṣṇava saint'.

to Viṣṇu—in the *Mahābhārata* and the *Rāmāyaṇa* she is said to be Viṣṇu's tongue:[10]

She is worshipped for the sake of learning and is one of the deities worshipped in the temples together with Śrī, with whom she is also identified. She is sometimes said to be Viṣṇu's wife. . . . It is understandable, that this goddess, after having been identified with Vāc, that is with Eternal Speech in its transcendent reality, tended to belong to the Highest God.[11]

Within the conventional picture of the poet as a 'type' it is very difficult to establish specific details about the life of Jayadeva—the few autobiographical hints in the *Gītagovinda* provide questions rather than answers. He describes himself as like 'the moon born from the ocean of Tindubilva (or Kindubilva or Kendubilva)' (III. 10); as the moon arose from the ocean when it was churned by the gods, he arose from Tindubilva village (and by suggestion, as the moon is filled with the nectar of immortality, he is filled with the nectar of devotion or of poetry). This is taken as a reference to Kenduli village on the Ajaya river in the Birbhum district of West Bengal by the inhabitants of that village and by Bengali scholars. To the Oriyas this is clearly a reference to Kenduli village on the Pracī river in the Purī district of Orissa; the village has also been said to be in Gujarat, Maharashtra and Bihar.[12]

It is probable that the *Gītagovinda* was written under the patronage of Lakṣmaṇasena of Bengal (*c.* 1179–1209);[13] in an

Date ob ?

[10] As cited by E. W. Hopkins, *Epic Mythology*, Indian edn (Varanasi and Delhi, 1968), p. 207.

[11] Gonda, op. cit., p. 228.

[12] Monmohan Chakravarty, 'Sanskrit Literature in Bengal during the Sena Rule', *Journal of the Asiatic Society of Bengal*, vol. II, no. 5, 1906, argues the Bengali case; K. N. Mahapatra, 'New Light on Poet Jayadeva, the Author of the Gītagovinda', *Orissa Historical Research Journal*, vol. VII, pp. 191 ff., argues the Orissan case. The commentator Lakṣmīdhara (cited by Krishnamachariar, op. cit., p. 337) asserts that Jayadeva was from Gujarat. Colebrook (cited by Chakravarty, p. 163) reports various claims: 'Jayadeva is said by the Maithilas to be their countryman. In Tirhoot [Bihar], a town on the Belan river near Jenjhapur bears the name Kendoli, supposed to be the same as Kenduli.'

[13] A verse is said to have been seen by Rūpa and Sanātana Gosvāmin inscribed on King Lakṣmaṇasena's assembly hall at Navadvipa: 'Govardhana, Śaraṇa, Jayadeva, Umāpati, and Kavirāja [Dhoyī] are the jewels of Lakṣmaṇasena's assembly-hall (*govardhanas ca śaraṇo jayadeva umāpatiḥ kavirājas ca ratnāni samitau lakṣmaṇasya*)' (see S. C. Mukherji, op. cit., p. 71).

entirely conventional[14] prefatory verse he acknowledges other Sena poets:

Umāpatidhara causes words to blossom; only Jayadeva knows the regular coherence of words; Śaraṇa is praiseworthy for the complex, quick [verse]; no one rivals Ācārya Govardhana for compositions which are predominantly of the erotic-mood; Śrutidhara is famous; Dhoyī is a king of poets. [I. 4]

The verse provides a frame of reference—he places himself among writers of erotic and heroic *kāvya*, poetry describing various phases of carnal love, poetry celebrating various gods and praising the exploits of kings and princes.[15] Among them, at least Umāpatidhara and Śaraṇa wrote verses about the love-play of Rādhā and Kṛṣṇa:

Who had been fancied secretly on the way by the one girl with the moving creeper-brows, by another with the winking of the eye, by yet another with the sprinkled moonlight-smile—the gazes of the foe of Kaṃsa that fall in fear and supplication on Rādhā's face which is bright in the splendour of modesty and in disregard owing to outbursting pride, may [those glances] be victorious![16]

Overcome with regret as he recalls the Kālindī with the streams of water along the banks, the mountain slopes dark like lotuses, the caves fragrant with *kadamba* flowers, and Rādhā lovely in her first secret adventure, let Dāmodara, the lord of Dvāravatī, be the joy for the three worlds![17]

[14] Ingalls (*Anthology* . . . , p. 439), explains the impulse behind such verses: '. . . the author speaks of the virtues of former authors in order to express his diffidence of undertaking the work in hand. The author thus anticipates criticism and by his modesty would deprecate it.'

[15] On these authors and their work see Pischel, op. cit., M. Chakravarty, op. cit., and De, *Bengal's Contribution to Sanskrit Literature*.

[16] Umāpatidhara, *Saduktikarṇāmṛta* 273 (Banerji edn) (Hardy trans.):
bhrū-vallī-calanaiḥ kayāpi nayanonmeṣaiḥ kayāpi smita-
jyotsnā-vicchuritaiḥ kayāpi nibhṛtaṃ saṃbhāvitasyādhvani
garvodbheda-kṛtāvahelavinaya-śrī-bhāji rādhānane
sātaṅkānunayaṃ jayanti patitāḥ kaṃsa-dviṣo dṛṣṭayaḥ.

[17] Śaraṇa, ibid., 302:
kālindīṃ anukūla-komala-rayām indīvara-śyāmalāḥ
śailopānta-bhuvaḥ kadamba-kusumair āmodinaḥ kandarān
rādhāṃ ca prathamābhisāra-madhurāṃ jātānutāpaḥ smarann
astu dvāravatī-patis tribhuvanāmodāya dāmodaraḥ.

Several verses attributed to King Lakṣmaṇasena himself also deal with the love of Rādhā and Kṛṣṇa, and one in particular echoes, in its last line, the opening verse of the *Gītagovinda*:

> 'Kṛṣṇa, along with your forest garland I found in the grove this peacock-feather-wreath for the hair of a cowherdess which someone has made; take it!' When the cattle boy said this with an innocent face, the gazes of Rādhā and Mādhava, downcast in embarrassment, languid with a crooked smile, are victorious![18]

Lakṣmaṇasena is described in the court records by the title *Parama-Vaiṣṇava*,[19] but although he was officially a devotee of Viṣṇu, worship of, devotion to, other gods seems to have been encouraged. It was a time of syncretic trends—Jayadeva's inclusion of the Buddha in the list of *avatāras* (I. 13) is indicative of the attitude. Lakṣmaṇasena sponsored Vedic ritualistic writing; his court records refer to his 'gifts to Brāhmaṇas, proficient in Vedic lore, and to his performance of orthodox ritualistic ceremonies';[20] his court scholar (*rāja-paṇḍita*), Halāyudha, seems to have written works on Vedic ritual, on both Vaiṣṇava and Śaiva ritual.[21] In the *Saduktikarṇāmṛta*, which was compiled under his patronage, many gods are invoked—Viṣṇu, Śiva, Brahmā, Gaṇeśā, Sūrya, and the Goddess by various names. It was a period of literary revival and Sanskrit learning:

> During the eleventh and twelfth centuries a general revival of Sanskrit learning is noticeable in Hindustan . . . *Paṇḍits* and their students travelled in numbers from one court to another, from one *tol* to the other. All this encouraged the study of Sanskrit in Bengal, where it had been not much attended to up to that time, presumably on account of Buddhistic influence.[22]

Lakṣmaṇasena's patronage was liberal; Dhoyī, in the *Pavana-*

[18] Ibid., 272:

kṛṣṇa tvad-vana-mālayā saha kṛtaṃ kenāpi kuñjāntare
gopī-kuntala-barha-dāma tad idaṃ prāptaṃ mayā gṛhyatām;
itthaṃ mugdha-mukhena gopa-śiśunā'khyāte trapānamrayo
rādhā-mādhavayor jayanti valita-smerālasā.

[19] S. C. Mukherji, op. cit., p. 70.

[20] De, '*Bengal's Contribution . . .*', p. 68.

[21] S. C. Mukherji, op. cit., p. 87 and M. Chakravarty, op. cit., p. 176.

[22] M. Chakravarty, op. cit., p. 157.

15

dūta, tells of being given gold-handled fly-whisks and elephants by the king.[23]

Luxuries were chiefly manifested in fine clothes, palatial buildings, costly furniture and sumptuous feasts . . . Wealth, luxury and extravagance are hardly compatible with a strict code of morality. Evidences, both literary and epigraphic, testify to the immorality and sensual excesses in ancient Bengal. An idea of the moral laxity of the fashionable young men and women of Gauḍa [Bengal] may be formed by the vivid descriptions of their amorous activity in . . . *Pavanadūta*. The language of Dhoyī seems to imply that these were not merely tolerated but regarded as a normal part of social life. . . . Courtesans were familiar, and presumably not unwelcome features of city-life, for appreciative references are made to them not only in the *Pavanadūta* and *Rāmacarita* but also in the official records of the Sena Kings.[24]

The delights of sensual love, the glories of martial conquest, the sanctity of religious practice—these seem to have been the ideals of the period, ideals clearly reflected in the *Gītagovinda*.

Five of the verses from the *Gītagovinda* (vi. 11; xi. 35; xii. 10, 12, 14) are included in the *Saduktikarṇāmṛta*, compiled in 1205–6 under Lakṣmaṇasena's patronage, and there are also in that anthology several erotic verses attributed to Jayadeva—a description of the moon,[25] praise of a loving-woman's lower-lip,[26] a verse on the line of hair on a beautiful woman's stomach,[27] and a depiction of playful wag-tail birds in autumn.[28] The verses are conventional evocations of the erotic sentiment, the mood of love in a purely carnal sense, not sacralized by any reference to Kṛṣṇa or any other god. The majority of poems attributed to Jayadeva in the anthology are panegyrics praising his royal patron's generosity, extolling his virtue and fame, hailing his noble brow and foot, his glorious sword and fierce arm.[29]

The king, like the Lord, is praised as a lover, as a warrior, and as a god—in a spirit of service and devotion.

[23] *Pavanadūta* 101.

[24] R. C. Majumdar, *History of Ancient Bengal* (Calcutta, 1971), pp. 464–5.

[25] *Saduktikarṇāmṛta* i. lxxxv. 5 (Śarma ed.).

[26] Ibid., ii. lxxii. 4.

[27] Ibid., ii. lxxvii. 5.

[28] Ibid., ii. clxx. 5.

[29] Ibid., iii. v. 4; ix. 4, 5; x. 4; xi. 5; xv. 5; xix. 5; xx. 5; xxiii. 4; xxix. 5; xxxiv. 3, 4, 5; xxxviii. 3; xxxix. 4; xl. 5; lii. 5; v. xvii. 2.

THE SAINT

> . . . Vishnu said, 'Vyas, what you have said in the *Purana* is not understood in the *Kali Yuga* by ignorant men because of its difficulties. I wish you to become an *avatar* in the form of Jayadeva, and bring about the salvation of mankind.' Having thus heard the wish of Vishnu, Vyas prostrated himself on the ground before Him. In accordance with Vishnu's wish, Vyas became a full *avatar* in a Brahman family, in a town by the name of Tundubilva, near the sacred city devoted to Jagannath, Lord of the universe. . . .[30]

In the early seventeenth century Nābhādās composed a *Bhaktamāla* in Hindi, a collection of stories about various Vaiṣṇava 'saints'—Jayadeva, Nāmdev, Tukarām, Kabīr, Tulsīdās, Rāmdās, and so forth.[31] The hagiography is presumably a compilation and embellishment of stories from an oral folk tradition. In the several centuries after Nābhādās the stories were translated into many vernacular languages, rewritten, elaborated, retold; and they remain very much a part of current Indian popular lore.

The stories are meant to inspire devotion to Viṣṇu through descriptions of the ideal behaviour of Vaiṣṇava devotees, saints who are themselves to be worshipped. The theological concept, the philosophical idea, the Upaniṣadic revelation, all that is ascetical or ritualistic or contemplative is avoided in favour of the simple tale which hyberbolically describes the wondrous devotional acts, the exalted state of *bhakti*, of various historical people. The devoted are always rewarded in the stories by a compassionate God who watches over his flock. The stories are meant for people who are striving simply to cope with and endure *saṃsāra*; there is nothing mystical or arcane in the legends; they are meant for the householder. Mahīpati assures

[30] *Bhaktavijaya* of Mahīpati (trans. from Marathi by Abbot and Godbole), op. cit., II. 1–3.

[31] A critical edition and English translation of the *Bhaktamālā* by G. A. Pollet is forthcoming from Louvain. I have used the Sanskrit translation by Candradatta, of which the *sargas* dealing with the life of Jayadeva are republished in the Nirṇaya-Sāgar Press edn of the *Gītagovinda*, pp. 3–17.

his readers of the simplicity and directness which they will find in his tales:

I am mentally dull and ignorant. I have never read the books known as *Puranas*. I do not know the divine (Sanskrit) language. But here there is one special thing that Hari is fond of, the stories of His *bhaktas*.[32]

It is perhaps an irony that the stories in the *Bhaktavijaya* about Jayadeva are told by and for people who cannot necessarily read the *Gītagovinda*.[33] The poet's life, as illustrative of *bhakti*, takes precedence over his work. The stories are meant, above all, for the specific problems of the Kali Age—the traditional, orthodox, religious texts are deemed no longer pertinent to the age of degeneration—the problems are practical. Only devotion can help.

Now let us bow to saints and good men, whom Hari loves with all his heart, and who in this *Kali Yuga* came to save the dull, the fools and the ignorant.[34]

Viṣṇu sends *avatāras* in the form of great devotees, saints, to save mankind by teaching them *bhakti*.

. . . we Thy *bhaktas* are purified by Thee, Shri Hari. Or as trees develop from water, or as clusters of stars arise in the heaven, so we Thy *bhaktas* are (dependent) on Thee, Shri Hari. As cloth comes from threads, as light comes from lamps, so we Thy *bhaktas* are purified by Thee, Shri Hari. As fishes develop from water, as honey comes from flowers, so we Thy *bhaktas* seem important because of Thee. As the bracelet is made of gold, as the wind comes out of the sky, so we Thy *bhaktas* have become *avatars* on earth with human qualities.[35]

[32] *Bhaktavijaya* I. 8–9.

[33] When I was in Purī (August, 1974) I found that almost everyone knew some stories about the life of Jayadeva, some piece of hagiography which they accepted as historically true; many people could sing the second song of the *Gītagovinda*; but no one I met, with the exception of Paṇḍit Sadashiv Rath Sharma, had actually read the work. The stories were always told with enthusiasm and always connected with some bit of local geography—Jayadeva was said to have spent an evening with Padmāvatī in a particular park, or on the particular spot where now a certain shop stands. The stories are felt to be auspicious in themselves; a tour conductor said to me, 'I will tell you stories about Swāmi Jayadeva and the beautiful Padmāvatī, oh golden stories! Tell them to your superiors in England and they will give you a Doctorate, oh golden Doctorate!'

[34] *Bhaktavijaya* I. 5.

[35] *Bhaktavijaya* I. 78–82.

1 *Portrait of the artist as a young saint*[36]

Jayadeva qualifies as a saint because he sang the praises of Kṛṣṇa, the best way to express devotion in the Kali Age. The stories are meant, perhaps, to give solace to the poor, to suggest that they might endure their suffering more easily by devotion to Kṛṣṇa:

To meet a debt his parents had to surrender their house to a neighbor named Niranjan. One day it caught fire, Jayadeva at once ran inside and the fire was at once extinguished. Niranjan prostrated himself to the boy and became his life-long devotee. Left an orphan when still a child, Jayadeva lived in rags and had only water in hand but spent most of his time worshipping God and singing His holy Names.[37]

The legends portray Jayadeva as a mendicant, a holy man, a *sādhu*, a *sannyāsin*:

He is described by the author of the *Bhagat Mal* [i.e., Nābhādās] as an incarnation and treasury of melody, on which, however, he, owing to his ascetic habits, long preferred to feast his own soul rather than communicate to the world the splendid gifts he possessed. He wandered in several countries, provided with only a water-pot and dressed in the patched coat of a mendicant. Even pens, ink, and paper, generally so indispensable to literary men, were luxuries which he did not allow himself. Such was his determination to love nothing but God, that he would not sleep for two nights in succession under the same tree, lest he should conceive an undue preference for it and forget his Creator.[38]

In order to lure him out of this asceticism, to induce him to write the *Gītagovinda*, the Lord arranged for him to marry Padmāvatī—she, according to the legends, taught him the way of human love, gave him the experience of love which inspired his poem about Rādhā and Kṛṣṇa. Human love, rather than

[36] For the legends about the life of Jayadeva I have relied chiefly upon oral communication from Paṇḍit Sadashiv Rath Sharma, Research Scholar of the Jagannāth Temple in Purī; upon Macauliffe, *The Sikh Religion*, vol. VI, op. cit. (who has culled stories from 'the writings of Nabhaji, Uddhava, Chidghan, Mahipati, Ganesh Dattatre, Maharaja Raghuraj Sinha, Dahyabhai Ghelabhai Pandit and others in different Indian languages' [p. 2]); Candradatta's *Bhaktamālā*, op. cit.; the *Bhaktavijaya*, op. cit.; and several other sources cited below.

[37] Greenlees, op. cit., p. ix. He bases his account upon 'Sri Gurudatta Sarma's *Sri Gitagovinda Radhavinodanca*, Hanuman Prasad Poddar's *Premibhakta* and Munshi Haribhakshaji's *Haribhaktaprakasa*'.

[38] Macauliffe, op. cit., p. 5.

obstructing Jayadeva, led him closer to the Lord. This attitude, that the bondage of love is not really bondage at all, typifies the popular milieu in which these stories were told—the world is *māyā*, but the Lord 'arranges for those who love Him that His *māyā* cannot bind His *bhaktas* to the world'.[39] Devotion makes everything else permissible. '*Ama et fac quod vis!*'

As a result of his devotion Jayadeva has visions of Kṛṣṇa:

> Jayadeva left his village with his neighbor Parāshara, making a pilgrimage to Purī. On the way he fell swooning with thirst; Sri Krishna Himself, in the form of a cowherd, brought him fresh water and cool milk and fanned him with His cloth; then He led the two devotees as far as Purī, and there vanished. . . . One day he saw a vision of Sri Krishna smiling at him with His flute in hand; this led afterwards to the beginning of his great poem.[40]

By means of *bhakti* Jayadeva is able to be a householder and yet have the religious stature of a man who has renounced the world. He is depicted as a *guru* initiating various disciples, among them a rich, but devoted, merchant—just as the stories are meant to give solace to the poor by showing that poverty can be made bearable by devotion, they are also meant to give solace to the rich, to assure them that they may experience God's grace despite their wealth if they are devoted, and so that there is no need for them to renounce their riches.

The merchant took Jayadeva to his home where the poet 'performed services of praise. . . . He helped the dull-minded,

[39] S. N. Das Gupta, *Hindu Mysticism*, p. 145.

[40] Greenlees, loc. cit. The reference to 'Parāshara' is apparently a projection onto *Gītagovinda* XII. 30. The slightest hints become material for elaborate legends. This kind of groundless projection is prevalent also in literature which pretends to be scholarly. The Marxist critic D. D. Kosambi, for example, finds in Jayadeva a proletarian writer—Jayadeva, he claims in his 'Introduction' to his critical edition of the *Subhāṣitaratnakoṣa* (p. liv): '. . . was an author trained in music as well as the brahmin's craft, but who wandered far to gain real insight into the minds of his people. His life and art were one; he has romantically wooed and wedded a beautiful wife; they continued to travel, he singing verses of his own while she danced. This was not on the same level as the ordinary country jongleur, for in their performance they exemplified the sublime love of Rādhā and Kṛṣṇa, a theme he developed in his *Gītagovinda* for Lakṣmaṇsena's court. But in his country songs, he used a more popular idiom. . . . A festival is still annually celebrated at Jayadeva's birthplace Kenduli, not in memory of his *Gītagovinda* but because he introduced a new and joyous life, with faith in a personal god who is close to rich and poor.'

the ignorant and the doers of evil to a life of devotion to God.'[41]

When Jayadeva left the merchant to return home to his wife he was followed by two 'thugs' who, assuming that the merchant must have given the poet a great deal of money, stopped Jayadeva in the forest to rob him. Jayadeva merely bowed to them and told them to take as much of his wealth as they needed or wished. After taking all of the poet's money they cut off his hands and feet and threw him into a deep pit, fearing that if they freed him he would tell the merchant of the robbery.

In the meantime Jayadev was thinking of God, remembering Him with feelings of love. Thinking of this subject with true perception, he said to himself, 'It is only the body that can be killed. The soul is not touched thereby. It is quite separate from happiness and pain. The body is the home of all diseases. The body is the root of lust and anger. The unhappy changes of the three *guna*s arise out of the body.' Therefore Jayadev quickly becoming unconscious of his body thought of God, who finds His pleasure in the Ocean of milk.[42]

A royal party happened to pass near the pit some days later and the king heard Jayadeva singing the *Gītagovinda* and rapturously chanting the names of the Lord. Having ordered his servants to remove Jayadeva from the pit, the king asked him how he came to be there and who had amputated his hands and feet. Bearing neither regret nor hatred for his assailants, Jayadeva said that he had been born that way, and the king then fell prostrate upon the ground, realizing that he was in the presence of a very holy man. The king asked the poet if he could become his disciple. Jayadeva instructed the monarch in the way of service, service to the Lord through service to his *bhakta*s. The king made obeisance to every devotee, every wandering *sādhu*, who came to his palace. Hearing of the king who served holy men, the 'thugs' who had robbed Jayadeva decided to dress like *sādhu*s and go to the king and ask him for money.

As the two came forward Jayadev recognized them from afar. As soon as their eyes met they understood one another. Jayadev naturally lived without enmity towards every creature, as forgiveness and peace were personified in him. Enemies to him were like friends. He ignored the faults or merits of others. All mankind was to him a form of Brahma. Such is the characteristic of saints.[43]

[41] *Bhaktavijaya* II. 77. [42] Ibid., II. 102–105. [43] *Bhaktavijaya* II. 145–147.

Jayadeva asked the king to be especially devoted to these men and to give them anything that they requested; but the 'thugs' were suspicious, fearing that Jayadeva would have them executed, and so they told the king that they had known Jayadeva years before in the court of another king and that his hands and feet had been cut off by royal decree for evils he had committed in that court. The Lord could not bear such lies against his *bhakta* and so the earth trembled, quaked, opened and swallowed the 'thugs' up. When Jayadeva heard what had happened he began to weep and to beseech Kṛṣṇa to deliver the men from hell, to seat them in the heavens, to allow them to enjoy the highest spiritual state.

Listening to this moving plea, the Life of the World was delighted. At once Vishnu came to Jayadev. God Narayan became pleased with him and appeared before Jayadev in His *sagun* form with four arms. Immediately afterwards, Jayadeva's hands and feet sprouted from him like the branches of a tree . . . the pretending saints were taken to Vaikuntha (the heaven of Vishnu). Such is the power of *bhakti* which even Brahmadev and the other gods cannot understand. The husband of Rukmini said to Jayadev, 'Blessed is your devotion . . . '.[44]

Wishing to keep Jayadeva with him forever, the king sent a palanquin to Kenduli village to have Padmāvatī brought to the court. For a while Jayadeva and Padmāvatī resided in the court where Jayadeva served as court-poet. But ultimately

Jaidev and his wife . . . had sufficient experience of regal life. They were glad to abandon all state and return to their lowly home at Kenduli, where they enjoyed the society of saints and transferred their idolatrous devotion to the love and homage of the one true God.[45]

In his old age it became hard for Jayadeva to fulfil a vow he had made—to walk every day to bathe in the Ganges—and so the Goddess Gaṅgā entered the well near his house to save him the pain of the journey.[46]

The stories implicitly interpret the *Gītagovinda*; a work by a man so utterly devoted to Kṛṣṇa must necessarily be a very sacred text:

[44] *Bhaktavijaya* II. 193–196.

[45] Macauliffe (retelling Nābhādās), op. cit., p. 15.

[46] Story told by Paṇḍit Rath Sharma, in Greenlees and in the *Bhaktamālā*.

... the author of the *Bhagat Mal* [Nābhādās] ... states that the love scene and rhetorical graces of the poet are not to be understood in the sense that persons of evil minds and dispositions attach to them. Rādhikā the heroine is heavenly wisdom. The milkmaids who divert Krishan from his allegiance to her, are the senses of smell, sight, touch, taste, and hearing. Krishan represented as pursuing them is the human soul, which attaches itself to earthly pleasures. The return of Krishan to his first love is the return of the repentant sinner to God, which gives joy in heaven.[47]

The ideal presented in these legends is of the husband and wife, Jayadeva and Padmāvatī, worshipping Viṣṇu together. They are at once in the world, participants in society, affirming the social order by their marriage, and at the same time they are able to transcend the world and all social fetters by the loyalty to, and love for, Kṛṣṇa. The devotion which Kṛṣṇa feels for Rādhā in the *Gītagovinda* is felt to express the love of God for the devotee, and to reflect the love of the author for his wife.

2 *Padmāvatī*

Jayadeva describes himself as '*padmāvatī-caraṇa-cāraṇa-cakravartin*, an emperor of wandering-singers at Padmāvatī's feet' (I. 2), a phrase which has been understood as meaning both that he is supreme among the pilgrims making obeisance to the Goddess Lakṣmī and that he is a master of singers for the dancing-feet of his wife Padmāvatī. Śaṅkara Miśra explains that the name refers to both the goddess and the woman and thus to both the sacred activity of venerating the consort of Viṣṇu and the profane activity of playing music for his wife's dancing; the commentator maintains that Jayadeva suggests both the erotic mood and devotion to Kṛṣṇa.[48] Jayadeva declares that his song makes Padmāvatī abundantly happy (XI. 21) and again the commentators gloss the name as Lakṣmī and/or the wife of the poet; there is a variant reading of the last verse of the nineteenth song (X. 9) which adds another reference to Padmāvatī: '*padmā-*

[47] Macauliffe, op. cit., p. 10. I suspect that Macauliffe is elaborating on Nābhādās.

[48] '*kṛṣṇa-bhaktyādhikyaṃ śṛṅgāritvaṃ ca dhvanitaṃ*'. Mānāṅka glosses the name first as Lakṣmī but then adds that it refers to a '*brāhmaṇī*' skilled in the arts of dancing and secondarily as Rādhā. Kumbha rejects the idea that the name refers to Jayadeva's wife as 'speculative' (*vicāra-cāru*).

vatī-ramaṇa-jayadeva-kavi-bhāratī-bhaṇitam'—either the song is the composition of the poet Jayadeva who is the husband of Padmāvatī or the song which is the composition of the poet Jayadeva is the delight of Padmāvatī (as either his wife or the goddess).

From these vague references arose the notion of a dancer to whom Jayadeva was married—in the Bengali tradition she is a dancer in the court of Lakṣmaṇasena; in the Orissan tradition she is a dancer in the temple of the Jagannāth in Purī. 'Beautiful Padmāvatī', Paṇḍit Rath Sharma explained, 'performed the dance of the ten *avatāras*, based on her husband's famous song, for Lord Jagannāth.'[49] In both descriptions she is a paragon of beauty, a storehouse of wisdom, an ideal wife, and an ardent devotee of Kṛṣṇa—she and Jayadeva spent all of their time 'dancing, singing, eagerly engaged in praising Śrī Kṛṣṇa'.[50] She is considered the very incarnation of Lakṣmī, metaphorically or actually—'this dutiful wife was a mine of beauty, the *avatar* of Indira. . . . The king bowed with reverence before this honoured mother.'[51] And as she is the incarnation of Śrī, so Jayadeva himself becomes for Mahīpati an incarnation of Viṣṇu—when the father of Padmāvatī offers her to Viṣṇu the Lord says in a dream:

> You offered your daughter to me, but I live in the *Kali Yuga* in the form of Buddha [i.e., as an ascetic]. Now the poet Jayadev is a portion of me, an *avatar* of mine. You should give her to Jayadev with all her ornaments.[52]

Jayadeva and Padmāvatī then have a relationship parallel to that of Viṣṇu and Lakṣmī, and the link between them, between the human and the divine, is the relationship between Kṛṣṇa and Rādhā. The cosmic is made domestic in the legends, and in the process the home is invested with wider possibilities of sacredness—the devotee with his wife might reflect on the marriage of Jayadeva and Padmāvatī to discover the meaning of marriage in terms of the supernal and eternal union of Viṣṇu

[49] The dance is still performed in the Jagannāth temple, a tradition said to have been begun by Padmāvatī.

[50] *Bhaktamālā*, p. 5: '*nṛtyantau cāpi gāyantau śrī-kṛṣṇārcana-tat-parau*'.

[51] *Bhaktavijaya* II. 200–201.

[52] Ibid., II. 54–55.

and Lakṣmī. The portrayal of the poet's marriage to Padmāvatī in the legends reconciles the life of the householder with the life of the devotee; the ideal of fulfilling one's social position and function, one's duty within the world, is reconciled with the ideal of liberation, one's emancipation from the world. The legends about Padmāvatī represent a particular interpretation of the *Gītagovinda* as a religious poem, i.e., as a poem written by a man dedicated not to the sensual delights of the court but rather to the transcendental delights of Kṛṣṇa's eternal Vṛndāvana, delights which he savours with his wife. The legends justify the eroticism of the poem—Padmāvatī does not bind Jayadeva with selfish and fleshly desire; rather she inspires him to liberating devotion. She is the link between the sacred and the profane, the heavens and the home. The stories suggest an interpretation of the poem as neither a celebration of sexuality nor as an allegorical glorification of chastity—rather it is a 'middle-way' that is idealized—sexual love in the context of familial devotion. Married love is acclaimed. Nothing could be farther from the sentiments expressed in the poem with its apparent sanction, if not advocacy, of adulterous love.[53] Paṇḍit Rath Sharma, who accepts the legends as historically true, has dedicated his life to service to Lord Jagannāth, and is a dedicated householder, interprets the *Gītagovinda* in these terms:

> From the ninth to the eleventh century in India people were *vairāgya* [passionless and aloof from the material world]. Śaṅkarācārya and Rāmānujācārya and other *yogī*s of the period attest to that. These *yogī*s said that if you want liberation you must leave your family: go to the forest. But *bhakti* teachers like Jayadev held that you can get liberation in the home. Devotion is all that matters. Devotion is liberation. The family should be based on devotion and then all the people in the family will learn devotion. Jayadev describes the *āsana*s from the *Kāmaśāstra* [i.e., the sexual positions described in the canons of the erotic sciences] and all the joyful aspects of *maithuna* [sexual union] to attract human nature to the pleasures of family life. Through the *Gītagovinda* a family holder can be inspired to *bhakti*. *Bhakti* will lead him from *rati* [sexual pleasure] to divine pleasure.

The legends reflect a modification of the traditional doctrine of the four stages of life (the *āśrama*s), the doctrine that in youth

[53] See above, pp. 113–20.

one should study the scriptures, and then with maturity become a householder, pursuing *artha* and *kāma*; and when this is done become a hermit, and then with old age, in order to gain liberation, become a *sannyāsin*, a *bhikṣu*, a wandering beggar, a holy man. The stories about Jayadeva and Padmāvatī suggest that the liberation traditionally felt to be achieved in the fourth stage might be achieved in the second stage, by the householder. The young *sannyāsin* becomes a more mature householder.

A devoted *brāhman*, Devaśarman, longing for progeny, made offerings and prayers to Lord Jagannāth, promising to give his first-born daughter to the Lord that she might serve Him in His temple. A daughter was soon born and Devaśarman named her Padmāvatī because 'her beautiful form seemed like that of Lakṣmī'.[54] In fulfilment of his vow Devaśarman took the girl to the temple of Jagannāth to offer her to the Lord when she came of age. But the Lord appeared to the *brāhman* in a dream and ordered him to take the girl instead to Jayadeva who was living alone in a small hut near Kenduli village:

> He was poor, without care, zealously studying the scriptures; his face was happy and he was contemplating the Lord with his mind.[55]

At first he refused to accept Padmāvatī, for he was dedicated to a life of celibacy as a *sannyāsin*, but 'it pleased God, with the object, it is stated [in the *Bhaktamālā*], of saving the human race, to withdraw Jaidev from his ascetic life'.[56] And so the Lord appeared to him in a dream and instructed him to marry Padmāvatī. In some accounts he marries her out of compassion when he sees the sorrow she feels over being rejected by him (i.e., he selflessly renounces renunciation); in other accounts he accepts her when astrologers affirm the auspiciousness of the

[54] *Bhaktavijaya* II. 50.

[55] *Bhaktamālā*, p. 4:
daridraṃ nirapekṣaṃ ca śāstraṃ paśyantam ādarāt
manasā tu jagannāthaṃ dhyāyantaṃ muditānanaṃ.

[56] Nābhadās paraphrased by Macauliffe, op. cit., pp. 5–6.

betrothal. 'After their marriage they went to Kenduli village where they worshipped Rādhā-Mādhava in the Viṣṇu temple there.'[57]

When Devaśarman gave his daughter to Jayadeva he instructed her to be obedient:

This one is your husband now, my child, to be always worshipped by you. The woman who is intent upon service to her husband attains eternal happiness.[58]

The ideal was an ancient and traditional one:

> She is a wife who is skilful in the home;
> She is a wife who bears children; she is a
> wife whose life is her husband; she is a wife
> who is devoted to her husband. The wife is
> half the man; the wife is the best friend; the
> wife is the root of the three ends of life. . . .[59]

Those women who love their husband whether he lives in the forest or the city, whether he is good or bad—they have the worlds of high happiness. For women who are noble the husband is the supreme god even if he has a bad character, is lustful or if he has no wealth.[60]

The ideal wife is totally devoted to her lord, her husband; the ideal *bhakta* is totally devoted to his lord, the Lord. Thus the *bhakta* may learn devotion from his wife. The motif is common in the popular legends, medieval and modern, of the *bhakti*

[57] Paṇḍit Rath Sharma.

[58] *Bhaktamālā*, p. 5:
ayaṃ tu patiḥ putri tvayā pūjyaś ca sarvadā
pati-sevāparā nārī sukham akṣayam aśnute.

[59] *Mahābhārata* I. 68. 39–40:
sā bhāryā yā gṛhe dakṣā, sā bhāryā yā prajāvatī;
sā bhāryā yā pati-prāṇā, sā bhāryā yā pati-vratā.
ardhaṃ bhāryā manuṣyasya, bhāryā śreṣṭhatamaḥ sakhā;
bhāryā mūlaṃ trivargasya. . . .

[60] *Rāmāyaṇa* II. 109. 23–24:
nagarastho vanastho vā pāpo vā yadi vā śubhaḥ
yāsām strīnāṃ priyo bhartā tāsāṃ lokā mahodayāḥ
duḥśīlaḥ kāma-vṛtto vā dhanair vā parivarjitaḥ
strīṇām ārya-svabhāvānāṃ paraṃ daivataṃ patiḥ.

saints: there is a story about Tulsīdās, for example, about his longing for his beloved wife Ratnā—he goes to her in a storm:[61]

Padmāvatī, desiring to make a very sacred offering to Lord Jagannāth, decided to copy out the entire text of her husband's

[61] *Tulsidas* by Suresh Chandra Sharma (Bombay, n.d.), p. 17. On the cover of this 'comic book' the editor writes: 'His [Tulsīdās'] works and teachings stressed the importance of the life of householder and weaned the people away from the tantric cults.'

Gītagovinda on *tulsī* leaves with sandal paste. Determined to honour her lord and her Lord, she vowed that she would commit suicide at the end of the month if the work was not complete. Time was running out and the copying was far from complete, but then, one night, two mysterious *sādhus* (in reality Jaya and Vijaya, the gods who guard the Jagannāth temple) came and helped her finish the task. She went to the temple to offer the *Gītagovinda* on *tulsī* leaves to Lord Jagannāth, to adorn him with those leaves; but the offering was rejected by the temple priests. In despair Padmāvatī took the leaves and scattered them in the sea. But, the next morning, when the priests opened the temple doors they found the Jagannāth idol adorned with those leaves. It was proclaimed that the *Gītagovinda* is a sacred work and that Padmāvatī is a true and great devotee.[62]

Another story demonstrating the devotion of Padmāvatī is set in the royal court, variously that of Lakṣmaṇasena or that of the King of Assam:

Now it happened on a certain day that the king's wives were seated together. Padmāvatī, a mine of virtues, shone from her 'dais'. . . . Like musk among all other fragrant perfumes, like a necklace of glass beads among ornaments, so Padmāvatī seemed when sitting on the chief seat. Suddenly a message arrived announcing that the queen's brother Sujan had passed away. Thinking of her husband in her heart, Sujan's wife started to burn herself on his funeral pyre. . . .[63]

The women of the court began to debate the merits of the action. Padmāvatī declared:

When one hears of the death of one's husband, one should immediately commit suicide; just as when the sun sets, the rays disappear.[64]

The queens, doubting Padmāvatī's sincerity, arranged to trick her into believing Jayadeva had been killed by a tiger. A garment dipped in blood was brought to Padmāvatī and she was told that her husband was dead.

Just as the moon's phases wane in succession after the full moon, or as

[62] This story was told to me by various people in Purī. It is customary in the Jagannāth temple to offer *tulsī* leaves with lines from the *Gītagovinda* written on them in sandal paste to the idol.

[63] *Bhaktavijaya* II. 205–209.

[64] Ibid., II. 214.

flowers lose their fragrance when crushed between the hands, so as soon as she heard of the death of Jayadev she gave up her life.[65]

When Jayadeva learned what happened he approached the corpse of Padmāvatī and began to sing the *Gītagovinda.* Hearing the songs, Kṛṣṇa came at once to Jayadeva's side, embraced him and said:

Whoever sings your verses, or listens to them with fondness, the Husband of Lakshmī (Vishnu) stands at his side with hands joined palm to palm.[66]

And at that moment Padmāvatī came back to life. The power of Viṣṇu, a power vast enough to raise the dead, may be tapped by devotion, the devotion which is the essence of the *Gītagovinda* and which is the essence of Padmāvatī's relationship with Jayadeva. Perhaps the *Gītagovinda* and the legends about Jayadeva and Padmāvatī are both about devotion, but there is a difference in the nature of the devotion—the devotion displayed in the former is passionate; the devotion displayed in the latter is dutiful. The wife burns herself on the husband's pyre because it is her duty to do so; the lover burns with passion, with the 'fever of love' because she cannot help doing so. Padmāvatī as an ideal wife differs from Rādhā as an ideal lover. Padmāvatī is a desexualized Rādhā, a domesticated, 'house-broken', Rādhā. Rādhā is the ideal established in the tradition of Sanskrit court poetry—the playful, sulking, passionate woman with swelling breasts and thighs, prone to anger, jealousy, and overwhelmed with longing. Padmāvatī is the ideal established in the *dharma-śāstra*—she obeys her husband; she is beautiful but the physical details of her beauty are never described; she neither sulks nor shows any anger. Passionate carnal love is the key-note of the *Gītagovinda* and Sanskrit court poetry generally, but it has no place in the popular legends—Jayadeva and Padmāvatī are never described as burning with the passionate fires of sexual love; never are there references to physical manifestations of their love. The *Gītagovinda* was written in Sanskrit for the *rasika,* for the courtier, the man of taste, and perhaps also for the ardent devotee; the legends were written, and are told, in the vernacular languages for the 'common man', the householder, the farmer, the servant. Passion is meaningful in the context of

[65] Ibid., II. 229–230. [66] Ibid., II. 242.

the courtly life and in the context of the ecstatic life of the ardent devotee, but for the householder passion undermines the very ideal upon which the household, the marriage, is based. The wife must never be passionate:

She should always be cheerful, and skilful in her domestic duties with her household vessels well cleansed, and her hand tight on the purse-strings. . . .[67]

The hagiographies also depict the husband's devotion to his wife, the God's devotion to the human being. Troubled over a line of the *Gītagovinda* which would not come to him, Jayadeva decided to go for a bath. While he was gone Kṛṣṇa assumed the bodily appearance of the poet and, appearing before Padmāvatī, asked her to give him his writing book. In the place where Jayadeva had left off Kṛṣṇa wrote:

Place the noble sprout of your foot as an ornament upon my head, it dispels the poison of love [in separation]; the pitiless sun which is the destruction by love burns in me—may [your foot] take away the symptom produced. [x. 8]

When Jayadeva returned from his bath and asked for his writing book he was astonished to see the new verse there. He asked Padmāvatī who wrote it and when she replied, 'you', he immediately realized what had occurred, that Kṛṣṇa himself had written the verse. Jayadeva made obeisance to Padmāvatī: 'You are truly blessed to have had a *darśana* [vision] of the Lord.'[68] Kṛṣṇa is of course simply saying in the verse that in union with Rādhā he will no longer feel the pain of separation—he wants to enjoy her sexually; but Paṇḍit Rath Sharma interpreted the line as indicating that sexual love is a burning poison and that only through devotion is it dispelled.

That the *Gītagovinda* has lines in it written by the Lord himself sacralizes it; it is repeatedly described as particularly dear to Kṛṣṇa.

3 *The sacred text*

A king of Orissa wrote his own version of the *Gītagovinda*, similar to Jayadeva's work. The king ordered his own to be sung, for-

[67] *Manu-smṛti* v. 150 (Bühler trans.).

[68] Paṇḍit Rath Sharma. This legend is given by most of the hagiographers.

bidding anyone to sing Jayadeva's poem. The *paṇḍits* at the temple of Lord Jagannāth argued which poem was superior, and finally it was decided to let the Lord himself decide—the two manuscripts were placed at the feet of the Jagannāth and the doors of the temple closed. In the morning Jayadeva's poem was still at the feet of the idol and the king's work had been thrown out of the temple. The king, overcome with despair, wished to end his life. But the Lord appeared and told him that his book, because it was full of *bhakti*, was also dear to Him:

God . . . took from the king's book twenty-four verses. The life of the world then wrote them down in Jayadeva's book.[69]

In the traditional accounts of the Jagannāth temple at Purī seven texts were placed before the Jagannāth: Jayadeva's work, the king's, and also the *Bhāgavata Purāṇa*, the *Puruṣa-sūkta*, *Śrī-sūkta*, *Nārāyaṇa-sūkta*, and the *Nṛsiṃhatapinī Upaniṣad*. And in the morning the *Gītagovinda* of Jayadeva was on top of the pile, preferred by the Lord even to the Vedic texts. The account reflects the anti-orthodox trends in the *bhakti* movement, the feeling that God is to be experienced in spontaneous devotion rather than in Vedic study or Brāhmanic ritual.

O servant, where dost thou seek Me?
Lo! I am beside thee.
I am neither in temple nor in mosque: I am neither in
  Kaaba nor in Kailash:
Neither am I in rites and ceremonies, nor in Yoga and
  renunciation.
If thou art a true seeker, thou shalt at once see Me:
  thou shalt meet Me in a moment of time.
Kabīr says, 'O Sādhu! God is the breath of all breath.'[70]

The anti-orthodox trends encompassed a reaction against the caste system, a reaction also reflected in the legends. One night a female vegetable vendor was gathering egg plants and as she worked she was singing the deeds of Kṛṣṇa as written by Jayadeva. When the Lord heard the song he followed her in order to enjoy her devotion and the devotion of the poet. In the morning the priests of the Jagannāth temple found the garments of the idol torn and they could not understand how that could have

[69] *Bhaktavijaya* II. 47–48.
[70] Kabīr translated by Tagore in Alphonso-Karkala, op. cit., p. 552.

happened. The Lord appeared to the *Rājā* of Purī in a dream to explain to him that the garments had been torn as He followed the vegetable vendor; when the *Rājā* revealed this, all the *brāhmans* gathered to honour the low-caste woman. Candradatta ends his account of the story with the exclamation, 'The devotion of men accomplishes all things!'[71] In some versions of the legend the verse which the vegetable vendor was singing, the verse so particularly dear to Kṛṣṇa, is given:

He has gone to the tryst which is the essence of sexual delight; he has the beautiful appearance of [the god of] love; O fair-hipped-girl! don't delay going to him! follow him, the lord of your heart! Where the wind is gentle, on the banks of the Yamunā, he, the one who wears a forest garland, dwells in the forest, his hands are ever-moving in squeezing the cowherdesses' swollen breasts. [v. 8]

The whole song is intensely erotic but the last verse makes the devotional ramifications clear:

While Jayadeva sings in service to Hari most delightfully, bow, delighted in heart, to Hari who is very compassionate, who is to be desired for his virtues . . . . [v. 15]

Most of the legends about Jayadeva are set in Purī and have to do with the temple of the Jagannāth; the songs of the *Gītagovinda* are sung there each evening and have been at least since the beginning of the sixteenth century—there is an inscription in the temple:

On Wednesday the 10th *tithi* of *Kakaḍā*, bright half in the 9th *aṅka* of the warrior, the elephant Lord, King over Gauḍa, and ninety millions of Karṇāta and *Kalabarga* [i.e., July 17, 1499 A.D.], the mighty Pratāparudradeva Mahārāja orders: Dancing will be performed thus at the *Bhoga* time of Balarāma and *Gītagovinda* Thākura [i.e., Jagannātha] . . . the female dancers . . . all will learn no other song than *Gītagovinda*. . . . No other kind of dancing should be performed before the god. Besides the dancing, there are four Vaiṣṇava singers: they will sing only *Gītagovinda*. Hearing in one tone from them, those who are ignorant will learn the *Gītagovinda* song; they should not learn any other song. That Superintendent who knowingly allows other songs to be sung, and other dancings to be performed, rebels against Jagannātha.[72]

[71] *Bhaktamālā*, p. 9: '*bhaktir hi nṛṇaṃ sarvārtha-sādhinī*.'

[72] Ed. and trans. M. N. Chakravarti, cited by K. C. Mishra, *The Cult of Jagannātha* (Cuttack, 1971), pp. 54–5.

Legends about Jayadeva and the *Gītagovinda* have developed in Orissa to explain the meaning of the poem for the cult of the Jagannāth, the place of the work in the history of that cult. One such legend deals with the 'sacred cloth of Lord Jagannāth', twelve pieces of cloth, each one woven with a canto of the *Gītagovinda*, pieces of cloth still used in temple ritual. Kenduli village at the time of Jayadeva was particularly renowned for its weavers. Wanting to present his work to the Lord in a form that would cause Him the greatest delight, Jayadeva asked the weavers of Kenduli to weave the *Gītagovinda* in cloth. They consented on the condition that while they worked he would sing his songs.

One day while they were on the seventh *sarga* Jayadeva became overwhelmed with emotion and went into a state of *taṭastha* [a spontaneous loss of consciousness, an ecstasy induced by his poem]. The weavers stopped their work but at that time a cheerful *sādhu* came along playing the coconut *karatāla* [cymbals] and he said to the weavers, 'Do not stop your work. I am Jayadeva's younger brother and I know the *Gītagovinda* by heart. I'll sing it so you can finish your sacred work.' The *sādhu* left before the poet woke up. When he finally regained consciousness the weavers told him what had happened. Jayadeva said to them, 'I have no brother. Only Padmāvatī and I know the *Gītagovinda*'. The weavers showed Jayadeva the line the *sādhu* was singing for they had woven it into the cloth:

kaṃsārir api saṃsāra-vāsanā-bandha-sṛnkhalāṃ
rādhām ādhāya hṛdaye tatyāja vraja-sundarīḥ

[Moreover, the Enemy of Kaṃsa, having placed Rādhā in his heart as the chain binding him with desire for the world, abandoned the beauties of Vraja.] [III. 1]

Jayadeva said to the weavers: 'Only Lord Jagannāth could have sung this to you. I salute you, true *bhaktas*.'[73]

The traditional accounts of the temple of the Jagannāth explain that the twelve pieces of cloth were not immediately accepted into the temple. The question as to whether the *Gītagovinda* is a sacred or profane work was discussed by the *paṇḍits* of the temple:

Paṇḍit Nīlādrimiśra insisted that devotees should be concerned with

[73] Paṇḍit Rath Sharma. He brought one piece of cloth from the temple to show me.

ritual functions and prayers and that a song which tempted them with desire and pleasure was bad. But Jayadeva answered him politely, '*Kāma* is the basis of human nature. It is the *ādi-rasa* of the world. Without *kāma* the mind cannot be attracted to anything, the devotee cannot love the Lord. This is supported by the *Veda*. This is especially true in the *Kali Yuga*. Puruṣa enjoys Māyā, Śiva enjoys Pārvatī, Viṣṇu enjoys Lakṣmī. It is universal.' Then the Rājā declared, 'If the Lord wants the cloth He will tell us. Have Jayadeva throw the cloth into Mārkaṇḍeya tank and if it does not sink we shall know that Lord Jagannāth does want the *Gītagovinda* cloth in His Temple.' Everyone gathered to watch the test. Jayadeva sang the first *sarga* of the *Gītagovinda* with love and then, in tears, he threw the cloth into the tank. Through the mercy of Lord Jagannāth the cloth did not sink. All the people of Purī declared that Jayadeva was no ordinary poet but a great *sādhu* and an excellent *bhakta* of the Lord. He was known from then as '*Sādhu-pradhāna-Jayadeva-Gosvāmin*'. . . . Although the cloth was admitted into the temple, the songs of the *Gītagovinda* were not allowed to be sung there during Jayadeva's lifetime. The poet was very disappointed so he left Purī and went to Bengal where he was engaged for some time as the court *paṇḍit* of Rājā Lakṣmaṇasena.[74]

The last incident in the story seems to have been developed to reconcile the Orissan account of Jayadeva with the generally accepted notion of Jayadeva as a court poet under the patronage of Lakṣmaṇasena. A fabulous account of Lakṣmaṇasena's court is given in Halāyudha's *Sekaśubhodaya*[75]—the work describes the visit of a Moslem saint to Lakṣmaṇasena's court before the coming of the Turks. One anecdote concerns Jayadeva: he is portrayed not so much as a devotee or holy man, but as a kind of musical magician. A wandering singer, Buḍhana Miśra, came to the court of Lakṣmaṇasena and challenged the court musicians to a singing contest. To demonstrate his skill he sang, and as he did the leaves on a nearby tree fell to the ground. Everyone admitted that this was a marvellous display, and Lakṣmaṇasena ordered that he be given a 'Writ of Victory'. But then Padmāvatī spoke up and said that only her husband deserved the kudos; she began to sing one of his songs and as she did all the boats on the Ganges, though un-manned, began to come toward her, drawn by the magic of the song. The

[74] Idem.

[75] Compiled early in the sixteenth century. Ed. and trans. Sukumar Sen (Calcutta, 1963).

poets and *paṇḍits* of the court began to debate which feat was the more miraculous when Jayadeva suddenly arrived. He sang his *vasanta-rāga* (presumably the third song of the *Gītagovinda*) and leaves immediately began to sprout on the tree which had been made barren by Buḍhana Miśra's song. Jayadeva was unanimously declared the greatest singer in the world. In the Orissan tradition the magic is possible, not because of Jayadeva's musical skill, but because of the purity of his devotion. Lakṣmaṇasena offers Jayadeva and Padmāvatī great sums of gold; Paṇḍit Rath Sharma explained:

> The devoted *sādhu* said, 'I cannot accept your gift—my wife is a *devadāsī* in the service of Lord Jagannāth and I am his devotee. We are not interested in money or wealth. You should give all honour to the Lord, not to me. He made the tree bloom. Devotion made the tree bloom.'... From that time on King Lakṣmaṇasena was fond of making *pūjā*s to Lord Jagannāth and toward the end of his reign he came to Purī to worship in the Temple of Lord Jagannāth.

Paṇḍit Rath Sharma was 'not interested in money or wealth' either; he spent a great deal of time telling me these legends, and he asked for nothing in return. 'Telling the stories about the life of the great *bhakta* of Lord Jagannāth, Jayadeva Gosvāmin, is good,' he said, 'it is *bhakti*.'

7

THE *GĪTAGOVINDA*: A TRANSLATION

A translation remains perhaps the most direct form of commentary. . .[1]

CONFESSIONS OF A TRAITOR: A PREFACE TO THE TRANSLATION

'*Traditore = traduttore*', 'the translator is a traitor', is the old Italian equation; all translation involves distortion. Information is inevitably lost in the abyss between the source and receptor languages and the greater the differences between the two languages, the two cultures, the greater and deeper the abyss, the more perilous the crossing. The language produces and reflects the culture as the culture produces and reflects the language. They are inextricable. A word cannot be transplanted from one field of occurrence to another without changing. The word 'love' is the lexical equivalent of *amour* and *Liebe* and عشق and אהבה and любовь and ἔρως and 愛 and प्रेम, but outside the dictionaries equivalences dissolve. The words mean different things, elicit different associations and responses in different social and psychological fields. Each word has its own history in each cultural context and the word defies translation by being fixed in that context. Different languages segment experience, classify the world, in different ways; a text and a translation of it into a different language are expressions of two different conceptions of experience, two different relationships between two selves and one world, two subjects and one

[1] D. G. Rossetti, cited by Achilles Fang, 'Some reflections on the difficulty of translation', in Reuben A. Brower, *On Translation* (New York, 1966), p. 111.

object. The objective world is subjectified differently in each language. And the more extreme the subjectivity the more extreme is the difficulty in translating the word. The word 'elephant' is more easily translated than the word 'love' because it is less subjective. But if the word 'elephant' occurs in an English poem, its emotive value, beyond its referential value, its force, its suggestive impact, its meaning, is quite different from that of its lexical equivalent in a Sanskrit poem. With poetry the subjectivity of the words increases, their cultural and social implications, their personal and psychological associations, their assonance and resonance become more crucially a part of their meaning. 'Poetry', Samuel Johnson said, 'cannot be translated.'[2]

There are two basic approaches to this attempt to do what cannot be done—the translator can be a traitor to the 'spirit' of the poem by attempting to be literal or a traitor to the 'letter' by attempting to be literary. In the literary approach the translator attempts to write a poem in the receptor language which elicits in the readers of the translation a response comparable to the one experienced by readers of the original. 'Better a live sparrow', Edward Fitzgerald wrote in defence of this approach, 'than a stuffed eagle.'[3] But for studying the eagle, an eagle stuffed may be better than a live sparrow. And there is, of course, the danger of producing a stuffed sparrow.

In the literal approach, the translator attempts to reproduce meanings in terms of the source context, to preserve grammatical units. The literary translator tries to *do* what the poet did, the literal translator tries to *show* what the poet did. 'The clumsiest literal translation', Vladimir Nabokov wrote in defence of this approach, 'is a thousand times more useful than the prettiest paraphrase.'[4]

The literary approach to Sanskrit *kāvya* seems futile and vain: the ideals, aims, subtleties, constructions, standards of *kāvya* are so utterly different from those of English poetry.

The Kāvya literature . . . is essentially untranslatable; German poets like Rückert [who translated the *Gītagovinda* into German] can, indeed, base excellent work on Sanskrit originals, but the effects produced are achieved by wholly different means, while English efforts at verse

[2] Cited by B. Q. Morgan, 'Bibliography', in Brower, op. cit., p. 271.
[3] Cited by Eugene A. Nida, *Toward a Science of Translating* (Leiden, 1964), p. 2.
[4] Cited by Morgan, op. cit., p. 291.

translations fall invariably below a tolerable mediocrity, their diffuse tepidity contrasting painfully with the brilliant condensation of style, the elegance of metre, and the close adaption of sound to sense of the originals.[5]

Neither the literary nor the literal approach can convey the sensual impact of the poem, the sense-sound of the original; nor can they convey that intensity in *kāvya* which arises from the compactness and condensity achieved by compounding groups of words into monolithic units.[6] *Kāvya* is untranslatable because it is not only *in* Sanskrit, it *is* Sanskrit.

But the literal approach can, I believe, convey at least the intellectual impact of the poem, the sense-content of the original, particularly if it is supplemented with notes and discussion to make explicit information which would have been implicit to the original audience. This is the method that I have used in translating the *Gītagovinda*, an approach quite different from the approaches of the other English translators of the text. The differences of method, as well as some of the particular translational problems which the *Gītagovinda* poses, are apparent in a comparison of *all* the other English translations[7] as exemplified in a single verse (I. 49):

> *rāsollāsa-bhareṇa vibhrama-bhṛtām ābhīra-vāma-bhruvām*
> *abhyarṇaṃ parirabhya nirbharam uraḥ premāndhayā rādhayā|*
> *sādhu tvad-vadanaṃ sudhā-mayam iti vyāhṛtya gīta-stuti-*
> *vyājād udbhaṭa-cumbitaḥ smita-manohārī hariḥ pātu vaḥ||*

Sir William Jones's translation, one of the first translations of

[5] Cited by Keith, *History of Sanskrit Literature*, pp. vii–viii.

[6] Occasionally it is possible to mimic the original, e.g. *Gītagovinda* I. 28:

> *lalita-lavaṅga-latā-pariśīlana-komala-malaya-samīre*
> *madhu-kara-nikara-karambita-kokila-kūjita-kuñja-kuṭīre*

might be rendered:

> when sandal-scented-southern-currents clutch climbing-cloves
> when bee-bands buzz as cozy-cuckoos coo-coo in coppice-coves.

[7] The comparison does, I believe, show a justification for my undertaking a new, a literal, translation. Barbara Stoler Miller has been working for several years on another verse translation. The text has been translated into Latin by C. Lassen (Bonn, 1836); into French by H. Fouche (Paris, 1850) and G. Courtillier (Paris, 1904); into German by F. H. von Dalberg (Erfurt, 1802), F. Rückert (1829; revised and reprinted, Berlin, 1920) and R. Wogen (Halle, 1907); into Dutch by B. Faddegon (Santpoort, 1932); and into many Indian languages including Assamese, Bengali, Hindi, Kannada, Malayalam, Marathi, Nepali, Oriya, Tamil and Telugu.

any Sanskrit work directly into English,[8] is heroic and elegant, full of an eighteenth-century charm and grace; but it is fragmentary and often paraphrastic:

Thus wanton Hari frolicks, in the season of sweets, among the maids of *Vraja*, who rush to his embraces, as if he were Pleasure itself assuming a human form; and one of them, under a pretext of hymning his divine perfections, whispers in his ear, 'Thy lips, my beloved, are nectar'.[9]

Edwin Arnold's *The Indian Song of Songs* is more of a Victorian idyll inspired by the *Gītagovinda* than a translation. All meaning, nuance, form, structure, all the elements of the original, become secondary to Arnold's attempt to maintain English rhyme within a pedestrian metrical scheme:

> But may He guide us all to glory high
> Who laughed when Radha glided, hidden, by,
> And all among those damsels free and bold
> Touched Krishna with a soft mouth, kind and cold;
> And like the others, leaning on his breast,
> Unlike the others, left there Love's unrest;
> And like the others, joining in his song,
> Unlike the others, made him silent long.[10]

George Keyt's translation often stays quite close in general purport to the Sanskrit, but again the effort to be literary, to compose English poetry, often obscures the meaning of the original and obstructs an understanding of what Jayadeva wrote:

> May smiling captivating Hari protect you, whom Radha, blinded by love,
> Violently kissed as she made as if singing a song of welcome saying,
> 'Your face is nectar, excellent,' ardently clasping his bosom
> In the presence of the fair-browed herd-girls dazed in the sport of love![11]

[8] The first was Charles Wilkins's *Bhagavad Gītā* (1784) followed by his *Hitopadeśa* (1787) and then Jones's *Śakuntalā* (1789) followed by his *Gītagovinda* in 1792.

[9] *Asiatick Researches*, vol. III (Calcutta 1792), p. 187.

[10] *Indian Poetry* (London, 1881), pp. 20–1.

[11] *Sri Jayadeva's Gita Govinda: The Loves of Kṛṣṇa and Radha* (Bombay 1940), cited in Alphonso-Karkala, *Anthology of Indian Literature*, p. 510. Some of Keyt's errors distort the poem drastically: in the Eighth *sarga* Rādhā, angered, offended

Duncan Greenlees's *The Song of Divine Love* is a Theosophical rendering of the *Gītagovinda* heavily laden with the language of Christian mysticism. He criticizes other translators:

Too often they have been mere Orientalists, scholars of Sanskrit, or dilettantes of erotism, and so they have spoiled the sweetness. . . .[12]

and he explains that his translation has been undertaken 'with a reverential love for Sri Krishna . . . the all-perfect universal God of Love'.[13]

This mystic Dance gives rapture, bliss untold,
And overwhelms the Gopi devotees
With dainty eyebrows. Radha, set aflame
By madness of desire, with sudden cry:
'Ah, Love, how sweet! Your honeyed mouth
Is filled with nectar!' hurls herself upon
His bosom, and with burning thirst assails
Those smiling lips with fiery kiss of love.
O crafty Radha! You pretend to praise
His song; what sweets of ecstasy you steal!
May this delightful Hari by His smile
Of flowing grace protect you from all harm![14]

Monika Varma describes her method of rendering the poem:

. . . the translator can enter into the philosophical feel of the poem, and transcreate and interpret. . . . It is then not merely a translation of the language but a translation of the *bhava*: the inherent meaning. . . .[15]

Her own interpretation of the poem as a metaphysical treatise on 'cosmic consciousness' is imposed without restraint:

Even in the sight of the distracted lovers,
the infatuated Srimati, blinded by love,
forgetting and throwing aside all modesty,
holds close to her, her Madhava.
Pretending to be chanting words of worship,
Kesava whispers: Immortal nectar lies in your mouth;
and then drinks deep, clinging lip to lip.

---

and jealous, tells Kṛṣṇa to go away, to follow some other cowherdess and this display of 'pique' [*māna*] is crucial to Jayadeva's portrait of Rādhā; but Keyt has the friend speaking to Kṛṣṇa, telling him to follow Rādhā!

[12] Greenlees, op. cit., p. vi.

[13] Ibid., p. vii. [14] Ibid., p. 15. [15] Varma, op. cit., p. v.

> Satisfied, Kesava smiles,
> and may that smile lighten the burden
> of all your griefs, and sorrowed days.[16]

Dr Bankey Behari's paraphrase is so loose and free that it is difficult to know whether or not he has tried to translate this verse at all. What does in fact come at the end of his rendition of the first *sarga* typifies his approach:

> Man! learn ye the mystery of Divine Love and the glory of Surrender. Cast aside the sensual veil of attachment to earthly objects that warp the vision to the Divine Realm of Beatitude. . . [*etc.*].[17]

I have attempted to be much more literal and to impose as little of myself as possible:

> In the presence of the cowherds' beautiful-browed-women
> who were carried-away by whirling-about with a
> great-amount of {joy / jumping} in the *rāsa* [dance], [Hari]
> was passionately kissed by Rādhā who was blind with
> love, having ardently embraced his breast, having said,
> 'Wonderful! your {mouth / voice} consists of nectar!' under
> the pretext of praising his song; may Hari, ravishing
> with his smile, protect you!

By following the Sanskrit syntax as closely as the lack of inflection in English permits, by use of braces to indicate double meanings, by the use of brackets to include words understood but not present in the original, by the use of hyphens to indicate single Sanskrit words rendered by several English words, and by risking awkward English construction for the sake of greater faithfulness to the Sanskrit idiom, my intent has been to present the sense-content of the *Gītagovinda*; my intent has been to *show* in English what *is* in Sanskrit. And I hope that, as in the judgement of bigamists and murderers, 'intent' is taken into account in the judgement of the translator as traitor.

The translation follows the Nirṇaya-Sāgar Press edition (Bombay, 1949). Abbreviations used in the notes to the translation:

[16] Ibid., p. 16.
[17] Behari, op. cit., p. 18.

M. King Mānāṅka: his commentary (*tippaṇikā*) on the *Gīta-govinda* (fourteenth century?).

M.W. Monier-Williams, *Sanskrit-English Dictionary*.

R.K. [Raṇa] Kumbha: his commentary, *Rasikapriyā*, on the *Gītagovinda* (fifteenth century).

Ś.M. Śaṅkara Miśra: his commentary, *Rasamañjarī*, on the *Gītagovinda* (sixteenth century).

THE GĪTAGOVINDA

THE FIRST CANTO

JOYFUL DĀMODARA[1]

1. 'The sky is densely clouded, the forest grounds are dark with *tamāla*[2] trees; at night he [Kṛṣṇa] is afraid. Rādhā, you alone must take him home.' This is Nanda's[3] command. [But] Rādhā and Mādhava[4] stray to a tree in the grove by the path and on the bank of the Yamunā[5] their secret love-games prevail.[6]

2. [He is] an abode of thoughts elaborated by the deeds of the Word Goddess,[7] an emperor of wandering-singers at Padmāvatī's[8] feet—the poet Jayadeva composes this work comprised of tales about the love-play of Śrī and Vāsudeva.[9]

3. If your mind is passionate[10] in remembrance of Hari, if it is curious about the amatory arts, then listen to Jayadeva's eloquence: a sweet, tender, lovely string of verses.[11]

4. Umāpatidhara causes words to blossom; only Jayadeva

[1] Dāmodara: 'he who has a rope around his waist'—in an attempt to keep the infant Kṛṣṇa from mischief, Yaśodā, his foster-mother, the wife of Nanda, put a rope around Kṛṣṇa's waist and tied him to a large mortar.

[2] *tamāla*: the tree has a straight trunk, fragrant leaves, white blossoms, a very dark bark and it grows near rivers; it is associated with night and darkness generally but particularly with nocturnal or autumnal trysts, with the erotic sentiment and Kṛṣṇa.

[3] Nanda: chief of the cowherds at Gokula and foster-father of Kṛṣṇa.

[4] Mādhava: the epithet of Kṛṣṇa also means 'honey-like' and 'vernal'.

[5] Yamunā: the Jumna river—associated with Kṛṣṇa in both his heroic (e.g., the quelling of the serpent Kāliya) and his erotic (e.g., sporting with the bathing cowherdesses) aspects.

[6] 'prevail' [*jayanti*]: or 'are victorious'—used in the sense of 'long-live' or 'hail' the love-games.

[7] 'An abode . . .': or 'Jayadeva's mind [i.e., the abode of his thought] is adorned by the deeds of the Word Goddess . . .'. The Word Goddess is Vāc—speech personified, deified, identified with Sarasvatī or Bhāratī.

[8] Padmāvatī: Viṣṇu's consort (Lakṣmī, Padmā, Kamalā, Śrī, etc.) but traditionally taken to be the name of a dancer to whom Jayadeva was married.

[9] 'Śrī and Vāsudevā: or *Śrī* as the honorific prefix; the epithet of Kṛṣṇa designates him as the son of Vasudeva of the Yadu family.

[10] 'passionate' [*sa-rasa*]: 'if your mind has (or experiences) *rasa* (relish, bliss, the [erotic] sentiment) . . .'.

[11] 'a sweet . . .': or 'a string of verses about the sweet and tender lover [*kānta*]'.

knows the regular coherence[12] of words; Śaraṇa is praiseworthy for the complex, quick [verse];[13] no one rivals Ācārya Govardhana for compositions which are predominantly of the erotic-mood; Śrutidhara[14] is famous; Dhoyī is a king of poets.[15]

The First Song (*mālava-gauḍa rāga, rūpaka tāla*):

5. In the oceanic waters of destruction you supported the *Veda* without fatigue, performing the function of a boat [for it]; O Keśava who bore the form of the Fish! O Lord of the World! Victory! O Hari![16]

6. The earth stands on your expansive back which is
{ most venerable for / very thickened with } the rings of callus [caused by] supporting the world; O Keśava who bore the form of the Tortoise! O Lord of the World! Victory! O Hari!

7. The earth abides on the point of your tusk, fixed like the dark-spot upon the hare-marked-moon;[17] O Keśava who bore the form of the Boar! O Lord of the World! Victory! O Hari![18]

8. The nail on your fair lotus-hand has a marvellous point which tore-open the bee-like body of Hiraṇyakaśipu; O Keśava

[12] 'regular coherence' [*saṃdarbha-śuddhi*]: following M.W.; there is purity, correctness, clarity [*śuddhi*] in his literary-compositions [*saṃdarbha*].

[13] 'complex' [*durūha*] or 'difficult to be inferred or understood' (M.W.) and 'quick' (following Ś.M.'s gloss of *druti* as *śīghra-vacana*)—he can compose difficult verses quickly.

[14] Taking Śrutidhara as the name of a poet (following R.K.)—but other commentators gloss the word as an adjective qualifying Dhoyī as 'well-remembering' or 'observing the *Veda*'.

[15] The poets were members of literary courts of the Sena dynasty. M. Chakravarty (op. cit., p. 161), following various commentators, has noted criticism inherent in each flattery: 'Umāpatidhara sprouts words . . . i.e., lengthens verses by the addition of adjectives, etc.' and Dhoyī's 'good memory' implies 'that he was not original, probably alluding to his fondness for imitation'. Along the same line of interpretation Śaraṇa would be described as overly hasty and obscure.

[16] At the end of the past cosmic era [*kalpa*], in the dissolution of the universe, the demon, Hayagrīva, stole the *Vedas* whereupon Viṣṇu took the form of a fish to slay the demon and recover the sacred text from the depths of the cosmic sea.

[17] 'fixed like . . .': or possibly his tusk is the bright crescent of the moon and the earth is the dark phase of the moon fixed upon that crescent.

[18] Viṣṇu took on the form of a boar to rescue the earth from the sea where it had been thrown by the demon Hiraṇyākṣa.

who bore the form of the Man-lion! O Lord of the World! Victory! O Hari![19]

9. You outwit Bali in your striding, O marvellous Dwarf, you purify men with the water [which flows down] from your toe-nails; O Keśava who bore the form of the Dwarf! O Lord of the World! Victory! O Hari![20]

10. In fluid consisting of the Kṣatriya's blood you cleanse the world so that evil is expelled, the burning-pain of existence is calmed; O Keśava who bore the form of the Bhṛgu Chieftain! O Lord of the World! Victory! O Hari![21]

11. In combat you bestow an agreeable offering, the crowns from the ten heads [of Rāvaṇa], in every direction, [an offering which is] desirable to the Lords of the Quarters; O Keśava who bore the form of Rāma! O Lord of the World! Victory! O Hari![22]

12. On your spotless body you wear a garment resembling a rain-cloud, resembling the Yamunā which came [to you] for fear of the blow of your plough; O Keśava who bore the form of the Plough-holder! O Lord of the World! Victory! O Hari![23]

[19] Hiraṇyakaśipu, enraged over the death of Hiraṇyākṣa, his brother, was determined to kill his son, Prahlāda, for worshipping Viṣṇu. Prahlāda sought Viṣṇu's aid. Hiraṇyakaśipu had obtained a boon ensuring that he could not be harmed by either man or beast, either at night or during the day; Viṣṇu took on a form, half lion and half man, and killed the demon at sunset.

[20] Bali had taken control of the three worlds (Earth, Sky, Heaven); Viṣṇu, appearing in the form of a dwarf before the demon, asked him if he could have as much land as he could cover in three steps. Bali agreed to the dwarf's proposal whereupon Viṣṇu assumed cosmic proportions and in three steps regained the universe. The purificatory water of the Ganges flowed down from his foot, from the place where his toe-nails broke through the world-egg, at the end of his third stride.

[21] The Bhṛgu Chieftain or Paraśu-Rāma (Rāma with an Axe): the Kṣatriyas, the military caste, had taken the power from the Brahmins, the priestly caste; Viṣṇu as the Bhṛgu Chieftain reinstated the priests and thereby re-affirmed and re-established the social order.

[22] Rāma killed Rāvaṇa to rescue his wife, Sītā, from the demon. There are ten heavenly quarters, ten directions (*viz.*, the four cardinal and four intermediary directions plus up and down).

[23] The Plough-holder or Balarāma—traditionally the brother of Kṛṣṇa noted for his exuberant, excessive drinking; his fair body is clothed with a garment that is dark (like a rain-cloud and like the Yamunā) in contrast to Kṛṣṇa's dark, yellow-clad, body. The plough-holder ordered the Yamunā to come to him and when it refused he dug a furrow with his plough to divert it to himself.

13. You censure, because of the precept of sacrifice, ah, ah, the collection of *Veda*s, [the texts in which] the slaughter of tethered-animals [cattle] is taught; O you who are compassionate at heart! O Keśava who bore the form of the Buddha! O Lord of the World! Victory! O Hari![24]

14. In the destruction of the barbarian hordes you carry a sword like a comet [but] even more terrible; O Keśava who bore the form of Kalki! O Lord of the World! Victory! O Hari![25]

15. Hear this noble proclamation of the poet, Śrī Jayadeva, which gives happiness, gives prosperity, which is the epitome of existence; O Keśava who bore ten forms! O Lord of the World! Victory! O Hari!

* * *

16. Lifting up the *Veda*s, supporting the world, raising-up the globe, tearing the demon [Hiraṇyakaśipu] to pieces, outwitting Bali, destroying the Kṣatriyas, conquering Paulastya [Rāvaṇa], bearing a plough, extending compassion, deranging the barbarians, creator of ten forms, homage to you, Kṛṣṇa!

The Second Song (*gurjarī rāga, niḥsāra tāla*):

17. You lay upon the roundness of Kamalā's breasts, wore ear-rings, wore a fair forest-garland! Victory! Victory! O God! Hari![26]

18. You whose ornament is the day-jewel's disk [the sun], you break-up the world, O [Lake] Mānasa gander (mental

[24] The notion of the Buddha as an incarnation of Viṣṇu reflects a conception of Hinduism incorporating and absorbing Buddhism. Mānāṅka, however, reads the verse without the caesura between *sadaya-hṛdaya* and *darśita* (as it would be written in manuscript) and he takes the former, 'compassionate at heart', to refer not to the Buddha but to the Brahmin priests, emphasizing the heretical rather than the compassionate aspect of the Buddha; this notion reflects a Hindu conception that Buddhists, those lured away from the authority of the *Vedas*, are being tricked and punished for their sins by Viṣṇu.

[25] Kalki—the *avatāra* ('descent') who is to come at the end of the present age, the Kali Era, the age of strife, to destroy the world for the creation of a new age. The comet [*dhūma-ketu*] (lit., 'smoke-bannered') is an omen of destruction. Here, given the historical context in which the song was written, the reference seems to be to a belief or hope that Kalki would save India from the invading Moslems.

[26] The refrain, *jaya jaya deva hare*, when read without caesuras plays on the poet's name: 'Victory! Jayadeva and Hari!'

17

supreme-spirit)[27] of the sages! Victory! Victory! O God! Hari!

19. O queller of the venom-bearing serpent Kāliya,[28] O delighter of men, O lord-of-day [sun] to the lotus which is the Yadu[29] family! Victory! Victory! O God! Hari!

20. O annihilator of Madhu, Mura and Naraka,[30] O rider of Garuḍa,[31] O primary-cause of games in the community of the gods! Victory! Victory! O God! Hari!

21. Your eye is a spotless lotus-petal, O liberator from phenomenal-existence, O support of the mansions of the three worlds! Victory! Victory! O God! Hari!

22. You made ornaments[32] for Janaka's daughter [Sītā], you slew Dūṣaṇa [one of Rāvaṇa's generals], you destroyed the ten-necked-one [Rāvaṇa] in battle! Victory! Victory! O God! Hari!

23. You are beautiful like a young rain-cloud, you supported [Mount] Mandara,[33] O *cakora* [bird][34] to the moon which is the face of Śrī! Victory! Victory! O God! Hari!

24. We bow down to your foot—for this reason cherish us! Cause prosperity among us who are bent down [in obeisance]! Victory! Victory! O God! Hari!

[27] *mānasa*: mental or spiritual (as opposed to carnal) *or* Lake Mānasa, a sacred pilgrimage place on Mount Kailāsa to which *haṃsa* birds migrate; *haṃsa*: the Indian goose but also used to refer to the soul or spirit which is migratory like the goose and white (or pure) like the goose—it may indicate either the individual or universal spirit.

[28] Kāliya: living in a tide-pool of the Yamunā, the serpent-demon defiled its waters with his poison; Kṛṣṇa subdued the serpent, made Kāliya worship him.

[29] Kṛṣṇa is a descendant of Yadu and is known as Yadunandana (Son-of-the-Yadus), also Best-of-the-Yadus, Support-of-the-Yadus, etc.

[30] Demons killed by Viṣṇu/Kṛṣṇa; Madhu stole the *Vedas*; Naraka kidnapped the celestial maidens, stole the umbrella of Varuṇa, and the ear-rings of Aditi and took these to Prāgjyotiṣa, a city of demons guarded by Mura. Kṛṣṇa is known as the Enemy or Destroyer-of-Madhu and of Mura (Madhusūdana, Madhuripu, Murāri, etc.).

[31] Garuḍa: 'the devourer'—the sun-bird, the vehicle of Viṣṇu; Kṛṣṇa attacked Prāgjyotiṣa riding upon Garuḍa.

[32] 'You made ornaments': i.e., he drew designs on Sītā's cheeks and breasts, etc., *or* 'You *are* an ornament made for Sītā', i.e., he is her beloved lover.

[33] Mount Mandara: In his incarnation as the Tortoise, Viṣṇu dived to the bottom of the sea that his back might be the support for Mount Mandara which was used as a churning stick to churn the nectar of immortality from the ocean.

[34] *cakora*: a kind of nocturnal partridge which, according to poetic convention, is supposed to subsist solely by drinking the rays of light from the moon.

25. This splendid song of the poet Śrī Jayadeva causes happiness and joy! Victory! Victory! O God! Hari!

* * *

26. The chest of the Killer-of-Madhu is marked with saffron, fixed [there] by embraces on the surface of Padmā's breasts;[35] [his chest] is flooded with the sweat of sexual-fatigue from [making] agitated love,[36] [a flood] like passion made visible; May [his chest] fulfil your pleasure!

27. In spring, with limbs delicate like *vāsantī*[37] blossoms, [Rādhā] wanders in the forest, searching hard[38] for Kṛṣṇa; intensely her distress increases with the mental confusion which is produced by the fever of love; passionately a companion said this to Rādhā:

The Third Song (*vasanta rāga, yati tāla*):

28. When the tender Malayan wind[39] touches the lovely clove creeper, when the hut of the grove is filled with the sounds of the cuckoo[40] intermingled with [the sounds of] swarms of honey-making-bees, Hari plays, now, in the amorous springtime, endless for separated-lovers, he dances with the young-girls, O friend!

29. When travellers' brides moan from being out of their minds with love, when the cluster of *bakula*[41] [trees] is constantly

[35] 'breast' [*payo-dhara*]: lit., 'milk-carrier'.

[36] 'love' [*anaṅga*]: lit., 'the bodiless'—Kāma-deva, the love-god, fired one of his flower-tipped arrows at Śiva to arouse the great ascetic from his meditation and to overwhelm him with desire for Pārvatī; awakened in anger, Śiva consumed the body of Kāma-deva with the fire from his third eye. The various epithets of Kāma-deva (e.g., Remembrance [*smara*], Intoxicator [*madana*], Mind-born [*manasi-ja*], Five-arrowed-one [*pañca-bāṇa*], Mind-churner [*manmatha*], etc.) are used ambiguously as either proper or common nouns.

[37] *vāsantī*: a type of spring jasmine.

[38] 'searching hard': lit., 'her search was much-done [*bahu-vihita* . . .]'.

[39] Malayan wind: the wind from the Western Ghāts, a low mountain range in western India, along the western margin of the Deccan plateau and bordering on the Arabian Sea; the mountains are said to abound in sandal-wood trees and the wind from those mountains is heavily scented with sandal fragrance.

[40] 'cuckoo' [*kokila*]: Koïl bird or Indian cuckoo—a dark bird frequenting the mango groves and crying out in spring. Its song resounds with the 'fifth note' which by convention is the amorous note of the Indian scale.

[41] *bakula*: a fragrant tree said to blossom only when sprinkled with nectar from a young woman's mouth.

covered[42] with blossoms which are full of swarms of sting-possessing-bees, Hari plays, now, in the amorous springtime . . . .

30. When the *tamāla* [tree], garlanded with fresh leaves, is overcome by the violent-passion of the fragrance of musk, when a mass of *kiṃśuka*[43] [blossoms] has the appearance of the nails of the love-god who lacerates the hearts of the young, Hari plays, now, in the amorous springtime . . . .

31. When the opening-up of the *keśara*[44] flower has the appearance of the golden parasol[45] of the emperor of passion, when [the god of] love's quiver makes its appearance as *pāṭala*[46] clusters full of { bees / arrows }, Hari plays, now, in the amorous springtime . . . .

32. When young *karuṇa* [trees] laugh at the sight of people whose modesty has disappeared,[47] when the quarters are bristled with *ketaka*[48] [leaves] shaped like the spear-heads which pierce separated-lovers, Hari plays, now, in the amorous springtime . . . .

33. [Spring is] lovely on account of the fragrance of *mādhavikā* [flowers], fragrant on account of the groups of fresh jasmine,[49] [spring] produces infatuation even in the minds of

[42] 'constantly covered with' (*nirākula*]: or 'pervaded with' following M.'s gloss (*nirantaraṃ ākulā vyāptāh*) (contrary to M.W.).

[43] *kiṃśuka*: or 'flame-tree' has brilliant crimson flowers.

[44] *keśara*: the safflower or crocus or *bakula*.

[45] 'golden parasol' [*kanaka-daṇḍa*]: lit., 'golden stick'—love's sceptre, the symbol of his sovereignty. The stick and quiver of love are possibly euphemisms for the male and female genitals respectively.

[46] *pāṭala*: a begonia with large, deep-red or purple, trumpet-shaped flowers.

[47] 'when young *karuṇa* . . .': the *karuṇa* is a kind of citrus tree with small, white blossoms; 'to show the whiteness [of one's teeth]' is an expression meaning 'to laugh'—hence when the *karuṇa* tree shows the whiteness of its blossoms, i.e., when it blooms, it may be said to laugh. Furthermore *karuṇa*, meaning pitiful, sad, mournful, is a technical term for the 'tragic' sentiment in Sanskrit poetics and *hāsya* is the name of the 'comic' sentiment [*rasa*]—the line, then, might also be rendered: 'comedy becomes tragedy for the young at the sight of people whose modesty has disappeared', i.e., young separated-lovers, seeing lovers enjoying themselves in spring, cease to laugh and begin to feel sad.

[48] *ketaka* (or *ketakī*): the screw-pine, a highly fragrant evergreen with large, spinous leaves. The flowers are worn in a woman's hair to attract a lover.

[49] '*mādhavikā* . . . jasmine [*mālatī*]': spring creepers with white, heavily perfumed flowers. The juxtaposition of the two plants plays on the names of the lovers in Bhavabhūti's drama, *Mālatī-Mādhava*.

sages, [spring is] spontaneously favourable to the young; Hari plays, now, in the amorous springtime....

34. When the mango is bristled-with-delight and [made to] close-its-buds[50] in the embraces of the trembling *atimukta*[51] creeper, when the Vṛndāvana[52] forest is purified by the waters of the Yamunā which flow around its edge, Hari plays, now, in the amorous springtime....

35. This song of Śri Jayadeva shines,[53] [it is] the essence of remembrance of the feet of Hari, [it is] a description of the forest during the amorous spring season, imitating the symptoms of love; Hari plays, now, in the amorous springtime....

* * *

36. [The wind] perfumes the woods with fragrant powders of pollen shaken [loose] from the slightly opened [flowers of the] jasmine creepers; here, the wind, which is like the breath of [the god of] love[54] advancing, which is favourable to *ketakī* [flower] fragrances, inflames the mind.

37. Fevers in the ears [of travellers] are caused by the soft-sweet-toned murmurs of the cuckoos playing in the mango shoots which are shaken by honey-drinking-bees greedy for the fragrance of the emerging honey; with great difficulty these days are passed by travellers who delight in the relish of the union with their lovers which is obtained [only] in moments of meditational attentiveness.[55]

[50] 'bristled ... [*pulakita*] and closed ... [*mukulita*]': *pulakita*, i.e., horripilated conventionally describes the person rapt with love and *mukulita* describes the closing of the eyes with that rapture.

[51] *atimukta*: a type of jasmine; lit., 'beyond the pearl [in whiteness]', but also 'entirely liberated, free from sensual desire' (ironically here).

[52] Vṛndāvana: the wood near Gokula where Kṛṣṇa spent his youth. In later Vaiṣṇava theology the conception arose of the earthly Vṛndāvana as the mere reflection of the Eternal Vṛndāvana, the site of the eternal love-play of Rādhā and Kṛṣṇa.

[53] 'shines' [*udayati*]: following M.'s gloss (*rājate*).

[54] 'love' [*asama-bāṇa*]: lit., 'he who has an unequal [number, i.e., five] of arrows'.

[55] 'travellers ...': men separated from their beloveds find the joy of union only in their imaginations—the vocabulary used is ambivalently sexual and ascetical: *rasollāsa*, the delight of the relish, might also refer to the yogic practice of raising the semen up the spinal column; an *āgama* is a Tantric text which might describe such a process; *prāṇa-samāsamāgama* might be a phrase describing yogic breath control, 'the equal and unequal coming of the breath'; *dhyānāvadhāna* might refer either to intentness in thinking about the beloved, or attention in yogic meditation.

38. [The Enemy-of-Mura was] mind-stealing and trembling, eager for love-play in the excitement of the embraces of many women; that friend, being present, causing Rādhikā to see the Enemy-of-Mura from a distance, spoke again:

The Fourth Song (*rāmakarī rāga, yati tāla*):

39. [He has] a forest garland and a yellow garment on his sandal-smeared, blue body and laughter on his cheeks which are adorned with jewelled ear-rings shaking in play; here Hari plays in the coquettish and playful flock of artless[56] women.

40. Passionately embracing Hari with the massive weight of her swollen breasts a certain herdsman's wife sings a resounded fifth note;[57] here Hari plays . . . .

41. Another artless woman continually contemplates the lotus-face[58] of the Killer-of-Madhu, [the face-lotus] produced love[59] on account of the quivering of his eyes rolling about seductively;[60] here Hari plays . . . .

42. Another [woman] with beautiful buttocks, came-up to say something into his ear and sweetly kissed the beloved on the surface of his cheek which was obliging[61] with bristling-hairs; here Hari plays . . . .

43. And another, with eagerness for the arts of love-play, on the bank of the Yamunā's waters, with her hand on his robe, pulled him who was in the beautiful cane[62] grove; here Hari plays . . . .

44. While his sweet-toned flute is sounded, rows of bracelets shake on their clapping hands; in the enjoyment of the

[56] 'artless' [*mugdha*]: or lovely, beautiful, innocent; one type of woman as classified in the rhetorical texts.

[57] 'fifth note' [*pañcama-rāga*]: a love-cry made during or before coition to signify pleasure.

[58] 'lotus' [*saro-ja*]: lit., 'water-born'.

[59] 'love' [*mano-ja*]: lit., 'mind-born'.

[60] 'seductively' [*vilāsa*]: i.e., in love-play or with an amorous gesture.

[61] 'obliging' [*anukūla*]: obliging by showing his pleasure; this term, as a classification of the hero in the rhetorical texts, signifying the faithful lover, is ironic here.

[62] 'cane' [*vañjula*]: following R.K. and Ś.M.—grove of reeds or ratan or cane [*vetasa*].

rāsa[63] [dance] a deer-like young-girl, intent on dancing with him, praised him; here Hari plays . . . .

45. He embraces one, kisses one, sexually-pleases some sexually-pleasing one, he sees yet another beauty more charming still on account of her smiles and he chases her; here Hari plays . . . .

46. May this song of Śrī Jayadeva, the secret of marvellous Keśava's love-play, lovely and famous in the Vṛndāvana forest, spread-abroad prosperities! here Hari plays . . . .

47. Producing the joy of all-creatures by his love, initiating the festival of love[64] by his limbs which are dark and tender like bunches of [blue] lotuses, embraced by the beauties of Vraja[65] of their own free-will, entirely, all-over-his-body, he is like the erotic [sentiment] incarnate, O friend, Hari plays in spring.

48. [In spring:] as if from discomfort on account of the bites of the serpents which dwell permanently in the hollows [of sandalwood trees], with a desire to bathe in the snow, the sandal-mountain wind seeks the Mountain of the Lord;[66] moreover, because of a rising of their joy at seeing the buds on the tops of the {oily / lovely} mango [trees] the voices of the cuckoos come-forth loud and sweet: '*kuhū-kuhū*'.

49. In the presence of the cowherds' beautiful-browed-women who were carried-away[67] by whirling-about with a great-amount of {joy / jumping} in the *rāsa* [dance], [Hari] was passionately kissed by Rādhā who was blind with love, having ardently embraced his breast, having said, 'Wonderful! your {mouth / voice} consists of nectar!', under the pretext of praising his song; may Hari, ravishing with his smile, protect you!

[63] *rāsa*: a sportive, pastoral, circular dance which became a prominent feature of the Kṛṣṇa mythology and theology.

[64] 'initiating . . .' [*upanayann . . . anaṅgotsavam*]: or 'bringing the merriment of love'.

[65] Vraja: or Braj, the district around Agra and Mathurā where Kṛṣṇa spent his youth.

[66] Mountain of the Lord: Mount Kailāsa, the abode of Śiva, in the Himālayas.

[67] 'who were carried-away': or 'who had lovely gestures on account of . . .'.

THE SECOND CANTO

CAREFREE[68] KEŚAVA

1. While Hari, loving them all-equally, roams-for-pleasure in the forest, Rādhā, on account of jealousy because she was no longer his favourite-beloved,[69] went in another direction; hiding somewhere in a grove of creepers which had crests that were noisy with circles of humming honey-bees, she was sad —she spoke privately to her friend:

The Fifth Song (*gurjarī rāga, yati tāla*):

2. His infatuating flute resounded with honied tones like the nectar from his quavering lower-lip; an ear-ornament was tremulous on his cheek as his head moved, [tremulous] at the corner of his quivering eye; my mind remembers Hari —he joked and played love-games here during the *rāsa* [dance].

3. His hair was surrounded with a ring of lovely peacock-tail-feathers with moon-eyes;[70] [in] his lovely robe [he] was like a dark cloud coloured with many rainbows;[71] my mind remembers Hari....

4. He had an eager desire for kisses from the mouths of the cowherds' fair-hipped-women;[72] the blossom of his honied lower-lip is like the *bandhujīva*[73] [flower], [he was] beautiful on account of his radiant smile; my mind remembers Hari....

5. A thousand cowherd girls were encircled by the shoots which are his very bristled arms; upon his chest and hands and

[68] 'carefree' [*a-kleśa*]: or 'without pain or affliction'; in the *Yogasūtra* a technical term for the five 'hindrances', *viz.*, ignorance [*avidyā*], egotism [*asmitā*], passion [*rāga*], aversion [*dveṣa*], tenacity [*abhiniveśa*].

[69] 'because she was no longer . . .': following Ś.M. *et al.*; lit., 'because her own excellence [over the others] was melted-away' [*vigalita-nijotkarṣād*].

[70] 'moon-eyes' [*candraka*]: i.e., the ocellated markings on the ends of peacock feathers.

[71] 'rainbow' [*puraṃdara-dhanur*]: lit., 'the bow of the Destroyer-of-Strongholds [Indra]'.

[72] 'He had an eager desire for . . .': or following M.—'his desire was aroused by the kisses . . .'.

[73] *bandhujīva*: lit., 'living in groups'—deep-red flowers that open at midday in autumn only to wither by the next morning.

feet, the darkness was dispelled by the rays of his pearl ornaments; my mind remembers Hari....

6. His forehead had a mark of sandal on it which surpassed the moon moving in a cluster of clouds; his broad chest[74] was cruel in crushing the regions of the swollen breasts [of the cowherdesses]; my mind remembers Hari....

7. His cheeks were adorned with enchanting ear-rings of *makaras*[75] made of jewels; he was noble; he had a yellow robe; he was followed by a great retinue of sages, men, spirits and gods; my mind remembers Hari....

8. He was met beneath the splendid *kadamba* [tree][76] allaying fear of the evil of the Kali [Era],[77] exciting me too somehow, in my imagination, with his gaze which is like the tremulous, restless love-god; my mind remembers Hari....

9. The song of Śrī Jayadeva has the beauty of the Enemy-of-Madhu, infatuating, so beautiful; it is exactly suitable for the virtuous for remembrance of the feet of Hari.[78]

* * *

10. [My mind] counts the multitude of his virtues, it does not think of his roaming-about even by mistake,[79] and it possesses delight, it pardons [him of his] transgressions from afar; even while fickle[80] Kṛṣṇa delights among the girls without me, yet again my perverse[81] mind loves him! What am I to do?

[74] 'his broad chest' [*hṛdaya-kapāṭam*]: lit., 'the door to his heart'.

[75] *makara*: a crocodile or sea-monster; painted on the woman's breasts or cheeks or worn as jewellery by the man; the insignia of the love-god, Kāma-deva and his vehicle.

[76] *kadamba* tree: bears orange, fragrant, spherical blossoms; blooms during the first rains; its flowers said to exude liquors; associated with Kṛṣṇa in his amorous aspect (e.g., he stole the clothes of the cowherdesses while they were bathing and climbed to hide from them in the *kadamba* tree) and in his heroic aspect (e.g., he dove from the branches of the *kadamba* tree into the Yamunā to fight Kāliya).

[77] Kali Era: the present age [*yuga*], the age of vice and degeneration; the last of the four ages.

[78] 'suitable for ...': or it is 'remembrance of the feet of Hari in accordance with the practice of the virtuous'.

[79] 'his roaming about ...': M. glosses it as, 'my mind does not roam from concentration on him even by mistake'.

[80] 'fickle' [*valat-tṛṣṇa*]: lit., he has 'an increasing [or turning] thirst'.

[81] 'perverse' [*vāma*]: i.e., she loves him despite herself; or she loves him and love [*kāma*] is cruel or sinister [*vāma*].

The Sixth Song (*mālava-gauḍa rāga, eka-tālī tā a*):

11. I went to his hut in the secret thick t; secretly at night he remained hiding; I looked fearfully i all directions; he laughed with an abundance of passion for e pleasure-of-love; O friend! Make the noble Slayer-of-Keśi [82] make-love to me passionately,[83] I am engrossed with desire r love!

12. I was shy at our first union; he vas obliging with hundreds of skilful flatteries; I spoke wi sweet and gentle smiles; he loosened the silk-garment on y hips; O friend! Make him make-love to me . . . .

13. I was laid-down on a bed of fresh-sh ots; for a long time he lay on *my* breast; I caressed him and kissed him; embracing me, he drank from my lower-lip; O friend! Make him make-love to me . . . .

14. My eyes were closed from sleepiness; his cheek was beautiful and bristling; my whole body was sweating;[84] he was very restless on account of the drunkenness of his great passion; O friend! Make him make-love to me . . . .

15. I cooed with the soft sound of the cuckoo; he mastered the procedures of the science of love; my tresses were strewn with loose flowers; the mass of my firm breast was scratched by his nails; O friend! Make him make-love to me . . . .

16. The jewelled anklets rang-out on my feet; he made-love to me in various ways;[85] my unfastened girdle jingled; he gave me kisses and pulled my hair; O friend! Make him make-love to me . . . .

17. I was languid with the taste of our happy union in love-pleasure, the lotuses which are his eyes were slightly closed; the creeper which is my body fell-down, limp; he declared his love,[86] he is the Killer-of-Madhu; O friend! Make him make-love to me . . . .

18. The speech of the longing cowherd's woman is this song

[82] Keśin: an evil-spirit [*asura*].

[83] 'passionately' [*sa-vikāram*]: lit., 'with true feeling'; or 'make him *savikāram*'—'make him change his feelings for me so that he loves me'.

[84] 'sweat' [*śrama-jala*]: lit., 'exertion-water'.

[85] 'various ways': i.e., in various coital postures—conventionally the ringing of her anklets indicates that they have adopted a posture in which she is beneath him; the jingling girdle indicates the woman positioned on top of the man.

[86] or: 'his love arose'.

of Śrī Jayadeva [about] the abounding erotic disposition of the Enemy-of-Madhu—may this [song] spread happiness with ease.

* * *

19. The sportive flute has fallen from his hand, the glance from the corner of his eye drives-on a flock of cowherdesses with curving creeper-like brows, the surface of his cheek is very sweaty; looking at me he is embarrassed, his face is charming with the nectar of his smile, in the forest, surrounded by a flock of the beauties of Vraja, I see Govinda and I rejoice!

20. The burgeoning of fresh *aśoka*[87] tendrils with small blossom-clusters, difficult to look at, and also the breeze from the grove by the pond distresses me; not even this pointed[88] blossoming of the mango [trees], charming on account of the humming of the swarming bees, makes me happy, O friend!

21. A meaningful smile, an ornamental-braid fallen loose and dishevelled, the creeper which is a splendid brow, a breast exposed by a hand raised over the shoulder in a pretence[89]—having looked secretly [at this sight] of the cowherdesses, [this sight which] captures the minds of fools,[90] for a long time thinking within-himself, his desire for them was dispelled; may young Keśava take away your distress!

THE THIRD CANTO

THE PERPLEXED[91] KILLER-OF-MADHU

1. Moreover, the Enemy-of-Kaṃsa,[92] having placed Rādhā

[87] *aśoka*: has a bright red flower said to bloom only when touched by the foot of a young woman.

[88] 'pointed' [*śikhariṇī*]: a pun—this verse is written in the *Śikhariṇī* metre (four times: ᴗ–– ––– ᴗᴗᴗ ᴗᴗ– –ᴗᴗ ᴗ–); thus superimposed upon Rādhā's statement is the poet's statement: 'this *Śikhariṇī* [metre] causes happiness'.

[89] 'a pretence': they pretend to have to fix their hair in order to expose their breasts to him.

[90] 'fools' [*mugdha*]: following R.K. *et al.*

[91] 'perplexed' [*mugdha*]: or 'bewildered'—the use of the adjective seems peculiar so close to its usage in the previous verse; when applied to a woman *mugdha* usually means artlessly beautiful, innocently charming, but when applied to a man it usually means foolish or inexperienced.

[92] Kaṃsa: a king of Mathurā; Kṛṣṇa's cousin; it was prophesied that he would

in his heart as the chain binding him with desire for the world,[93] abandoned the beauties of Vraja.

2. Having pursued Rādhikā here and there, his mind suffering from the wounds of love's arrows, repentant in the grove on the bank of the Kalinda-Nandinī,[94] Mādhava was despondent.

The Seventh Song (*gurjarī rāga, yati tāla*):

3. She left having seen me surrounded by the group of women, she was not stopped by me [for I was] truly guilty and very frightened; *Hari! Hari!*[95] Because her respect [for me] is destroyed, she is gone, apparently angry.

4. What will she do? What will she say [after] separation [from me for such] a long-time? What use have I for relatives[96] [or] wealth? for life [or] home? *Hari! Hari!* ... she is gone ....

5. I think of her face, her brow bent with an excess of anger, [her face] like a red lotus agitated by a bee flying-about above it; *Hari! Hari!* ... she is gone ....

6. She is united [with me] in [my] heart, incessantly and vehemently I make-love to her [there]; why do I search-for her here in the forest? Why do I lament in vain? *Hari! Hari!* ... she is gone ....

7. O slender-woman! I suppose your heart is distressed with jealousy—I cannot calm you for I do not know where you have gone; *Hari! Hari!* ... she is gone ....

8. You appear before me; you really do make me run around; why don't you give an eager embrace as before? *Hari! Hari!* ... she is gone ....

9. Forgive me—in the future I shall not do such things to you at any time; give me a vision [of you], O beautiful-woman! I am burning with passion; *Hari! Hari!* ... she is gone ....

10. This [song] is depicted by Jayadeva with devotion to

---

be killed by a child of Devakī (Kṛṣṇa was the son of Devakī and Vasudeva; his parents took him to Nanda and Yaśodā to prevent Kaṃsa from killing him). Kaṃsa tried to destroy Kṛṣṇa in a futile attempt to prevent the prophecy from being realized.

[93] 'the world' [*saṃsāra*]: empirical-existence, mundane illusion, the course of rebirth.

[94] Kalinda-Nandinī: 'the daughter of Mount Kalinda', i.e., the Yamunā river.

[95] *Hari!*: an indeclinable exclamation, 'alas!'

[96] 'relatives' [*jana*]: or 'people'.

Hari;[97] [Jayadeva is like] the moon[98] born from the ocean of Tindubilva.[99]

* * *

11. This is a lotus-tendril necklace on my chest, not the Lord-of-Serpents; this is a row of lotus petals on my neck, not the radiance of poison; this is sandal dust, not ashes, on me [for I am] deprived of my beloved; do not[100] attack me, mistaking me for Hara;[101] O Bodiless-love-god! why do you chase me angrily?

12. Do not hold that mango-arrow in your hand, do not string that bow! O you who conquer all [the world] for sport![102] What kind of bravery is it to strike people who are stupefied? O Mind-born-love-god! Torn apart by the rows of arrows which are the trembling, side-long glances of the deer-eyed-woman, of her alone, not even now does my mind even slightly recover.

13. The bow is the sprig of her brow, the arrows are her side-long glances, the bow-string is the tip of her ear: have [love's] weapons, which conquer the world, been transferred by love onto her, the living goddess who is the triumph of the Bodiless-love-god?

14. May the arrow of your side-long glance, placed on the bow of your brow, cause pain where I am vulnerable, also may the mass of your braids, dark natured and curly, make the strenuous-effort [to perform the function] of Māra;[103] mean-

[97] 'This song is . . .': or 'This song of Hari . . .' or '. . . by Jayadeva who is devoted to Hari'.

[98] moon' [*rohiṇī-ramaṇa*]: lit., 'the lover of Rohiṇī'.

[99] Tindubilva (or Kindubilva or Kendubilva): traditionally designated as the birthplace of Jayadeva; located in the Birbhum district of West Bengal according to the Bengali tradition, or in the Purī district of Orissa according to the Orissan tradition.

[100] 'not' [*na*]; grammatically incorrect—(*mā* should be used with the imperative tense) used here and elsewhere presumably for metrical considerations and sound-effect.

[101] Hara: Śiva—The Lord-of-Serpents, Vāsuki or Śeṣa, is represented binding up Śiva's hair, wrapped around his arm or curling around his neck; Śiva's throat is blue, stained by the *kālakūṭa* poison which he drank at the time of 'churning the ocean for nectar' to spare the gods and demons; his body is smeared with ashes from the cremation grounds.

[102] 'sport' [*krīḍā*]: or 'with love-play'.

[103] Māra: identified with Kāma, the love-god, but also with death, the Des-

while may this {impassioned / reddened} lower-lip, which is like the *bimba*[104] fruit, spread infatuation! O slender-woman! How does the well-rounded circle of your breast play with my life?

15. The pleasures of her touch and the tremulous, tender wandering of her eyes, the fragrance of the lotus which is her mouth, the cunning[105] flow of the nectar of her words, the mead from her *bimba*-like lower-lip—if thus, even in attachment to sense-objects, my mind is fixed in the highest-meditation[106] upon her, alas, how then can the sickness of love-in-separation increase?

16. [The waves of glances of him whose] ear-rings were shaking upon his tremulous head as his neck turned-to-the-side, were not noticed by the multitudes of loving-women whose attention was fixed on the place where the song was issuing from his flute; [the glance-waves] of the Enemy-of-Madhu were emitted-in-abundance upon the moon which is Rādhā's artless face, which is sweet and has the essence of nectar; may the waves of his side-long glances give you tranquillity for a long time.

THE FOURTH CANTO
THE AFFECTIONATE KILLER-OF-MADHU

1. Rādhikā's friend spoke to Mādhava, who was distracted by the burden of his love, who was dwelling languidly in a grove of reeds on the bank of the Yamunā:

The Eighth Song (*karṇāṭa rāga, eka-tālī tāla*):

2. She reviles sandal, thereafter she capriciously feels the

---

troyer; the command is then, 'may your hair kill me by making me love you, by making me attached to you'.

[104] *bimba*: a bright red, round fruit to which a woman's lower-lip is conventionally compared.

[105] 'cunning' [*vakriman*]: 'crooked'—refers to a kind of evasive speech conventionally used by lovers to say endearing things to each other in puns and ambiguities and to convey secret love-messages in the presence of others (see I. 49 above and VI. 12 below).

[106] 'highest-meditation' [*samādhi*]: complete concentration, ecstasy, integration, absorption, the unification of subject and object, the final stage of *Yoga*.

moon-beam [to be] a fatigue; she considers the Malayan wind to be like poison because of its contact with serpents' lairs. O Mādhava! as if from a fear of love's arrows she is merged[107] with you in her imagination—she is distressed in your absence.

3. She makes an armour out of a large mass of wet lotus leaves upon the vulnerable spot which is her own heart as if to be a protection for you [who dwell in her heart] from the incessantly falling arrow[s] of love; O Mādhava! . . . she is distressed in your absence.

4. She makes a bed of arrows[108] with the flower-shafts, [a bed] to be desired for [practising] the many amatory arts, [she makes] a bed of flowers as if [fulfilling] a religious-vow[109] for the happiness of your embrace; O Mādhava! . . . she is distressed in your absence.

5. And she has a noble lotus of a face which has a multitude of tears caused-to-flow from her eyes, [a face-lotus] like the moon which has a stream of nectar trickling from it on account of its being torn by the teeth of the monstrous moon-troubler;[110] O Mādhava! . . . she is distressed in your absence.

6. Secretly she draws you with musk as the love-god—she bows down [to you] placing the *makara* beneath you and the fresh mango arrow in your hand; O Mādhava! . . . she is distressed in your absence.

7. By her absorption in meditation you, who are so very difficult to be attained, are placed before her—she moans, laughs, grieves, weeps, wanders-about, lets-go of her grief; O Mādhava! . . . she is distressed in your absence.

8. At every step she proclaims this also: 'O Mādhava, I am fallen [in obeisance] at your foot—when you are turned-away [from me] immediately even the nectar-storing-moon spreads fire in my body.' O Mādhava! . . . she is distressed in your absence.

[107] 'merged' [*līna*]: she clings to, is devoted to, absorbed in, concealed in Kṛṣṇa.

[108] 'bed of arrows' [*śara-talpa*]: following M.W.—'a bed of arrows for a dead or wounded soldier'.

[109] 'religious-vow' [*vrata*]: i.e., making and lying upon the bed of arrows; ironically chastity is implicit in the word.

[110] 'moon-troubler' [*vidhuṃ-tuda*]: Rāhu who took the first sip of the nectar of immortality, after the churning of the ocean, for which Viṣṇu beheaded him; but having tasted the nectar he did not die; his head continued to chase after the moon for another taste; eclipses are described as Rāhu devouring the moon.

9. If this song of Śrī Jayadeva is to be performed repeatedly by the mind, the speech of the friend of the young cowherdess agitated by separation from Hari should be recited.

* * *

10. Her dwelling is like the forest and the garland of her dear friends is like the snare and her heated-sorrow with her sighing-breath is like the sheet of flames in the burning forest; and, alas, through your absence, she is like the doe [caught in the snare, in the burning forest]; Ah! How, moreover, is [the god of] love like [the god of] death [Yama] performing tiger's play?[111]

The Ninth Song (*deśākha rāga, eka-tālī tāla*):

11. O Keśava! In your absence Rādhikā, the slender-bodied one, considers even the exalted necklace placed upon her breasts to be like a burden.

12. O Keśava! In your absence Rādhikā fearfully regards the sandal unguent, truly passionate and smooth, to be like poison on her body.

13. O Keśava! In your absence Rādhikā bears her sighing-breath, incomparably long, burning like the scorching of love.

14. O Keśava! In your absence Rādhikā casts the lotus of her eye, which has a multitude of watery drops, in every direction, as if its stalk had fallen off.[112]

15. O Keśava! In your absence Rādhikā does not let her cheek go from the palm-of-her-hand—it is steady like the new moon in the evening.

16. O Keśava! In your absence Rādhikā considers the bed of sprouts which is right before her eyes to be made of its antithesis—fire.[113]

17. O Keśava! In your absence Rādhikā mutters[114] 'Hari,

[111] 'tiger's play' (*śārdūla-vikrīḍita*]: i.e., she is like a deer caught in the snare, easy prey for the tiger-like love-god. Again a pun on the name of the metre of the verse—'performing the *Śārdūla-vikrīḍita* [metre]' (four times: – – – ∪∪– ∪–∪ ∪∪– – –∪ – –∪ –).

[112] 'casts the lotus . . .': i.e., her eye, looking all about for Kṛṣṇa, is like a lotus moving all about (because its stalk no longer holds it still).

[113] 'fire' [*hutāśa*]: lit., 'oblation-eater'.

[114] 'mutters' [*japati*]: 'whispers' or as a devotional term, 'prays', invokes the name of Hari.

Hari' passionately, according to her desire, as if ordained to die of separation.

18. May this song, sung by Śrī Jayadeva, brought near the foot of Keśava, cause joy!

* * *

19. She bristles, makes love-cries, laments, trembles, gasps-for-breath, ponders, jumps-up, closes-her-eyes, falls, rises, and even faints; in such a great love-fever why should her beautiful body not live through your elixir[115] if you are pleased [to go to her], O you who are like a Heavenly Physician? Otherwise the suppliant is abandoned.[116]

20. O Beloved of the Heavenly Physicians! You don't help Rādhā [to be] rid of afflictions, [Rādhā] is sick with love and cured only by the nectar of contact with your body; O Upendra! You are more cruel than even the thunder-bolt.[117]

21. Her body long suffers from the fire of love's fever; O wonder! her mind suffers in thoughts of sandal, the moon and lotuses; but, because of fatigue, thinking of you who are the only one dear to her, [you who have] a cool body, waiting in private, she who is wasted somehow lives for a moment.

22. Formerly your separation was not endured even for a moment by her, suffering the closing of her eyes; how can she live through long separation, having seen the mango branch, its tip in flower?

23. Having lifted up Govardhana and bearing it from a desire to shelter Gokula[118] which was disturbed by rain, long

[115] 'elixir' [*rasa*]: the 'savouring' of the bliss of his love—i.e., only the 'relish' of his love can keep her alive for it is like the divine medicine of the Heavenly Physicians, the Aśvins, who avert disease and bring fortune.

[116] 'the suppliant' [*hastaka*]: lit., 'the hand', but taking *hastaka* as 'the one whose *hand* is out-stretched in supplication' following R.K.'s explanation of the general purport of the phrase: 'Unless you go to her you'll incur the fault [*dūṣaṇa*] of abandoning a suppliant [*āśrita*]'—i.e., it would be dishonourable for a god not to serve a needy devotee.

[117] 'O Upendra . . .': Upendra is an epithet of Kṛṣṇa qualifying him as the 'younger brother of Indra'; the thunder-bolt [*vajra*] is the weapon of Indra. Pun on the name of the metre of the verse, the *Upendravajrā* metre (four times: ᴗ–ᴗ ––ᴗ ᴗ–ᴗ ––).

[118] 'Govardhana . . . Gokula': Kṛṣṇa stole the food that had been offered to Indra; Indra, retaliating in anger, cast his thunder-bolt causing it to rain on Gokula, the village of the cowherds; Kṛṣṇa then lifted up Mount Govardhana and held it as an umbrella over Gokula.

kissed by the loving-women of the cowherds on account of their supreme joy, as if with pride, marked with vermilion imprints from the surfaces of their lower-lips which had been applied to it, may the arm of the Enemy-of Kaṃsa, who has a cowherd's body, spread your prosperity!

THE FIFTH CANTO

THE LOTUS-EYED-ONE IN A STATE-OF-LONGING

1. 'I'll stay here; you go to Rādhā; pacify her with my speech; bring her here!' said the Enemy-of-Madhu; the friend, so directed, went again to Rādhā herself and said this:

The Tenth Song (*deśivarāḍī rāga, rūpaka tāla*):

2. While the Malayan wind blows, bringing love near, while myriads of blossoms open to split the hearts of separated-lovers, he, the one who wears a forest garland, sadly-waits in your absence, O friend!

3. While the cool-rayed-moon burns him he seems to die; while the arrow of love falls he moans all the more distraught; he . . . sadly-waits . . . .

4. While the swarm of honey-drinkers hums he covers his ears; while his mind experiences separation night after night he falls ill; he . . . sadly-waits . . . .

5. He lives in the forest expanse, abandons his lovely house;[119] he tosses on his bed which is the ground; he often moans your name; he . . . sadly-waits . . . .

6. While the poet Jayadeva sings, may Hari appear, by [your] good deeds, in your mind which has an abundance of zeal because of the charm of love-in-separation; he . . . sadly-waits . . . .

* * *

7. Where he was together with you before the perfections of love[120] were attained; truly there in love's great-pilgrimage-place in the grove again Mādhava, meditating on you, con-

[119] 'lovely house' [*lalita-dhāman*]: or 'pleasure house', a hut reserved for trysting.

[120] 'love' [*rati-pati*]: Kāma-deva as the 'husband of Rati' (sexual-pleasure personified).

stantly chanting also a string of sacred-sounds as an invocation to you alone, desires again the nectar of the ardent embraces of the pitchers of your breasts.

The Eleventh Song (*gurjarī rāga, eka-tālī tāla*):

8. He has gone to the tryst which is the essence of sexual delight; he has the beautiful appearance of [the god of] love; O fair-hipped-girl! don't delay going to him! follow him, the lord of your heart! Where the wind is gentle, on the bank of the Yamunā, he, the one who wears a forest garland, dwells in the forest, his hands are ever-moving in squeezing the cowherdesses' swollen breasts.

9. He sounds his tender flute which makes the given signal together with your name: he very-much esteems even the pollen which is moved [to him] by the wind that touches your slender-body; where the wind is gentle . . . he dwells . . . .

10. When a feather falls [or] a leaf stirs, his trembling eye looks to the path [by which] your arrival is anxiously-expected and he prepares the bed; where the wind is gentle . . . he dwells . . . .

11. Abandon the noisy, capricious anklet which shakes-well during love-play—it is like an enemy;[121] Go, O friend, to the dense, dark grove; wear a dark-blue cloak; where the wind is gentle . . . he dwells . . . .

12. [You'll be] a necklace placed upon the chest of the Enemy-of-Mura, like a fluttering crane, like lightning, upon a [dark] cloud, O golden [woman]—you'll shine while making-love on-top-of-him as a reward for your good-deeds;[122] where the wind is gentle . . . he dwells . . . .

13. Place your hips, unclothed, the girdle unbound, uncovered,[123] upon bed of sprouts, O lotus-eyed [woman]—[Your hips are] like a treasure, a store-house of joy; where the wind is gentle . . . he dwells . . . .

[121] 'Abandon . . .': because of its jangling the anklet is desirable during love-making but it is taken off to go to the tryst clandestinely. The dark cloak is worn also to maintain that secrecy.

[122] '. . . fluttering crane . . .'; clouds are said to wear garlands of lightning; cranes are said to nest in clouds and the word 'crane' [*balākā*] can refer to a loving-woman.

[123] 'uncovered' [*apidhāna*]: following M. [*anāvṛta*] (contrary to M.W.).

14. Hari is full-of-pride; this night also is coming to an end now; heed my words at once—fulfil the desire of the Enemy-of-Madhu; where the wind is gentle . . . he dwells . . . .

15. While Jayadeva sings in service to Hari most delightfully, bow, delighted in heart, to Hari who is very compassionate, who is to be desired for his virtues; where the wind is gentle . . . he dwells . . . .

* * *

16. At one moment he scatters sighs, then looks in [all] directions, then resorts to the grove humming, then gasps-for-breath, then prepares the bed, then looks around bewilderedly —O lovely-woman, your beloved is wearied by the suffering of love.

17. Entirely along with your perversity the hot-rayed-sun has now set, along with the desire of Govinda the darkness has gained intensity; my long plea resembles the doleful cry of *koka* [birds];[124] so, O artless-woman, delay is fruitless, the moment of the tryst is delightful.

18. From an embrace, then from a kiss, then from scratching with their nails, then from love's rousing, then from shaking-about [in coition], then from sexual exertion, both are pleased —when a husband and wife who have gone to an affair with another [lover] come-together by mistake and [then] recognize [each other] by their speech here, in the darkness, their pleasure is mixed with embarrassment, isn't it? isn't it?

19. Fixing your vision, trembling in fear, upon the dark road, having stood at each tree for a moment, slowly stepping, somehow the secret-place was reached by your limbs which have a restless-motion on account of love—O lovely-faced-woman, seeing you, may the fortunate-and-charming-one attain his purpose!

20. [He is] the bee on the lotus which is Rādhā's artless face, the appropriate blue jewel for the adornment of the region which crowns the three worlds, the death of those [demons] whose descent is a burden to the world, spontaneously the beginning of a night of pleasure for the minds of the beauties of Vraja, the comet for the destruction of Kaṃsa—may he, the Son-of-Devakī, protect you!

[124] *koka* birds: following Ś.M.—the *cakravāka* bird, a kind of shelldrake said to mourn aloud each night when it must be separated from its mate.

THE SIXTH CANTO

ARDENTLY-LONGING VAIKUṆṬHA

1. Then, having seen her in the hut of creepers unable to go [to him] and passionate for a long time the friend related her behaviour to Govinda who was sluggish from love:

The Twelfth Song (*guṇakarī rāga*, *rūpaka tāla*):

2. Secretly she sees you everywhere, drinking the sweet honey of her lower-lip; O Lord Hari! Rādhā sadly-waits in her bed-chamber.

3. Rushing with eagerness for the tryst with you, [but] she falls down, moving just-a-few steps; O Lord Hari! . . .

4. She has a bracelet made of spotless lotus fibres and she lives henceforth [solely] by your skill in love-making;[125] O Lord Hari! . . .

5. Constantly seeing the play of her ornaments she is disposed to imagining: 'I am the Enemy-of-Madhu'; O Lord Hari! . . .

6. 'Why doesn't Hari come quickly to our tryst?' she asks her friend time-after-time; O Lord Hari! . . .

7. She embraces and kisses the great darkness which resembles a rain-cloud [saying] 'Hari has arrived'; O Lord Hari! . . .

8. While you delay her modesty is melted-away; she moans and weeps, ready in her chamber [for you, her beloved]; O Lord Hari! . . .

9. May this song of the poet Śrī Jayadeva spread great-joy to people of taste!

* * *

10. She has many rows of bristling-hairs, she utters loud love-cries[126] confused with emotional-tones[127] and senselessness

[125] 'skill in love-making' [*rati-kalā*]: or 'with a desire for love-pleasure' following Ś.M.'s gloss of *kalā* as *ākalana*.

[126] 'love-cries' [*sīt-kāra*]: lit., 'she makes the sound "*sīt*". . .', a sound made by drawing in the breath with the mouth almost closed to express intense delight in love-making.

[127] 'emotional-tones' [*kāku*]: M.W. gives 'a peculiar tone or change of voice resulting from distress or fear or anger . . .'.

is aroused within her; Cheat! having thoughts of intense passion about you, the deer-eyed-woman is immersed in an ocean of passionate-bliss, fixed in meditation.

11. She puts jewellery on her limbs; again-and-again also when a leaf stirs she believes that you have come; she spreads-out a bed; for a long time she meditates: Thus although engaged in hundreds of games of imagining the preparation of the bed and the choosing of finery, without you this beautiful body will not pass the night.

12. 'Why do you rest beneath the banyan[128] tree which is the abode of { black-snakes / Kṛṣṇa-the-enjoyer } ?[129] O Brother! Why don't you go to the joyful house of Nanda which is within sight's range from here?' Concealing a message for Rādhā, in the presence of Nanda, from the mouth of a traveller, Govinda's words, filled with excellence for evening guests, prevail!

THE SEVENTH CANTO
CUNNING[130] NĀRĀYAṆA

1. And at that time [the moon], beautiful with its distinct spot like a stain incurred by obstruction[131] of the path [by which] wanton-women [go to their lovers], illuminated the Vṛndāvana area with its rays; the moon was the sandal-spot on the face of a beautiful-woman of the Quarters.[132]

2. While the sphere which bears a hare [the moon] was rising and Mādhava was making a delay, the destitute-woman sorrowfully uttered aloud various lamentations:

[128] 'banyan' [*bhāṇḍīra*]: *nyag-rodha* or Indian fig tree; according to M.W. the name of a particular tree on Mount Govardhana in Vṛndāvana.

[129] 'black-snakes *and/or* Kṛṣṇa-the-enjoyer [*or*: the enjoyer(s) of Kṛṣṇa]' [*kṛṣṇa-bhogin*]: the black-snake is the deadliest of Indian cobras. The 'enjoyer' in the sense of the 'lover'. By the pun Kṛṣṇa both frightens the traveller away from the place where he wants to meet Rādhā *and* conveys to Rādhā the message of assignation.

[130] 'cunning' [*nāgara*]: lit., 'town-bred', i.e., sophisticated, clever.

[131] 'obstruction . . .': by shedding light the moon makes it difficult for women to go secretly to their lovers.

[132] 'beautiful woman of the Quarters': M.W., 'a quarter of the sky deified as a young virgin'.

The Thirteenth Song (*mālava rāga, yati tāla*):

3. Even at the time [which he] appointed, Hari, ah, ah, did not go to the forest; this youth of mine is fruitless even though it has a spotless body; to whom shall I, deceived by the words of my friends, go for refuge here?

4. For whose pursuit at night even this dark-thicket is frequented, because of him, this, my heart, is impaled by the five arrows [of love]; to whom shall I go for refuge . . . ?

5. My very death would be preferable—how can I, whose { tryst / body } is thus futile, who am senseless here, endure the flame of separation? To whom shall I go for refuge . . . ?

6. The sweet spring night depresses me, ah, ah; some fortunate[133] loving-woman enjoys Hari; to whom shall I go for refuge . . . ?

7. Ah, ah! I consider the jewelled ornaments, bracelets and so forth, very offensive because they carry the flame of separation from Hari; to whom shall I go for refuge . . . ?

8. Also the flower-garland upon my heart kills me, whose body is very delicate like a flower, by the play of love's arrows which has a very terrible[134] nature; to whom shall I go for refuge . . . ?

9. Here I stay, [I who have] not taken-account of the forest-tangles; the Enemy-of-Madhu does not even remember me with his mind; to whom shall I go for refuge . . . ?

10. May the { muse / speech } of the poet Jayadeva, [who goes for] refuge at the feet of Hari, dwell in your heart like a young girl who is skilled in the charming arts [of love].

* * *

11. Is he trysting-with some loving-woman or is he restrained by friends [who are playing] games of skill[135] [or] does he wander-about in the darkness near the forest? My lover, bored-with-me, is unable to set-out even a little [way] upon

[133] 'fortunate' [*kṛta-sukṛta*]: lit., 'done-good-deeds', i.e., she is fortunate in this life because she has done good deeds in a past life.

[134] 'terrible' [*viṣama*]: or 'unequal', punning on the unequal number of Kāma's arrows.

[135] 'by friends . . .': or 'by licentious friends'.

the road to the beautiful grove of creepers and cane as arranged —for he has not come.

12. Then, having seen that friend, silent with depression, return without Mādhava, apprehending that Janārdana[136] was sexually-pleased by some woman, she said this as if she had seen it:

The Fourteenth Song (*vasanta rāga, yati tāla*):

13. Her dwelling is appropriately arranged for the battle of love; her hair is slightly dishevelled with strewn flowers; some young-girl, who has so-many qualities,[137] plays-in-love with the Enemy-of-Madhu.

14. The emotions [of love] are stirred in the embraces of Hari; upon her pitcher-like breasts a necklace shimmers; some young-girl plays-in-love . . . .

15. The moon of her lovely face has curls shaking [over it];[138] she is made languid by the violent-passion of his lip-drinking-kisses; some young-girl plays-in-love . . . .

16. Her cheeks are scratched by jangling ear-rings; she shakes with the motions of her hips which have a jingling girdle; some young-girl plays-in-love . . . .

17. She laughs with embarrassment at the gaze of her lover; she cries out with the relish of love-pleasure in many varied moans; some young-girl plays-in-love . . . .

18. She undulates, trembles much and her hair is very bristled; her love blossoms as her eyes are closed in sighs; some young-girl plays-in-love . . . .

19. Her fortunate body bears drops of sweat; steadfast in the battle of love she falls upon his chest; some young-girl plays-in-love . . . .

20. May the delight of Hari as sung by Śrī Jayadeva cause the evil of the Kali [Era] to be destroyed!

* * *

21. Although by having the appearance of the lotus which is the face of the Enemy-of-Mura, pale from separation, it stops

[136] Janārdana: the epithet of Kṛṣṇa means 'tormenting people'.

[137] 'so many qualities': suggests, 'who is luckier than I am'.

[138] 'The moon . . .': the suggested simile is that her dark curls are like a cloud over her moon-like face.

my pain, as the friend of love alas, the moon greatly spreads the anguish of love in my heart.

The Fifteenth Song (*gurjarī rāga, eka-tālī tāla*):

22. Upon a delightful-woman's face, where love has arisen where a lower-lip is turned for a kiss, bristling-excitedly he draws a forehead-mark with musk, [a mark] like the animal upon the moon; in the forest on the bank of the Yamunā the triumphant Enemy-of-Mura delights now.

23. In her hair which resembles a mass of clouds, which sets-shaking the faces of young-men, he places a *kurubaka*[139] flower, splendid as lightning, in the forest of the deer who is love; . . . the Enemy-of-Mura delights now.

24. He places a string of jewels, a spotless multitude of stars, upon the { very firm / beautifully clouded } firmament which is her pair of breasts smeared with musk, adorned with a moon in nail-marks;[140] . . . the Enemy-of-Mura delights now.

25. He puts an emerald bracelet, a swarm of honey-makers, upon the delicate pair of arms which surpass pieces of lotus-fibre having lotus leaves as the palms of her hands, cool as snow:[141] . . . the Enemy-of-Mura delights now.

26. He spreads a girdle made of jewels that mocks a festal-arch upon her hips which possess the abode of pleasure, upon the voluptuous portion [of her body] which is the golden seat of love, which is perfumed;[142] . . . the Enemy-of-Mura delights now.

27. He puts lac as an outer-covering upon the sprout which is her foot which is made to rest upon his chest,[143] the dwelling place of Kamalā, adorned by the jewels which are her nails; . . . the Enemy-of-Mura delights now.

[139] *kurubaka*: a red amaranth which is said to bloom only when touched by a young woman's foot or by her gaze.

[140] Her breast with the jewels, musk and nailmark is like the sky with its stars, clouds and the moon [lit., night-maker]; 'the moon' is a technical term in the erotic texts for a particular kind of nailmark; musk powder was used like talcum powder particularly by women after love-making.

[141] Her arm with the bracelet is like a lotus tendril with bees upon it.

[142] The girdle over her vulva (euphemistically described as the 'abode of pleasure' or 'throne of love') is like the decorative arch over a doorway.

[143] 'upon his chest' [*hṛdi*]: or 'in his heart'.

28 While the brother of the mischievous Plough-holder very-much delights some-woman who has beautiful eyes, tell me, O friend, why have I stayed here in this thicket, fruitlessly, for so long, unpleasantly?[144] . . . the Enemy-of Mura delights now.

29. Let not the evil, which is produced by the Kali Age, remain here, in Jayadevaka, the king of poets—he has a song of taste,[145] he recounts Hari's qualities, he is a servant of the feet of the Enemy-of-Madhu; . . . the Enemy-of-Mura delights now.

* * *

30. O friend, if my pitiless false-lover[146] has not come, O messenger, why should you burn-with-sorrow? He who has many lovers sports there of his-own-free-will—is that your fault? Look now, drawn to the tryst with the dear-one by the beloved's qualities, bursting as if from an excess of longing's pain, this soul will go itself.

The Sixteenth Song (deśavarāḍī rāga, rūpaka tāla):

31. O friend, she who is being sexually-pleased by the one who wears a forest-garland, by the one whose eyes are like lotuses[147] moving in the breeze—she does not burn on account of the bed of blossoms.

32. O friend, she who is being sexually-pleased by the one who wears a forest-garland, by the one whose lovely mouth is an opened lotus—she does not break-apart on account of love's arrows.

33. O friend, she who is being sexually-pleased by the one who wears a forest-garland, by the one whose sweet speech is more gentle than the nectar-of-immortality—she does not blaze in the Malaya-born breeze.

34. O friend, she who is being sexually-pleased by the one who wears a forest-garland, by the one whose feet and hands

144 'unpleasantly' [*virasa*]: 'without *rasa*', without enjoying the erotic sentiment particularly.

145 'a song of taste': 'a song of *rasa*', a song which has the erotic sentiment particularly.

146 'false-lover' [*śaṭha*]: a type of hero as classified in the rhetorical texts.

147 'lotus' [*kuvalaya*]: specifically a dark lotus or water-lily that opens at night and closes during the day.

resemble lotuses on the bank[148]—she does not toss on account of the cool-rayed [moon.]

35. O friend, she who is being sexually-pleased by the one who wears a forest-garland, by the one who is beautiful like a mass of rain-clouds—she does not break in her heart on account of long separation.

36. O friend, she who is being sexually-pleased by the one who wears a forest-garland, by the one who in his shining robe resembles a touchstone of gold[149]—she does not sigh on account of the mocking-laughter of her companions.

37. O friend, she who is being sexually-pleased by the one who wears a forest-garland, by the youth who is better than all the people of the world—she does not bear splitting-pain with great sorrow.[150]

38. Through this speech as sung by Śrī Jayadeva, may Hari too enter [your] heart!

* * *

39. O joy of [the god of] love! Sandal breeze! Be calm, oh, O southern [wind] stop [your] perversity! For a moment, O life-breath of the world, having put Mādhava [in my thoughts] formerly, you will be the taking of my life-breath.[151]

40. This house of my friend hurts like an enemy, the cool breeze [hurts] like fire, the nectar-rayed [moon hurts] like poison, while he is on my mind; even when he is cruel again my heart is attached to him by force—perverse and totally-unfettered[152] is lotus-eyed-women's love!

41. Make torment O Malayan breeze! O Love! Take my life! I shall not seek refuge at home again! Why should you

[148] 'lotuses on the bank' [*sthala-jala-ruha*]: lit., 'land-water-born'; perhaps a kind of hibiscus sometimes known as a 'land-lotus' [*sthala-kamala*].

[149] 'who in his shining robe . . .': his dark body with his yellow robe is like a black touch-stone streaked with gold.

[150] 'sorrow' [*karuṇa*]: suggests that Rādhā suffers the tragic sentiment [*karuṇa-rasa*] while the cowherdess with Kṛṣṇa enjoys the erotic sentiment [*śṛṅgāra-rasa*].

[151] 'O southern . . .': playing on the several meanings of *dakṣiṇa* (southern, right, good) and *vāma* (perverse, left, cruel); the wind is being perverse or paradoxical in that, although it is the life-breath [*prāṇa*] of the world, it is taking away her individual life-breath.

[152] 'totally-unfettered' [*nikāma-niraṅkuśa*]: or, following M.W., perverse love 'freely rules over' the deer-eyed-women.

have mercy on me? O Sister-of-Death![153] Moisten my limbs with waves! Let the fire in my body cease!

42. At dawn, while the circle of friends was laughing freely having timidly looked at Rādhā's breast, clad in the yellow garment, at Acyuta[154] in the blue garment, may he, having held the edge-of-the-garment,[155] shaking with embarrassment, to the eyes of Rādhā's face, the Son-of-Nanda, this one whose mouth has a sweet smile, be for the joy of the world![156]

THE EIGHTH CANTO

LAKṢMĪ'S MASTER VEXED

1. Then, somehow having passed the night, although truly torn apart by the arrows of love, at dawn, she spoke indignantly to her lover although he was bent-down [in obeisance] before her, uttering a speech of conciliation:

The Seventeenth Song (*bhairavī rāga*, *yati tāla*):

2. Red with the passion of long wakefulness produced during the night, sluggishly settling, your eye displays, as if your passion were made visible,[157] your attachment to the savour-of-love which has arisen; *Hari! Hari!* Go Mādhava! Go Keśava! Don't speak your deceitful speech [to me]! Follow her, O lotus-eyed-one, her who dispels your depression!

3. Having an appearance of darkness caused by kissing eyes dark with mascara, your red lip, O Kṛṣṇa, conforms [in colour now] to your [dark] body; O *Hari! Hari!* Go . . . .

[153] 'Sister-of-Death': the Yamunā river.

[154] Acyuta: the epithet of Kṛṣṇa means 'firm' or 'imperishable'.

[155] or: 'having directed a side-long glance at Rādhā's face . . .'.

[156] Rādhā is wearing Kṛṣṇa's yellow garment and he is wearing her blue garment. The implication is that, dressing together in the dark after love-making, they mistakenly put on each other's clothes. The friends laugh at dawn when they discover the mistake, when they see this evidence of the love-making. In later poetry about the love-play of Rādhā and Kṛṣṇa the exchange of clothes became a conventional theme and the lovers do so consciously. This verse refers back to their previous union.

[157] 'passion made visible . . .': *rāga* can mean both 'red' and 'passion' and the double meaning is frequently used. The word for 'red' used here [*kaṣāya*] suggests 'stain, impurity'.

4. Your body, which has lines of wounds from hard curved-nails [inflicted] in the battle of love, resembles a record[158] of [her] victory in love-pleasure from an inscribing on gold done with an emerald chip; *Hari! Hari!* Go . . . .

5. Streaked with lac[159] dissolved from the lotus of [her] foot, this, your noble chest seems to show outwardly a covering-hedge of the young shoots of the tree of love; *Hari! Hari!* Go . . . .

6. The mark of teeth on your lower-lip produces pain in my mind—it says, 'how can this body of yours still be one-and-the-same as mine?' *Hari! Hari!* Go . . . .

7. Like your outer [self], O Kṛṣṇa, your inner-self will become very dark indeed—how then can you cheat the person who follows you, [people] afflicted with the fever of love? *Hari! Hari!* Go . . . .

8. You wander in the forest to devour young-women—what's wonderful in that? Pūtanikā[160] indeed reveals your character as the cruel infant who is a lady killer! *Hari! Hari!* Go . . . .

9. Listen to the lamentation of a young-woman abandoned and deceived in love-pleasure as sung by Śrī Jayadeva, O learned-men! It is sweet like nectar, unobtainable even from the abode of the gods.

* * *

10. Seeing this, your chest with its red colour, sprinkled with lac from the foot of your beloved, as if your spreading passion were on the outside, with the humiliation of great affection [being] made-public now, O Cheat, looking at you causes me even more shame than anguish.

11. Of the gazelle-eyed women [the sound of the flute] is a great spell for delight, [for making them] rave, for stupefaction, [for making] fall *mandāra*[161] [flowers] quivering on account of

[158] 'lines . . . record' [*rekhā* . . . *lekhā*]: the two words can indicate either a line, as of writing, or a scratch, as of nails.

[159] 'streaked with lac . . .': following Ś.M., the lac on his chest indicates that they have adopted a particular coital posture in which the woman's feet are placed against the man's chest (the 'position of anger' [*krodha-bandha*]).

[160] Pūtanikā: in an attempt to kill the infant Kṛṣṇa Pūtanikā offered him her poisonous breast to suckle; but Kṛṣṇa, immune to the venom, killed her, sucked her life from her.

[161] *mandāra*: the coral tree; one of the five trees said to grow in heaven. Kṛṣṇa stole the *mandāra* tree from Indra's pleasure garden.

the shaking of their heads with inner infatuation; it is the ruin of the irrepressible sorrow and distress of the gods when they are afflicted by haughty demons; may the sound of the flute of the Enemy-of-Kaṃsa remove your misfortunes!

THE NINTH CANTO
ARTLESS MUKUNDA

1. To her [who was], then, exhausted with passion, hurt by her violent-desire for love-pleasure, possessed of despair, [who] had pondered Hari's behaviour, [to the one] separated [from her lover] on account of a quarrel, the friend spoke privately:

The Eighteenth Song (*gurjarī rāga, yati tāla*):

2. Hari goes-to-the tryst while the sweet [spring] wind blows; what other great joy is there, O friend, in the world? O piqued-woman! don't direct your pique against Mādhava, oh!

3. The pitcher of your breast is more heavy and {passionate / full-of-juice} than coconuts—why make it fruitless? O piqued-woman!...

4. How many times has this not been said, word for word, recently: 'Don't despise Hari who is so very splendid'? O piqued-woman!...

5. Why do you sink-down in weeping, helpless? The whole assembly of your young-women [friends] laughs [at you]; O piqued-woman!...

6. Upon a cool bed of wet lotus sprouts behold Hari, make your eyes fruitful; O piqued-woman!...

7. Why do you produce this heavy sorrow in your mind? Listen to my speech about displeasing separation; O piqued-woman!...

8. May Hari come! May he speak very sweetly! Why make your heart so very lonely? O piqued-woman!...

9. Sung by Śrī Jayadeva very delightfully, may the story-of-the-deeds of Hari cause happiness for people of taste!

* * *

10. When the beloved is tender you are rough, when he

bends-down [in obeisance] you are unbending, when he is passionate you are hostile, when he has his face-raised [in expectation] you have your face-turned-away [in aversion]; it is appropriate then, O perverse-woman, that your sandal unguent is poison, the cool-rayed-moon is the burning-sun, frost is fire, the pleasures of play[162] are tortures.

11. By the intensely joyous hosts of gods, Puraṃdara and the others, who bow down [to the lotus-foot] on account of their great reverence, having the blue jewels of Indra [sapphires] in their crowns, [the foot-lotus] has the appearance of having bumble-bees [upon it]; [the foot-lotus is] viscid with the Mandākinī which freely flows [from it] beautifully like nectar; we venerate, for the destruction of evil, the lotus of Śrī Govinda's foot![163]

THE TENTH CANTO

THE CLEVER FOUR-ARMED-ONE

1. On that occasion, having approached her who was governed by harsh anger, her whose mouth was weak from boundless sighs, the fair-mouthed-woman, to her who had seen the face of her friend bashfully, stammering his words on account of his joy, in the evening Hari spoke thus:

The Nineteenth Song (*deśavarāḍī rāga*, *aṣṭa-tālī tāla*):

2. If you speak, even a little-bit, the splendid moonlight of your teeth rips-apart the very dreadful darkness of fear; may the moon which is your face cause the *cakora* [bird] which is my eye to long for the intoxicating-nectar from your quavering lower-lip; beloved! sweet-natured-woman! stop being piqued with me—it is unjustified! Suddenly the fire of love burns my mind—give me a drink of the mead from the lotus of your mouth!

[162] 'the pleasures of play' [*krīḍā-mudaḥ*]: Ś.M. explains it as the games of her female friends.

[163] 'By the intensely . . .': i.e., his foot is like a lotus and as nectar flows from a lotus so the Heavenly Ganges [the Mandākinī] flows from his foot and as bees appear on a lotus, so the emeralds in the crowns of the gods appear on his foot as the gods make obeisance to him. (This reading following R.K.)

3. If you are *truly* angry, O lovely-toothed-woman, give me a wound with the arrows that are your sharp nails, bind me with the fetters which are your arms or bite me, by which all-the-conditions of pleasure come to be; beloved! sweet-natured-woman! stop being piqued with me . . .!

4. You are my adornment, you are my life, you are my jewel in the ocean of existence! May you be here, constantly compliant to me—for that my heart is making great-efforts! beloved! sweet-natured-woman! stop being piqued with me . . .!

5. Your eye, although it is like a blue lotus, O slender-woman, bears the appearance of a red lotus; if you {redden / impassion} me, who am {Kṛṣṇa / blue-black} with the emotion which is the arrow of love, then this [condition] matches that [your eye]; beloved! sweet-natured-woman! stop being piqued with me . . .!

6. May the garland of jewels glitter upon the pitchers of your breasts—may it {redden / impassion} the region of your heart, also may the girdle upon the circle of your firm hips ring-out—may it proclaim the command of love! beloved! sweet-natured-woman! stop being piqued with me . . .!

7. Speak, O soft-voiced-woman, and I'll make your feet red with passionate, shining lac, [your feet which] excel the land-lotus [hibiscus], delighting my heart, having produced excellence on the stage of love-pleasure; beloved! sweet-natured-woman! stop being piqued with me . . .!

8. Place the noble sprout of your foot as an ornament upon my head, it dispels the poison of love [in separation]; the pitiless sun which is the destruction by love burns in me—may [your foot] take away the symptom produced; beloved! sweet-natured-woman! stop being piqued with me . . .!

9. Pleasing and clever in sweet flatteries, the words of the Enemy-of-Mura to Rādhikā, embellished with the eloquence of the poet Jayadeva, which has caused joy in piqued-women —may [that speech] prevail!

* * *

10. O afflicted-woman! put-aside doubt about my heart which is overcome by you always, by you whose breasts and

hips are firm: there is no-room for another, not any lucky-person other than the bodiless [god of love] enters my heart; O beloved-lady, be disposed to undertaking embraces![164]

11. O artless-woman! give me squeezes against your ample[165] breasts and bind me with the creepers of your arms and bite me with cruel teeth! O fierce-passionate-woman, rejoice! my life-breaths leave me because of the splitting wound from the arrow of that outcast, love!

12. O moon-faced-woman! your curved brow [your frown] looks like a terrible black-snake, [the cause of] infatuation; for young-men only the liquorous nectar from your lower-lip is an effective charm for dispelling the danger which arises from it [your serpent-brow].

13. Your silence unnecessarily disquiets me,[166] O slender-woman, dwell upon the fifth [note], O young-woman, with sweet conversations, with your glances dispel my fever, O lovely-faced-woman, give-up so much aversion [to me], don't let me go! Your beloved, so very tender, O artless-woman, is near-and-ready!

14. This lower-lip is akin in splendour to the *bandhūka*[167] [flower]; your shiny cheek has the complexion of the *madhūka*[168] [flower], O fierce-passionate-woman! Your eye, which emits the lustre of a blue-lotus, shines; your nose resembles the sesame flower;[169] your teeth are like jasmine; O beloved, above all, by employment of [the features of] your face the flower-weaponed [god of love] conquers all.[170]

15. Your eyes are languid with [passion's] drunkenness, your face shines like the moon, your gait delights people's minds, your two thighs surpass the plantain; your love-making is skilful, your brows are beautiful shining lines, oh, you, on

[164] 'be disposed . . .' [*parirambhārambhe vidhehi vidheyatāṃ*]: lit., 'make the state of disposability in the beginning of embraces'.

[165] 'ample' [*nibiḍa*]: lit., her breasts are so large there is 'no space' between them.

[166] 'your silence': i.e., your silence is unjust.

[167] *bandhūka* = *bandhujīva*.

[168] *madhūka*: a velvety, pale spring flower; its blossoms and leaves were distilled to produce a sweet intoxicant.

[169] 'resembles': lit., 'Your nose follows the path of the sesame flower . . .'.

[170] R.K. explains that these five features of her face, each like a flower, act as the five flower-tipped arrows of the love-god.

earth, lead the gods' young-women, O slender-woman.[171]

16. May Hari grant delight! In battle he came into conflict with Kuvalayāpīḍa [the war-elephant of Kaṃsa] whose forehead reminded him of Rādhā's swollen breasts; wherein, while he was perspiring and closing his eyes for but a moment, quickly, in confusion at seeing him [like that], 'Victory! Victory! Victory!' was the uproarious-cry of Kaṃsa.

THE ELEVENTH CANTO
BLISSFUL DĀMODARA

1. While Keśava, for a long time having pleased the deer-eyed woman with supplication, adorned, was going to the bed in the grove, while the twilight, the robber of sight, was glimmering, [to her whose] splendid ornaments were arranged, to Rādhā who was not sunken-down [in the sadness of separation any longer] some-woman said:

The Twentieth Song (*vasanta rāga*, *yati tāla*):

2. He has composed flattering speeches bowed-down-in-reverence at your foot; now he has gone to the bed of love-play on the edge of the beautiful cane grove, O artless-woman! the Killer-of-Madhu has pursued [you]! Follow him, O Rādhikā!

3. O you who bear the burden of firm breasts and thighs, approach [him], wandering-pleasurably with slightly slow steps, with jewelled anklets jingled, act like a *marāla*[172] [bird]; O artless-woman! . . .

4. Listen to the most pleasing song of the honey-drinking-bees which is the infatuation of young girls; while a throng of cuckoos is herald for the behest of [love] the flower-bowed-one, enjoy the emotion-of-love; O artless-woman! . . .

5. With a hand which is a mass of shoots shaking in the wind, a mass of creepers seems to be urging [you], O you-whose-thighs-are-like-elephant-trunks! stop your delay in going; O artless-woman! . . .

[171] 'on earth': i.e., she's a living goddess. *Pṛthivī* also plays on the name of the metre of this verse (four times: ⏑–⏑ ⏑⏑– ⏑–⏑ ⏑⏑– ⏑–– ⏑–).

[172] *marāla*: a kind of goose or flamingo; supposed to have a lovely gait and a voice appropriate for trysting.

6. Consult that pitcher which is your breast, throbbing as if because of the waves of love, indicating the embrace of Hari, having a pure stream of water which is your beautiful necklace; O artless woman! . . .[173]

7. This is understood by all your friends—that your body too is ready for the battle of love-pleasure; O fierce-passionate-woman, go to him noisily with the battle-drum-uproar of your jangling girdle, passionately, not-bashfully! O artless-woman! . . .

8. Go playfully to Hari, clinging to a friend, with your hand, its nails beautiful as the arrows of love; with the jingling of your bracelets inform him also of your approach;[174] O artless-woman! . . .

9. May this song of Śrī Jayadeva eternally abide in the throats of those whose minds are fixed-upon Hari—[this song on the necks of devotees] puts to shame a necklace [or] a beautiful woman![175]

* * *

10. 'She will look at me, she will speak love-talk, by embraces of my entire body she will attain delight, she'll be pleased, O friend, in uniting with me', with these thoughts he is disturbed; in the grove, of which the mass of darkness is steady, the beloved sees you, trembles, bristles, rejoices, sweats, advances-to-meet-you, faints.

11. Putting mascara on the eyes, clusters of *tāpiccha*[176] flowers on the ears, a wreath of dark lotuses on the head, a musk-design on the breasts of the crafty-women enjoying the excitement of the tryst, everywhere in the grove, O friend, the darkness, beautiful as a black cloak, embraces every limb of the lovely-eyed-women.

12. Streaked all-over with the clusters of the light of the women-going-to-their-lovers whose bodies are yellow with saffron,[177] this night, most black like *tamāla* leaves, manifests [itself] as a touchstone for the gold which is love for him.

[173] 'indicating . . .': the waves in the pitcher are an omen of love; the stream of water, indicating that the pitcher is overflowing, suggests the fulness of her breast (following R.K.).

[174] 'inform . . .': the suggestion (following Ś.M.) is 'be a fair-fighter', i.e., give your enemy (lover) warning of your approach for battle (love-making).

[175] Following Ś.M. [176] *tāpiccha* = *tamāla.*

[177] 'saffron' [*kāśmīra*]: specifically that which comes from Kashmir.

13. Having seen Hari in the door-way of his grove-retreat which was shining with the splendour of the jewels in his bracelets, his arm-bands, the golden cord of his girdle, the central gem of his pearl-necklace, she [the messenger] then said to her bashful friend [Rādhā]:

The Twenty-first Song (*varāḍī rāga, rūpaka tāla*):

14. In the hut of love-play within the loveliest grove, play-in-love, O you whose face laughs from your violent-desire for love-pleasure; enter here, O Rādhā, into Mādhava's presence!

15. [In the place which is] excellent on account of the bed of fresh and shimmering *aśoka* leaves, play-in-love, O you whose necklace quivers on the pitchers of your breasts; enter here, O Rādhā, into Mādhava's presence!

16. In the radiant bed-chamber which is arranged with floral bouquets, play-in-love, O you whose body is as delicate as a flower; enter here, O Rādhā, into Mādhava's presence!

17. [In the place which is] cool and fragrant on account of the gentle, wafting Malayan wind, play-in-love, O you who are fearful of love's arrows; enter here, O Rādhā, into Mādhava's presence!

18. [In the place which is] thick with fresh sprouts and many outspread creepers, play-in-love for a long while, O you whose swollen hips are languorous; enter here, O Rādhā, into Mādhava's presence!

19. [In the place where] humming is made by swarms of bees delighted by the honey, play-in-love, O you who have the passionate emotions of the sentiment of love; enter here, O Rādhā, into Mādhava's presence!

20. [In the place which is] noisy with the cries of flocks of very sweet cuckoos, play-in-love, O you who have splendid rubies as your shining teeth;[178] enter here, O Rādhā, into Mādhava's presence!

21. While the king of poet-kings sings [this song which] makes an abundance of happiness for Padmāvatī, bestow, O Enemy-of-Mura, hundreds of blessings!

* * *

[178] 'splendid rubies . . .': an inverted simile following Ś.M.; perhaps the redness of her teeth refers to having chewed betel.

22. This one, having you on his mind, for a very long time, very weary and greatly inflamed by love, wishes to drink [from] your lower-lip which is like a *bimba* [fruit] abounding in nectar; so adorn his lap for a moment, here: why this agitation about one who worships the lotus of your feet like a slave purchased with a fraction of the {wealth / beauty} of your frown?

23. She, joyfully and [yet] fearfully, her eyes longing for Govinda, with lovely anklets jingling, entered the pleasure-hut.

The Twenty-second Song (*varāḍī rāga, yati tāla*):

24. She saw Hari: he had manifestations of the various symptoms [of love] blossomed on account of the sight of Rādhā's face like the ocean which has its high waves set-in-motion at the sight of the lunar disk; [he had her as his] only love, he had long desired love-play, his face was overcome with his great joy, he was love's dwelling-place.

25. She saw Hari: he, having tarried far-off, bore a necklace of very pure pearls upon his chest—[he was] like the flow of the Yamunā's waters mixed with masses of very clear foam. . . .[179]

26. She saw Hari: the curve of his body, dark and tender, had a yellow robe put-upon it—[he was] like the blue lotus which has a root encircled with a mass of yellow pollen-veils. . . .

27. She saw Hari: the passion of love's pleasure was produced by his beautiful face on account of the quivering of his tremulous side-long-glances—[he was] like a pool in autumn which has a pair of wagtails[180] playfully-shaking in an open lotus [upon it]. . . .

28. She saw Hari: the radiant-beauty of his ear-rings equal to [that of] the sun brought into touch with the lotus of his face, and a longing for love-pleasure was caused by the sprout of his lower-lip splendidly gleaming with the splendour of her smile. . . .

29. She saw Hari: his hair, with flowers in it, was like a rain-cloud the inside of which is strewn with moon-beams and

[179] The waters of the Yamunā are conventionally depicted as dark, as 'Kṛṣṇa-coloured'.

[180] 'wagtails' [*khañjana*]: a small bird with a white breast and black wings which mates in rivers and ponds in autumn.

the form of his spotless sandal forehead-mark was the lunar disk risen in the darkness. . . .

30. She saw Hari: prickly with a mass of much bristled hair, excited on account of the art of play in love-pleasure, his body was beautiful on account of his ornaments radiant with a mass of light-beams from the multitude of his jewels. . . .

31. She saw Hari: the mass of his adornments is doubled by the greatness of Śrī Jayadeva's song; placing him in your heart, bow to Hari who is the essence of virtue's arising. . . .

* * *

32. Then, in the moment of beholding the most beloved-one, the multitude of Rādhā's joyful tears fell like the flood of sweat of her eyes made-to-go-beyond their corners, as if with an effort to go as-far-as the region of her ears, with their pupils made-to-quiver.[181]

33. While the entourage of watchful friends went outside the dwelling with their smiles concealed by the contrivance of [having to] scratch [their mouths], the embarrassment of the deer-eyed woman, of her who was going near his bed, of her who was looking at her beloved's face which was beautiful with the intention which is under-the-sway of love, [her embarrassment,] as if also embarrassed, went far-away.

34. Joyfully may the Son-of-Nanda bestow upon you exhilaration beyond measure! Slowly-slowly having taken Rādhā in the space between his arms and then squeezing her tightly on account of his delight: 'May the lofty breasts of the excellent-bodied-woman not break my back and quickly come out of it!'[182] [he said], looking-around with his neck bent.

35. As if honoured with *mandāra* flowers placed there by the Goddess of Victory, as if marked by himself with vermilion because of the delight of battle with the elephant, scattered with the blood-drops of the elephant, Kuvalayāpīḍa, which was killed by play with the crushing with his arms, the staff-like arm of the Conqueror-of-Mura prevails!

36. In the heart of Rādhā, [she who is] the one store of

[181] Her tears are like perspiration from her eyes, caused by the great effort which her eyes are making and (following R.K.) her eyes seem to be falling out with weariness.

[182] Alternately (following R.K.) the bodiless-love-god [*atanu*] takes on Rādhā's body to conquer the world.

beauty, who enjoys the lovely love-games which are the delight of love,[183] in that pool which is the one stage for the play of love, because he relishes the playful-shaking of the lotuses which are her delightful breasts he is proclaiming his similarity to the royal gander of [lake] Mānasa of the contemplative—May Mukunda give joy![184]

THE TWELFTH CANTO

THE VERY-DELIGHTED YELLOW-ROBED-ONE

1. When the flock of her friends were gone, having seen Rādhā whose mind was passionate, whose lower-lip was bathed in smiles bursting-forth with the overpowering intentions of love which was full of the burden of her sharp shame, whose eyes were cast-down constantly upon the bed of flowers and fresh sprouts, Hari spoke to his beloved:

The Twenty-third Song (*vibhāsa rāga*, *eka-tālī tāla*):

2. Upon the surface of the bed of sprouts put the lotus of your foot, O loving-woman—may this well-adorned [bed of sprouts] be defeated by its rival, the sprout which is your foot! For a moment now, follow Nārāyaṇa, [me, as I have] followed [you], O Rādhikā!

3. I'll do honour to your foot with the lotus of my hand for you have been made to come a long way—upon the bed, for a moment, help the anklet which like me has been heroic in following [you] . . . follow Nārāyaṇa . . . .

4. Compose an obliging speech like nectar trickling from the nectar-storing-moon which is your face; I'll remove the fine-garment on your chest which is the obstruction to your breasts as if [removing our] separation . . . follow Nārāyaṇa . . . .

5. Bristled as if full of a violent-passion for the embraces of the beloved, very-difficult-to-attain, the pitcher of your breast

[183] 'delight of love' [*anaṅga-lalanā*]: or 'delight of Love' comparing her to Rati, delight personified as the consort of Kāma-deva.

[184] Kṛṣṇa delighting upon Rādhā's breasts is like the royal gander delighting amidst the lotuses on Lake Mānasa and Kṛṣṇa is in Rādhā's heart as that gander is in the heart or mind of the contemplative, the devotee.

—place it upon my chest, extinguish the heated-pain of love . . . follow Nārāyaṇa . . . .

6. Bring the elixir of nectar from your lower-lip, O passionate-angry-woman! vivify me, your slave, as if I were dead; my mind is fixed upon you, my body is consumed by the fire of love-in-separation; I am without the pleasure-of-play . . . follow Nārāyaṇa . . . .

7. O moon-faced-woman! sound the cord of your jewelled girdle, which is harmonious with the sound of your voice, in my ears which have been disturbed by the cries of the cuckoo —calm my depression at last . . . follow Nārāyaṇa . . . .

8. This, your eye, as if embarrassed, now ceases to look at me who was maimed by your very futile anger—give up your very vain sexual frustration . . . follow Nārāyaṇa . . . .

9. May this song of Śrī Jayadeva, having the Enemy-of-Madhu's delight recited in its refrain, produce in people of taste an appreciation of the emotion of the sentiment of lovely love-pleasure! . . . follow Nārāyaṇa . . . .

* * *

10. A close embrace in which an obstacle arose on account of the rising bristling-hair, and the glances which inspire love-play in which [an obstacle arose] on account of blinking, the drinking of the nectar from lower-lips in which [an obstacle arose] on account of their playful talk also the battle which is the art of love in which [an obstacle arose] on account of the attainment of bliss—that, their undertaking of love-making was about to be enjoyed.

11. Held-captive by her arms, pressed by the weight of her breasts, pierced by her finger-nails, the cup of his lower-lip bitten by her teeth, crushed by the slope of her hips, bent-down by her hand on his hair, crazed by the trickling-flow of honey from her lower-lip, the lovely-beloved somehow obtained delight —so, oh! the way of love is paradoxical!

12. In the undertaking of the battle mixed with the play of love-pleasure, [an undertaking] having the mark of $\left\{\begin{matrix}\text{love}\\ \text{death}\end{matrix}\right\}$[185], something full of impetuosity was undertaken by her on-top [of him] for victory over her lover, from the flurry of which,

[185] Following M. the battle is marked by Māra as death and the play is marked by Māra as Love.

the surface of her hips was motionless, the creepers of her arms were loosened, her breast was shaken, her eyes were closed—how is the heroic sentiment of women demonstrated?[186]

13. Her bosom was branded by pink finger-nails, her eyes were reddish from [lack of] sleep, the redness from her lower-lip was washed-off, her hair had its garland fallen and disarranged, the cord of her girdle [allowed] the border of her shawl [to be] slightly loose—the next morning by these arrows of love fixed in his eyes, the mind of her lord was impaled—it was amazing![187]

14. The braided-mass of her hair was dishevelled, her curls had been shaken, her cheeks had an effusion of sweat, the radiant-beauty of her lower-lip like a *bimba* [fruit] was worn-off; the string of pearls, radiant on the pitchers of her breasts, was lost, the beauty of her girdle was hopelessly-destroyed—covering her foot, hips, breasts with her hand in the moment when she noticed that, she was ashamed; so, also when dishevelled, artless-charm pleases.

15. [Of her whose face had its] eyes slightly closed by force of the stream of artlessly echoing love-cries, [whose face had its] lower-lip bathed by the rays of her teeth from opening [her mouth] with indistinct, warbled cries of love-play [so that her] breast was still and tranquil on account of so much embracing, [whose] body was weak from the release of the excess of joy—fortunate is he to kiss the face of the deer-eyed-woman!

16. Then suddenly, very delightedly, after the love-making, she, her body so very wearied, Rādhā with respectful-affection said this with joy to Govinda:

The Twenty-fourth Song (*rāmakarī rāga*, *yati tāla*):

17. 'Put, O Yadu-nandana, with your hand, which is cool with sandal, a design of musk, here upon my breast which is akin to a pitcher for the festival of love', she said to him while Yadu-nandana, the joy of her heart was playing.

[186] That is, how can women succeed in the sentiment [*rasa*] of heroism, the natural sentiment of men, except in the battle of love, in trying to conquer their lovers?

[187] 'amazing' [*adbhuta*]: the marvellous or wondrous sentiment [*rasa*]; each of her features is one of the five arrows of love.

18. 'Make the mascara which puts-to-shame the bees, shine, O beloved, upon my eye, which has [had] its collyrium removed by the kisses of your lower-lip, [my eye] which releases the arrows of love', she said to him . . . .

19. 'Fasten, O beautifully-adorned-one, both ear-rings on my ears' circle which, possessing the appearance of love's snare, prevents the expansion of the bounding of the deer which are [my] eyes', she said to him . . . .

20. 'In my presence for a long while arrange a glossy curl which causes pleasure, creating a swarm of bees above my spotless face which surpasses a lotus', she said to him . . . .

21. 'Put a lovely mark drawn with musk essence, placed as a lunar spot, O lotus-faced-one, upon the moon which is my forehead which has ceased sweating', she said to him . . . .

22. 'Put flowers in my radiant hair, O giver-of-honour, [my hair] which is the fly-whisk and banner of love, fallen-loose from love-pleasure, lovely, extraordinary as a peacock's tail', she said to him . . . .

23. 'Put ornaments, clothes and the jewelled girdle, O good-hearted-one, upon my passionate hips which are firm and beautiful, which are the cave-dwelling of the elephant who is Love!'[188] she said to him . . . .

24. Put your compassionate heart in this splendid speech of Śrī Jayadeva, which is an adornment, which dispels the fever and impurity of the Kali [Era] for it is formed with the nectar which is remembrance of the feet of Hari.

* * *

25. 'Put a pattern on my breasts, make a design on my cheeks, fasten a girdle on my hips, fix the mass of my braids with artless garlands, put rows of bracelets on my arms and jewelled anklets on my feet'—thus directed the yellow-robed-one was pleased and he did so.

26. By the mingling of reflected-images joined in the multitude of jewels lining the hoods of the Serpent Lord who was made [Hari's] couch, [Hari] wishes to see with hundreds of eyes the lotus-footed daughter of the ocean [Śrī] who is the produc-

[188] 'the cave-dwelling of the elephant who is love' (*śambara-dāraṇa-vāraṇa-kandare*]: following Ś.M. who explains that Love (the Killer-of-Śambara) is like an elephant in fierceness and strength; her hips possess that cleft in which the powerful elephant of Love resides; alternately, her hips are the 'goad' of love.

tion of the sustaining Lord, performing emanations of his body as if in her honour; may Hari protect you very much.[189]

27. 'Not having obtained you who were intent upon your own-choice-of-a-husband, [who were intent] upon me, it seems to me, O beautiful-woman, that the infatuated Husband-of-Mṛḍānī [Śiva] drank the poison [which was] in the depths of the ocean of milk.' Thus with these prefatory, distracting,[190] words, throwing-aside Rādhā's upper-garment, may Hari, whose eyes were encountering the buds of her breasts, protect you!

28. Skill in the arts of the Gāndharvas,[191] meditation consecrated-to-Viṣṇu, playful-creation in poems which are literary-works on the truth of the discrimination in erotics—may wise-people joyfully purely-understand all that according to the *Śrī-Gītagovinda* of the poet and scholar Jayadeva whose soul is solely directed[192] to Kṛṣṇa.

29. May holy-men approve in this case of the devotion of aspirants [like me] truly of their own accord and having considered my labour in composition may wise-men respect it; I ask those who are occupied with listening to the literary-work of some other [author], having examined the one made thoroughly by me, let them announce the imperfection if [there is any] here—that idea will remain.

30. May the poetic-skill of the *Śrī-Gītagovinda* of Śrī Jayadeva, son of Rāmādevī, son of Śrī Bhojadeva, be in the throats of friends, Parāśara and others.[193]

31. O sweet-liquor! the thought of you is not good! O sugar! you are hard! O grape! who will see you? O nectar [of immortality]! you are dead! O milk! you taste like water! O

[189] The Serpent Lord is Śeṣa, a serpent with hundreds of heads upon which Viṣṇu sleeps in the intervals between creations; there is a jewel in each of Śeṣa's hoods. Hari looks into them to see himself multiplied, i.e., reflected in each jewel, and thereby each reflected Hari can look at Śrī. This verse as well as the ones which follow may be late interpolations—there is little consensus among various editions and manuscripts as to the arrangement or inclusion of these final verses. XII. 26, 27, 29 are given only in the footnotes of the Nirṇaya Sāgar Press edition and they are without commentary or gloss.

[190] 'distracting' for *anya-manasā*?

[191] 'arts of the Gāndharvas', the heavenly musicians, i.e., the musical arts.

[192] 'solely directed' [*eka-tāna*]: his soul or self [*ātman*] has Kṛṣṇa as its one object or only tone.

[193] Parāśara—not identifiable but in the biographical legends about Jayadeva said to be a friend of the poet who accompanied him on various pilgrimages.

mango! weep! O lover's lower-lip! bear no comparison! go! —because the clever words of Jayadeva bestow the emotion which, like a blessing, has the essence of the erotic [sentiment]!

32. Having thus played multitudes of love-games on the bank of the Yamunā together with Rādhā, where the juncture of her pearl necklace and the line-of-hair-on-her-stomach bore the appearance of the juncture [of the Yamunā and the Ganges at Prayāga], there [his hands] wished-to-take the *prayāga* fruits which are her joy-causing breasts—may the actions of Purusottama's hands give much joy and success![194]

[194] Following R.K.: The line of hair on her stomach is like the dark Yamunā and her pearl necklace is like the light Ganges; the Yamunā and Ganges meet at Prayāga (Allahabad) and her hair-line and necklace meet between her breasts which are like *prayāga* fruit (figs?).

APPENDIX: Transliterated text of the *Gītagovinda*

GĪTAGOVINDA-KĀVYAM

Prathamaḥ Sargaḥ
SĀMODA-DĀMODARAḤ

1. *meghair meduram ambaraṃ vana-bhuvaḥ śyāmās tamāla-drumair*
naktaṃ bhīrur ayaṃ tvam eva tad imaṃ rādhe gṛhaṃ prāpaya /
itthaṃ nanda-nideśataś calitayoḥ praty-adhva-kuñja-drumaṃ
rādhā-mādhavayor jayanti yamunā-kūle rahaḥ-kelayaḥ //
2. *vāg-devatā-carita-citrita-citta-sadmā*
padmāvatī-caraṇa-cāraṇa-cakravartī /
śrī-vāsudeva-rati-keli-kathā-sametam
etaṃ karoti jayadeva-kaviḥ prabandham //
3. *yadi hari-smaraṇe sarasaṃ mano*
yadi vilāsa-kalāsu kutūhalam /
madhura-komala-kānta-padāvalīṃ
śṛṇu tadā jayadeva-sarasvatīm //
4. *vācaḥ pallavayaty umāpatidharaḥ saṃdarbha-śuddhiṃ girāṃ*
jānīte jayadeva eva śaraṇaḥ ślāghyo durūha-druteḥ /
śṛṅgārottara-sat-prameya-racanair ācārya-govardhana-
spardhī ko'pi na viśrutaḥ śrutidharo dhoyī kavi-kṣmā-patiḥ //

mālava-gauḍa-rāgeṇa rūpaka-tālena gīyate / aṣṭapadī 1 //

5. *pralaya-payodhi-jale dhṛtavān asi vedaṃ*
vihita-vahitra-caritram akhedam /
keśava dhṛta-mīna-śarīra
jaya jagad-īśa hare // *dhruva-padam* //
6. *kṣitir ativipulatare tava tiṣṭhati pṛṣṭhe*
dharaṇi-dharaṇa-kiṇa-cakra-gariṣṭhe /
keśava dhṛta-kacchapa-rūpa
jaya jagad-īśa hare //
7. *vasati daśana-śikhare dharaṇī tava lagnā*
śaśini kalaṅka-kaleva nimagnā /
keśava dhṛta-śūkara-rūpa
jaya jagad-īśa hare //
8. *tava kara-kamala-vare nakham adbhuta-śṛṅgaṃ*
dalita-hiraṇyakaśipu-tanu-bhṛṅgam /

keśava dhṛta-nara-hari-rūpa
jaya jagad-īśa hare //

9. *chalayasi vikramaṇe balim adbhuta-vāmana*
pada-nakha-nīra-janita-jana-pāvana /
keśava dhṛta-vāmana-rūpa
jaya jagad-īśa hare //

10. *kṣatriya-rudhira-maye jagad apagata-pāpam*
snapayasi payasi śamita-bhava-tāpam /
keśava dhṛta-bhṛgu-pati-rūpa
jaya jagad-īśa hare //

11. *vitarasi dikṣu raṇe dik-pati-kamanīyaṃ*
daśa-mukha-mauli-baliṃ ramaṇīyam /
keśava dhṛta-rāma-śarīra
jaya jagad-īśa hare //

12. *vahasi vapuṣi viśade vasanaṃ jaladābhaṃ*
hala-hati-bhīti-milita-yamunābham /
keśava dhṛta-hala-dhara-rūpa
jaya jagad-īśa hare //

13. *nindasi yajña-vidher ahaha śruti-jātaṃ*
sadaya-hṛdaya darśita-paśu-ghātam /
keśava dhṛta-buddha-śarīra
jaya jagad-īśa hare //

14. *mleccha-nivaha-nidhane kalayasi karavālaṃ*
dhūma-ketum iva kim api karālam /
keśava dhṛta-kalki-śarīra
jaya jagad-īśa hare //

15. *śrī-jayadeva-kaver idam uditam udāraṃ*
śṛṇu sukhadaṃ śubhadaṃ bhava-sāram /
keśava dhṛta-daśa-vidha-rūpa
jaya jagad-īśa hare //

* * *

16. *vedān uddharate jagan nivahate bhūgolam udbibhrate*
daityaṃ dārayate baliṃ chalayate kṣatra-kṣayaṃ kurvate /
paulastyaṃ jayate halaṃ kalayate kāruṇyam ātanvate
mlecchān mūrcchayate daśākṛti-kṛte kṛṣṇāya tubhyaṃ namaḥ //

gurjarī-rāga-niḥsāra-tālābhyāṃ gīyate / prabandhaḥ 2 //

17. *śrita-kamalā-kuca-maṇḍala*
dhṛta-kuṇḍala e /

kalita-lalita-vana-māla
jaya jaya deva hare || dhruva-padam ||

18. *dina-maṇi-maṇḍala-maṇḍana*
bhava-khaṇḍana e |
muni-jana-mānasa-haṃsa
jaya jaya deva hare ||

19. *kāliya-viṣa-dhara-gañjana*
jana-rañjana e |
yadu-kula-nalina-dineśa
jaya jaya deva hare ||

20. *madhu-mura-naraka-vināśana*
garuḍāsana e |
sura-kula-keli-nidāna
jaya jaya deva hare ||

21. *amala-kamala-dala-locana*
bhava-mocana e |
tribhuvana-bhuvana-nidhāna
jaya jaya deva hare ||

22. *janaka-sutā-kṛta-bhūṣaṇa*
jita-dūṣaṇa e |
samara-śamita-daśa-kaṇṭha
jaya jaya deva hare ||

23. *abhinava-jala-dhara-sundara*
dhṛta-mandara e |
śrī-mukha-candra-cakora
jaya jaya deva hare ||

24. *tava caraṇaṃ praṇatā*
vayam iti bhāvaya e |
kuru kuśalaṃ praṇateṣu
jaya jaya deva hare ||

25. *śrī-jayadeva-kaver idaṃ*
kurute mudam e |
maṅgalam ujjvala-gītaṃ
jaya jaya deva hare ||

* * *

26. *padmā-payodhara-taṭī-parirambha-lagna-*
kāśmīra-mudritam uro madhu-sūdanasya |
vyaktānurāgam iva khelad-anaṅga-kheda-
svedāmbu-pūram anupūrayatu priyaṃ vaḥ ||

27. *vasante vāsantī-kusuma-sukumārair avayavair*
bhramantīṃ kāntāre bahu-vihita-kṛṣṇānusaraṇām /
amandaṃ kandarpa-jvara-janita-cintākulatayā
valad-bādhāṃ rādhāṃ sarasam idam ūce saha-carī //

vasanta-rāga-yati-tālābhyāṃ gīyate / prabandhaḥ 3 //

28. *lalita-lavaṅga-latā-pariśīlana-komala-malaya-samīre* /
madhu-kara-nikara-karambita-kokila-kūjita-kuñja-kuṭīre //
viharati harir iha sarasa-vasante
nṛtyati yuvati-janena samaṃ sakhi virahi-janasya durante // *dhr.*
29. *unmada-madana-manoratha-pathika-vadhū-jana-janita-vilāpe* /
ali-kula-saṃkula-kusuma-samūha-nirākula-bakula-kalāpe //
30. *mṛga-mada-saurabha-rabhasa-vaśaṃvada-nava-dala-māla-tamāle* /
yuva-jana-hṛdaya-vidāraṇa-manasija-nakha-ruci-kiṃśuka-jāle //
31. *madana-mahīpati-kanaka-daṇḍa-ruci-keśara-kusuma-vikāse* /
milita-śilī-mukha-pāṭala-paṭala-kṛta-smara-tūṇa-vilāse //
32. *vigalita-lajjita-jagad-avalokana-taruṇa-karuṇa-kṛta-hāse* /
virahi-nikṛntana-kunta-mukhākṛti-ketaka-danturitāśe //
33. *mādhavikā-parimala-lalite nava-mālati-jāti-sugandhau* /
muni-manasām api mohana-kāriṇi taruṇākāraṇa-bandhau //
34. *sphurad-atimukta-latā-parirambhaṇa-mukulita-pulakita-cūte* /
vṛndāvana-vipine parisara-parigata-yamunā-jala-pūte //
35. *śrī-jayadeva-bhaṇitam idam udayati hari-caraṇa-smṛti-sāram* /
sarasa-vasanta-samaya-vana-varṇanam anugata-madana-vikāram //

* * *

36. *dara-vidalita-mallī-valli-cañcat-parāga-*
prakaṭita-paṭa-vāsair vāsayan kānanāni /
iha hi dahati cetaḥ ketakī-gandha-bandhuḥ
prasarad-asama-bāṇa-prāṇavad-gandhavāhaḥ //
37. *unmīlan-madhu-gandha-lubdha-madhupa-vyādhūta-cūtāṅkura-*
krīḍat-kokila-kākalī-kalakalair udgīrṇa-karṇa-jvarāḥ /
nīyante pathikaiḥ kathaṃ katham api dhyānāvadhāna-kṣaṇa-
prāpta-prāṇa-samā-samāgama-rasollāsair amī vāsarāḥ //
38. *aneka-nārī-parirambha-saṃbhrama-*
sphuran-manohāri-vilāsa-lālasam /
murārim ārād upadarśayanty asau
sakhī samakṣaṃ punar āha rādhikām //

rāmakarī-rāga-yati-tālābhyāṃ gīyate / prabandhaḥ 4 //

39. *candana-carcita-nīla-kalevara-pīta-vasana-vana-mālī* /

keli-calan-maṇi-kuṇḍala-maṇḍita-gaṇḍa-yuga-smitaśālī //
harir iha mugdha-vadhū-nikare vilāsini vilasati kelipare // *dhr.* //
40. *pīna-payodhara-bhāra-bhareṇa hariṃ parirabhya sarāgam* /
gopa-vadhūr anugāyati kācid udañcita-pañcama-rāgam //
41. *kāpi vilāsa-vilola-vilocana-khelana-janita-manojam* /
dhyāyati mugdha-vadhūr adhikaṃ madhu-sūdana-vadana-sarojam //
42. *kāpi kapola-tale militā lapituṃ kim api śruti-mūle* /
cāru cucumba nitambavatī dayitaṃ pulakair anukūle //
43. *keli-kalā-kutukena ca kācid amuṃ yamunā-jala-kūle* /
mañjula-vañjula-kuñja-gatam vicakarṣa kareṇa dukule //
44. *karatala-tāla-tarala-valayāvali-kalita-kalasvana-vaṃśe* /
rāsa-rase saha-nṛtyaparā hariṇā yuvatiḥ praśaśaṃse //
45. *śliṣyati kām api cumbati kām api kām api ramayati rāmām* /
paśyati sa-smita-cāru-tarām aparām anugacchati vāmām //
46. *śrī-jayadeva-bhaṇitam idam adbhuta-keśava-keli-rahasyam* /
vṛndāvana-vipine lalitaṃ vitanotu śubhāni yaśasyam //

* * *

47. *viśveṣām anurañjanena janayann ānandam indīvara-*
śreṇī-śyāmala-komalair upanayann aṅgair anaṅgotsavam /
sva-cchandaṃ vraja-sundarībhir abhitaḥ pratyaṅgam āliṅgitaḥ
śṛṅgāraḥ sakhi mūrtimān iva madhau mugdho hariḥ krīḍati //
48. *nityotsaṅga-vasad-bhujaṅga-kavala-kleśād iveśācalaṃ*
prāleya-plavanecchayānusarati śrī-khaṇḍa-śailānilaḥ /
kiṃ ca snigdha-rasāla-mauli-mukulāny ālokya harṣodayād
unmīlanti kuhūḥ kuhūr iti kalottālāḥ pikānāṃ giraḥ //
49. *rāsollāsa-bhareṇa vibhrama-bhṛtām ābhīra-vāma-bhruvām*
abhyarṇaṃ parirabhya nirbharam uraḥ premāndhayā rādhayā /
sādhu tvad-vadanaṃ sudhā-mayam iti vyāhṛtya gīta-stuti-
vyājād udbhaṭa-cumbitaḥ smita-manohārī hariḥ pātu vaḥ //

Dvitīyaḥ Sargaḥ
AKLEŚA-KEŚAVAḤ

1. *viharati vane rādhā sādhāraṇa-praṇaye harau*
vigalita-nijotkarṣād īrṣyā-vaśena gatā'nyataḥ /
kva-cid api latā-kuñje guñjan-madhu-vrata-maṇḍalī-
mukhara-śikhare līnā dīnāpy uvāca rahaḥ sakhīm //

gurjarī-rāgeṇa yati-tālena gīyate / prabandhaḥ 5 //

2. *saṃcarad-adhara-sudhā-madhura-dhvani-mukharita-mohana-vaṃśam /*
calita-dṛg-añcala-cañcala-mauli-kapola-vilola-vataṃsam //
rāse harim iha vihita-vilāsaṃ
smarati mano mama kṛta-parihāsam // dhruva-padam //
3. *candraka-cāru-mayūra-śikhaṇḍaka-maṇḍala-valayita-keśam /*
pracura-purandara-dhanur-anurañjita-medura-mudira-suveśam //
4. *gopa-kadamba-nitambavatī-mukha-cumbana-lambhita-lobham /*
bandhujīva-madhurādhara-pallavam ullasita-smita-śobham //
5. *vipula-pulaka-bhuja-pallava-valayita-ballava-yuvati-sahasram /*
kara-caraṇorasi maṇi-gaṇa-bhūṣaṇa-kiraṇa-vibhinna-tamisram //
6. *jalada-paṭala-calad-indu-vinindaka-candana-tilaka-lalāṭam /*
pīna-payodhara-parisara-mardana-nirdaya-hṛdaya-kapāṭam //
7. *maṇi-maya-makara-manohara-kuṇḍala-maṇḍita-gaṇḍam udāram /*
pīta-vasanam anugata-muni-manuja-surāsura-vara-parivāram //
8. *viśada-kadamba-tale militaṃ kali-kaluṣa-bhayaṃ śamayantam /*
mām api kim api tarala-taraṅgad-anaṅga-dṛśā manasā ramayantam //
9. *śrī-jayadeva-bhaṇitam atisundara-mohana-madhu-ripu-rūpam /*
hari-caraṇa-smaraṇam prati samprati puṇyavatām anurūpam //
10. *gaṇayati guṇa-grāmaṃ bhrāmaṃ bhramād api nehate*
vahati ca paritoṣaṃ doṣaṃ vimuñcati dūrataḥ /
yuvatiṣu valat-tṛṣṇe kṛṣṇe vihāriṇi māṃ vinā
punar api mano vāmaṃ kāmaṃ karoti karomi kim //

mālava-gauḍa-rāgeṇa ekatālī-tālena ca gīyate / prabandhaḥ 6 //

11. *nibhṛta-nikuñja-gṛhaṃ gatayā niśi rahasi nilīya vasantam /*
cakita-vilokita-sakala-diśā rati-rabhasa-bhareṇa hasantam //
sakhi he keśi-mathanam udāraṃ
ramaya mayā saha madana-manoratha-bhāvitayā savikāram // dhr. //
12. *prathama-samāgama-lajjitayā paṭu-cāṭu-śatair anukūlam /*
mṛdu-madhura-smita-bhāṣitayā śithilī-kṛta-jaghana-dukūlam //
13. *kisalaya-śayana-niveśitayā ciram urasi mamaiva śayānam /*
kṛta-parirambhaṇa-cumbanayā parirabhya kṛtādhara-pānam //
14. *alasa-nimīlita-locanayā pulakāvali-lalita-kapolam /*
śrama-jala-sakala-kalevarayā vara-madana-madād atilolam //
15. *kokila-kala-rava-kūjitayā jita-manasija-tantra-vicāram /*
ślatha-kusumākula-kuntalayā nakha-likhita-ghana-stana-bhāram /
16. *caraṇa-raṇita-maṇi-nūpurayā paripūrita-surata-vitānam /*
mukhara-viśṛṅkhala-mekhalayā sakaca-graha-cumbana-dānam //

17. *rati-sukha-samaya-rasālasayā dara-mukulita-nayana-sarojam /*
niḥsaha-nipatita-tanu-latayā madhu sūdanam udita-manojam //
18. *śrī-jayadeva-bhaṇitam idam atiśaya-madhu-ripu-nidhuvana-śīlam /*
sukham utkaṇṭhita-gopa-vadhū-kathitaṃ vitanotu salīlam //

* * *

19. *hasta-srasta-vilāsa-vaṃśam anṛju-bhrū-vallimad-ballavī-*
vṛndotsāri-dṛganta-vīkṣitam atisvedārdra-gaṇḍa-sthalam /
mām udvīkṣya vilajjitaṃ smita-sudhā-mugdhānanaṃ kānane
govindaṃ vraja-sundarī-gaṇa-vṛtaṃ paśyāmi hṛṣyāmi ca //
20. *durāloka-stoka-stabaka-navakāśoka-latikā-*
vikāsaḥ kāsāropavana-pavano'pi vyathayati /
api bhrāmyad-bhṛṅgī-raṇita-ramaṇīyā na mukula-
prasūtiś cūtānāṃ sakhi śikhariṇīyaṃ sukhayati //
21. *sākūta-smitam ākulākula-galad-dhammillam ullāsita-*
bhrū-vallīkam alīka-darśita-bhujā-mūlordhva-hasta-stanam /
gopīnāṃ nibhṛtaṃ nirīkṣya gamitākāṅkṣaś ciraṃ cintayann
antar mugdha-manoharaṃ haratu vaḥ kleśaṃ navaḥ keśavaḥ //

Tṛtīyaḥ Sargaḥ
MUGDHA-MADHU-SŪDANAḤ

1. *kaṃsārir api saṃsāra-vāsanā-bandha-śṛṅkhalām /*
rādhām ādhāya hṛdaye tatyāja vraja-sundarīḥ //
2. *itas tatas tāṃ anusṛtya rādhikām*
anaṅga-bāṇa-vraṇa-khinna-mānasaḥ /
kṛtānutāpaḥ sa kalinda-nandinī-
taṭānta-kuñje viṣasāda mādhavaḥ //

gurjarī-rāga-yati-tālābhyāṃ gīyate / prabandhaḥ 7 //

3. *mām iyaṃ calitā vilokya vṛtaṃ vadhū-nicayena /*
sāparādhatayā mayāpi na vāritā'tibhayena //
hari hari hatādaratayā gatā sā kupiteva // dhruva-padam //
4. *kiṃ kariṣyati kiṃ vadiṣyati sā ciraṃ viraheṇa /*
kiṃ dhanena janena kiṃ mama jīvitena gṛheṇa //
5. *cintayāmi tad-ānanaṃ kuṭila-bhrū kopa-bhareṇa /*
śoṇa-padmam ivopari-bhramatākulaṃ bhramareṇa //
6. *tām ahaṃ hṛdi saṃgatām aniśaṃ bhṛśaṃ ramayāmi /*
kiṃ vane'nusarāmi tām iha kiṃ vṛthā vilapāmi //

7. *tanvi khinnam asūyayā hṛdayaṃ tavākalayāmi |*
tan na vedmi kuto gatāsi na tena te'nunayāmi ||
8. *dṛśyase purato gatāgatam eva me vidadhāsi |*
kiṃ pureva sasaṃbhramaṃ parirambhaṇaṃ na dadāsi ||
9. *kṣamyatām aparaṃ kadāpi tavedṛśaṃ na karomi |*
dehi sundari darśanaṃ mama manmathena dunomi ||
10. *varṇitaṃ jayadevakena harer idaṃ pravaṇena |*
kindubilva-samudra-saṃbhava-rohiṇī-ramaṇena ||

* * *

11. *hṛdi bisa-latā-hāro nāyaṃ bhujaṅgama-nāyakaḥ*
kuvalaya-dala-śreṇī kaṇṭhe na sā garala-dyutiḥ |
malayaja-rajo nedaṃ bhasma priyā-rahite mayi
prahara na hara-bhrāntyā'naṅga krudhā kimu dhāvasi ||
12. *pāṇau mā kuru cūta-sāyakam amuṃ mā cāpam āropaya*
krīḍā-nirjita-viśva mūrcchita-janāghātena kiṃ pauruṣam |
tasyā eva mṛgī-dṛśo manasija preṅkhat-kaṭākṣāśuga-
śreṇī-jarjaritaṃ manāg api mano nādyāpi saṃdhukṣate ||
13. *bhrū-pallavaṃ dhanur apāṅga-taraṅgitāni*
bāṇā guṇaḥ śravaṇa-pālir iti smareṇa |
tasyām anaṅga-jaya-jaṅgama-devatāyām
astrāṇi nirjita-jaganti kim arpitāni ||
14. *bhrū-cāpe nihitaḥ kaṭākṣa-viśikho nirmātu marma-vyathāṃ*
śyāmātmā kuṭilaḥ karotu kabarī-bhāro'pi mārodyamam |
mohaṃ tāvad ayaṃ ca tanvi tanutāṃ bimbādharo rāgavān
sad-vṛttaḥ stana-maṇḍalas tava kathaṃ prāṇair mama krīḍati ||
15. *tāni sparśa-sukhāni te ca taralāḥ snigdhā dṛśor vibhramās*
tad-vaktrāmbuja-saurabhaṃ sa ca sudhā-syandī girāṃ vakrimā |
sā bimbādhara-mādhurīti viṣayāsaṅge'pi cen mānasaṃ
tasyāṃ lagna-samādhi hanta viraha-vyādhiḥ kathaṃ vardhate ||
16. *tiryak-kaṇṭha-vilola-mauli-taralottaṃsasya vaṃśoccarad-*
gīta-sthāna-kṛtāvadhāna-lalanā-lakṣair na saṃlakṣitāḥ |
saṃmugdhe madhu-sūdanasya madhure rādhā-mukhendau sudhā-
sāre kandalitāś ciraṃ dadatu vaḥ kṣemaṃ kaṭākṣormayaḥ ||

Caturthaḥ Sargaḥ
SNIGDHA-MADHU-SŪDANAḤ

1. *yamunā-tīra-vānīra-nikuñje mandam āsthitam |*
prāha prema-bharodbhrāntaṃ mādhavaṃ rādhikā-sakhī ||

karṇāṭa-rāgaika-tālī-tālābhyāṃ gīyate / prabandhaḥ 8 //

2. *nindati candanam indukiraṇam anu vindati khedam adhīram /*
vyāla-nilaya-milanena garalam iva kalayati malaya-samīram //
mādhava manasija-viśikha-bhayād iva bhāvanayā tvayi līnā /
sā virahe tava dīnā // dhruva-padam //

3. *avirala-nipatita-madana-śarād iva bhavad-avanāya viśālam /*
sva-hṛdaya-marmaṇi varma karoti sajala-nalinī-dala-jālam //

4. *kusuma-viśikha-śara-talpam analpa-vilāsa-kalā-kamanīyam /*
vratam iva tava parirambha-sukhāya karoti kusuma-śayanīyam //

5. *vahati ca calita-vilocana-jala-bharam ānana-kamalam udāram /*
vidhum iva vikaṭa-vidhun-tuda-danta-dalana-galitāmṛta-dhāram //

6. *vilikhati rahasi kuraṅga-madena bhavantam asama-śara-bhūtam /*
praṇamati makaram adho vinidhāya kare ca śaraṃ nava-cūtam //

7. *dhyāna-layena puraḥ parikalpya bhavantam atīva durāpam /*
vilapati hasati viṣīdati roditi cañcati muñcati tāpam //

8. *prati-padam idam api nigadati mādhava tava caraṇe patitāham /*
tvayi vimukhe mayi sapadi sudhā-nidhir api tanute tanu-dāham //

9. *śrī-jayadeva-bhaṇitam idam adhikaṃ yadi manasā naṭanīyam /*
hari-virahākula-ballava-yuvati-sakhī-vacanaṃ paṭhanīyam //

* * *

10. *āvāso vipināyate priya-sakhī-mālāpi jālāyate*
tāpo'pi śvasitena dāvadahana-jvālā-kalāpāyate /
sāpi tvad-viraheṇa hanta hariṇī-rūpāyate hā kathaṃ
kandarpo'pi yamāyate viracayañ śārdūla-vikrīḍitam //

deśākha-rāgaika-tālī-tālābhyāṃ gīyate / prabandhaḥ 9 //

11. *stana-vinihitam api hāram udāram /*
sā manute kṛśa-tanur iva bhāram //
rādhikā tava virahe keśava // dhruva-padam //

12. *sarasa-masṛṇam api malayaja-paṅkam /*
paśyati viṣam iva vapuṣi saśaṅkam //

13. *śvasita-pavanam anupama-pariṇāham /*
madana-dahanam iva vahati sadāham //

14. *diśi diśi kirati sajala-kaṇa-jālam /*
nayana-nalinam iva vigalita-nālam //

15. *tyajati na pāṇitalena kapolam /*
bāla-śaśinam iva sāyam alolam //

16. *nayana-viṣayam api kisalaya-talpam /*
kalayati vihita-hutāśa-vikalpam //

17. *harir iti harir iti japati sakāmam /*
viraha-vihita-maraṇeva nikāmam //

18. *śrī-jayadeva-bhaṇitam iti gītam /*
sukhayatu keśava-padam upanītam //

* * *

19. *sā romāñcati sīt-karoti vilapaty utkampate tāmyati*
dhyāyaty udbhramati pramīlati pataty udyāti mūrcchaty api /
etāvaty atanu-jvare vara-tanur jīven na kiṃ te rasāt
svar-vaidya-pratima prasīdasi yadi tyakto'nyathā hastakaḥ //
20. *smarāturāṃ daivata-vaidya-hṛdya*
tvad-aṅga-saṅgāmṛta-mātra-sādhyām /
nivṛtta-bādhāṃ kuruṣe na rādhām
upendra vajrād api dāruṇo'si //
21. *kandarpa-jvara-saṃjvarātura-tanor āścaryam asyāś ciraṃ*
cetaś candana-candramaḥ-kamalinī-cintāsu saṃtāmyati /
kiṃtu klānti-vaśena śītala-tanuṃ tvām ekam eva priyaṃ
dhyāyantī rahasi sthitā katham api kṣīṇā kṣaṇaṃ prāṇiti //
22. *kṣaṇam api virahaḥ purā na sehe*
nayana-nimīlana-khinnayā yayā te /
śvasiti katham asau rasāla-śākhāṃ
cira-viraheṇa vilokya puṣpitāgrām //
23. *vṛṣṭi-vyākula-gokulāvana-rasād uddhṛtya govardhanaṃ*
bibhrad ballava-vallabhābhir adhikānandāc ciraṃ cumbitaḥ /
darpeṇeva tad-arpitādhara-taṭī-sindūra-mudrāṅkito
bāhur gopa-tanos tanotu bhavatāṃ śreyāṃsi kaṃsa-dviṣaḥ //

Pañcamaḥ Sargaḥ
SĀKĀṄKṢA-PUṆḌARĪKĀKṢAḤ

1. *aham iha nivasāmi yāhi rādhām*
anunaya mad-vacanena cānayethāḥ /
iti madhu-ripuṇā sakhī niyuktā
svayam idam etya punar jagāda rādhām //

deśīvarāḍi-rāgeṇa rūpaka-tālena gīyate / prabandhaḥ 10 //

2. *vahati malaya-samīre madanam upanidhāya /*
sphuṭati kusuma-nikare virahi-hṛdaya-dalanāya //
tava virahe vana-mālī sakhi sīdati // dhruva-padam //
3. *dahati śiśira-mayūkhe maraṇam anukaroti /*
patati madana-viśikhe vilapati vikalataro'ti //

4. *dhvanati madhupa-samūhe śravaṇam apidadhāti* |
manasi kalita-virahe niśi niśi rujam upayāti ||
5. *vasati vipina-vitāne tyajati lalita-dhāma* |
luṭhati dharaṇi-śayane bahu vilapati tava nāma ||
6. *bhaṇati kavi-jayadeve viraha-vilasitena* |
manasi rabhasa-vibhave harir udayatu sukṛtena ||

* * *

7. *pūrvaṃ yatra samaṃ tvayā rati-pater āsāditāḥ siddhayas*
tasminn eva nikuñja-manmatha-mahātīrthe punar mādhavaḥ |
dhyāyaṃs tvām aniśaṃ japann api tavaivālāpa-mantrāvalīṃ
bhūyas tvat-kuca-kumbha-nirbhara-parīrambhāmṛtaṃ vāñchati ||

gurjarī-rāgeṇa ekatālī-tālena gīyate / prabandhaḥ 11 //

8. *rati-sukha-sāre gatam abhisāre madana-manohara-veṣam* |
na kuru nitambini gamana-vilambanam anusara taṃ hṛdayeśam ||
dhīra-samīre yamunā-tīre vasati vane vana-mālī |
gopī-pīna-payodhara-mardana-cañcala-kara-yuga-śālī || *dhruva.* ||
9. *nāma-sametaṃ kṛta-saṅketaṃ vādayate mṛdu-veṇum* |
bahu manute'tanu te tanu-saṅgata-pavana-calitam api reṇum ||
10. *patati patatre vicalati patre śaṅkita-bhavad-upayānam* |
racayati śayanaṃ sacakita-nayanaṃ paśyati tava panthānam ||
11. *mukharam adhīraṃ tyaja mañjīraṃ ripum iva keli-sulolam* |
cala sakhi kuñjaṃ satimira-puñjaṃ śīlaya nīla-nicolam ||
12. *urasi murārer upahita-hāre ghana iva tarala-balāke* |
taḍid iva pīte rati-viparīte rājasi sukṛta-vipāke ||
13. *vigalita-vasanaṃ parihṛta-raśanaṃ ghaṭaya jaghanam apidhānam* |
kisalaya-śayane paṅkaja-nayane nidhim iva harṣa-nidhānam ||
14. *harir abhimānī rajanir idānīm iyam api yāti virāmam* |
kuru mama vacanaṃ satvara-racanaṃ pūraya madhu-ripu-kāmam ||
15. *śrī-jayadeve kṛta-hari-seve bhaṇati parama-ramaṇīyam* |
pramudita-hṛdayaṃ harim atisadayaṃ namata sukṛta-kamanīyam ||

* * *

16. *vikirati muhuḥ śvāsān āśāḥ puro muhur īkṣate*
praviśati muhuḥ kuñjaṃ guñjan muhur bahu tāmyati |
racayati muhuḥ śayyāṃ paryākulaṃ muhur īkṣate
madana-kadana-klāntaḥ kānte priyas tava vartate ||
17. *tvad-vāmyena samaṃ samagram adhunā tigmāṃśur astaṃ gato*
govindasya manorathena ca samaṃ prāptaṃ tamaḥ sāndratām |
kokānāṃ karuṇa-svanena sadṛśī dīrghā mad-abhyarthanā
tan mugdhe viphalaṃ vilambanam asau ramyo'bhisāra-kṣaṇaḥ ||

18. *āśleṣād anu cumbanād anu nakhollekhād anu svāntaja-*
prodbodhād anu saṃbhramād anu ratārambhād anu prītayoḥ /
anyārthaṃ gatayor bhramān militayoḥ saṃbhāṣaṇair jānator
dampatyor iha ko na ko na tamasi vrīḍā-vimiśro rasaḥ //
19. *sabhaya-cakitaṃ vinyasyantīṃ dṛśaṃ timire pathi*
pratitaru muhuḥ sthitvā mandaṃ padāni vitanvatīm /
katham api rahaḥ prāptām aṅgair anaṅga-taraṅgibhiḥ
sumukhi subhagaḥ paśyan sa tvām upaitu kṛtārthatām //
20. *rādhā-mugdha-mukhāravinda-madhupas trailokya-mauli-sthalī*
nepathyocita-nīla-ratnam avanī-bhārāvatārāntakaḥ /
svacchandaṃ vraja-sundarī-jana-manas-toṣa-pradoṣodayaḥ
kaṃsa-dhvaṃsana-dhūma-ketur avatu tvāṃ devakī-nandanaḥ //

Ṣaṣṭhaḥ Sargaḥ
SOTKAṆṬHA-VAIKUṆṬHAḤ

1. *atha tāṃ gantum aśaktāṃ ciram anuraktāṃ latā-gṛhe dṛṣṭvā /*
tac-caritaṃ govinde manasija-mande sakhī prāha //

guṇakarī-rāgeṇa rūpaka-tālena gīyate / prabandhaḥ 12 //

2. *paśyati diśi diśi rahasi bhavantam /*
tad-adhara-madhura-madhūni pibantam //
nātha hare sīdati rādhā'vāsa-gṛhe // dhruva-padam //
3. *tvad-abhisaraṇa-rabhasena valantī /*
patati padāni kiyanti calantī //
4. *vihita-viśada-bisa-kisalaya-valayā /*
jīvati param iha tava rati-kalayā //
5. *muhur avalokita-maṇḍana-līlā /*
madhu-ripur aham iti bhāvana-śīlā //
6. *tvaritam upaiti na katham abhisāram /*
harir iti vadati sakhīm anuvāram //
7. *śliṣyati cumbati jala-dhara-kalpam /*
harir upagata iti timiram analpam //
8. *bhavati vilambini vigalita-lajjā /*
vilapati roditi vāsaka-sajjā //
9. *śrī-jayadeva-kaver idam uditam /*
rasika-janaṃ tanutām atimuditam //

*   *   *

10. *vipula-pulaka-pāliḥ sphīta-sīt-kāram antar-*
janita-jaḍima-kāku-vyākulaṃ vyāharantī /
tava kitava vidhāyāmanda-kandarpa-cintāṃ
rasa-jala-nidhi-magnā dhyāna-lagnā mṛgākṣī //
11. *aṅgeṣv ābharaṇaṃ karoti bahuśaḥ patre'pi saṃcāriṇi*
prāptaṃ tvāṃ pariśaṅkate vitanute śayyāṃ ciraṃ dhyāyati /
ity ākalpa-vikalpa-talpa-racanā-saṅkalpa-līlā-śata-
vyāsaktāpi vinā tvayā vara-tanur naiṣā niśāṃ neṣyati //
12. *kiṃ viśrāmyasi kṛṣṇa-bhogi-bhavane bhāṇḍīra-bhūmī-ruhi*
bhrātar yāsi na dṛṣṭi-gocaram itaḥ sānanda-nandāspadam /
rādhāyā vacanaṃ tad adhvaga-mukhān nandāntike gopato
govindasya jayanti sāyam atithi-prāśastya-garbhā giraḥ //

Saptamaḥ Sargaḥ
NĀGARA-NĀRĀYAṆAḤ

1. *atrāntare ca kulaṭā-kula-vartma-ghāta-*
saṃjāta-pātaka iva sphuṭa-lāñchana-śrīḥ /
vṛndāvanāntaram adīpayad aṃśu-jālair
dik-sundarī-vadana-candana-bindur induḥ //
2. *prasarati śaśa-dhara-bimbe vihita-vilambe ca mādhave vidhurā* /
viracita-vividha-vilāpaṃ sā paritāpaṃ cakāroccaiḥ //

mālava-rāga-yati-tālābhyāṃ gīyate / prabandhaḥ 13 //

3. *kathita-samaye'pi harir ahaha na yayau vanam* /
mama viphalam idam amala-rūpam api yauvanam //
yāmi he kam iha śaraṇaṃ sakhī-jana-vacana-vañcitā // *dhruva.* //
4. *yad-anugamanāya niśi gahanam api śīlitam* /
tena mama hṛdayam idam asama-śara-kīlitam //
5. *mama maraṇam eva varam iti vitatha-ketanā* /
kim iha viṣahāmi virahānalam acetanā //
6. *mām ahaha vidhurayati madhura-madhu-yāminī* /
kāpi harim anubhavati kṛta-sukṛta-kāminī //
7. *ahaha kalayāmi valayādi-maṇi-bhūṣaṇam* /
hari-viraha-dahana-vahanena bahu-dūṣaṇam //
8. *kusuma-sukumāra-tanum atanu-śara-līlayā* /
srag api hṛdi hanti mām ativiṣama-śīlayā //
9. *aham iha nivasāmi na-gaṇita-vana-vetasā* /
smarati madhu-sūdano mām api na cetasā //

10. *hari-caraṇa-śaraṇa-jayadeva-kavi-bhāratī /*
vasatu hṛdi yuvatir iva komala-kalāvatī //

* * *

11. *tat-kiṃ kām api kāminīm abhisṛtaḥ kiṃ vā kalā-kelibhir*
baddho bandhubhir andhakāriṇi vanābhyarṇe kim udbhrāmyati /
kāntaḥ klānta-manā manāg api pathi prasthātum evākṣamaḥ
saṅketī-kṛta-mañju-vāñjula-latā-kuñje'pi yan nāgataḥ //

12. *athāgatāṃ mādhavam antareṇa*
sakhīm iyaṃ vīkṣya viṣāda-mūkām /
viśaṅkamānā ramitaṃ kayāpi
janārdanaṃ dṛṣṭavad etad āha //

vasanta-rāga-yati-tālābhyāṃ gīyate / prabandhaḥ 14 //

13. *smara-samarocita-viracita-veśā /*
galita-kusuma-dara-vilulita-keśā //
kāpi madhu-ripuṇā vilasati yuvatir adhika-guṇā // dhruva. //

14. *hari-parirambhaṇa-calita-vikārā /*
kuca-kalaśopari taralita-hārā //

15. *vicalad-alaka-lalitānana-candrā /*
tad-adhara-pāna-rabhasa-kṛta-tandrā //

16. *cañcala-kuṇḍala-dalita-kapolā /*
mukharita-raśana-jaghana-gati-lolā //

17. *dayita-vilokita-lajjita-hasitā /*
bahu-vidha-kūjita-rati-rasa-rasitā //

18. *vipula-pulaka-pṛthu-vepathu-bhaṅgā /*
śvasita-nimīlita-vikasad-anaṅgā //

19. *śrama-jala-kaṇa-bhara-subhaga-śarīrā /*
paripatitorasi rati-raṇa-dhīrā //

20. *śrī-jayadeva-bhaṇita-hari-ramitam /*
kali-kaluṣaṃ janayatu pariśamitam //

* * *

21. *viraha-pāṇḍu-murāri-mukhāmbuja-*
dyutir ayaṃ tirayann api vedanām /
vidhur atīva tanoti mano-bhuvaḥ
suhṛd aye hṛdaye madana-vyathām //

gurjarī-rāgaika-tālī-tālena gīyate / prabandhaḥ 15 //

22. *samudita-madane ramaṇī-vadane cumbana-valitādhare /*
mṛga-mada-tilakaṃ likhati sapulakaṃ mṛgam iva rajanī-kare //
ramate yamunā-pulina-vane vijayī murārir adhunā // dhruva. //

23. *ghana-caya-rucire racayati cikure taralita-taruṇānane |*
kurubaka-kusumaṃ capalā suṣamaṃ rati-pati-mṛga-kānane ||
24. *ghaṭayati sughane kuca-yuga-gagane mṛga-mada-ruci-rūṣite |*
maṇi-saram amalaṃ tāraka-paṭalaṃ nakha-pada-śaśi-bhūṣite ||
25. *jita-bisa-śakale mṛdu-bhuja-yugale kara-tala-nalinī-dale |*
marakata-valayaṃ madhu-kara-nicayaṃ vitarati hima-śītale ||
26. *rati-gṛha-jaghane vipulāpaghane manasija-kanakāsane |*
maṇi-maya-raśanaṃ toraṇa-hasanaṃ vikirati kṛta-vāsane ||
27. *caraṇa-kisalaye kamalā-nilaye nakha-maṇi-gaṇa-pūjite |*
bahir-apavaraṇaṃ yāvaka-bharaṇaṃ janayati hṛdi yojite ||
28. *ramayati subhṛśaṃ kām api sudṛśaṃ khala-hala-dhara-sodare |*
kim aphalam avasaṃ ciram iha virasaṃ vada sakhi viṭapodare ||
29. *iha rasa-bhaṇane kṛta-hari-guṇane madhu-ripu-pada-sevake |*
kali-yuga-racitaṃ na vasatu duritaṃ kavi-nṛpa-jayadevake ||

* * *

30. *nāyātaḥ sakhi nirdayo yadi śaṭhas tvaṃ dūti kiṃ dūyase*
svacchandaṃ bahu-vallabhaḥ sa ramate kiṃ tatra te dūṣaṇam |
paśyādya priya-saṃgamāya dayitasyākṛṣyamāṇaṃ guṇair
utkaṇṭhārti-bharād iva sphuṭad idaṃ cetaḥ svayaṃ yāsyati ||

deśavarāḍī-rāgeṇa rūpaka-tālena gīyate / prabandhaḥ 16 //

31. *anila-tarala-kuvalaya-nayanena |*
tapati na sā kisalaya-śayanena ||
sakhi yā ramitā vana-mālinā || dhruva-padam ||
32. *vikasita-sarasija-lalita-mukhena |*
sphuṭati na sā manasija-viśikhena ||
33. *amṛta-madhura-mṛdu-tara-vacanena |*
jvalati na sā malayaja-pavanena ||
34. *sthala-jala-ruha-ruci-kara-caraṇena |*
luṭhati na sā hima-kara-kiraṇena ||
35. *sajala-jalada-samudaya-ruciredṇa |*
dalati na sā hṛdi cira-viraheṇa ||
36. *kanaka-nikaṣa-ruci-śuci-vasanena |*
śvasiti na sā parijana-hasanena ||
37. *sakala-bhuvana-jana-vara-taruṇena |*
vahati na sā rujam atikaruṇena ||
38. *śrī-jayadeva-bhaṇita-vacanena |*
praviśatu harir api hṛdayam anena ||

* * *

39. *mano-bhavānandana-candanānila*
prasīda re dakṣiṇa muñca vāmatām /
kṣaṇaṃ jagat-prāṇa nidhāya mādhavaṃ
puro mama prāṇa-haro bhaviṣyasi //
40. *ripur iva sakhī-saṃvāso'yaṃ śikhīva himānilo*
viṣam iva sudhā-raśmir yasmin dunoti mano-gate /
hṛdayam adaye tasminn aivaṃ punar valate balāt
kuvalaya-dṛśāṃ vāmaḥ kāmo nikāma-niraṅkuśaḥ //
41. *bādhāṃ vidhehi malayānila pañca-bāṇa*
prāṇān gṛhāṇa na gṛhaṃ punar āśrayiṣye /
kiṃ te kṛtānta-bhagini kṣamayā taraṅgair
aṅgāni siñca mama śāmyatu deha-dāhaḥ //
42. *prātar nīla-nicolam acyutam uraḥ saṃvīta-pītāṃśukaṃ*
rādhāyāś cakitaṃ vilokya hasati svairam sakhī-maṇḍale /
vrīḍā-cañcalam añcalaṃ nayanayor ādhāya rādhānane
svādu-smera-mukho'yam astu jagad-ānandāya nandātmajaḥ //

Aṣṭamaḥ Sargaḥ
VILAKṢYA-LAKṢMĪ-PATIḤ

1. *atha katham api yāminīṃ vinīya*
smara-śara-jarjaritāpi sā prabhāte /
anunaya-vacanaṃ vadantam agre
praṇatam api priyam āha sābhyasūyam //

bhairavī-rāga-yati-tālābhyāṃ gīyate / prabandhaḥ 17 //

2. *rajani-janita-guru-jāgara-rāga-kaṣāyitam alasa-niveśam*
vahati nayanam anurāgam iva sphuṭam udita-rasābhiniveśam //
hari hari yāhi mādhava yāhi keśava mā vada kaitava-vādam /
tām anusara sarasī-ruha-locana yā tava harati viṣādam // *dhr.* //
3. *kajjala-malina-vilocana-cumbana-viracita-nīlima-rūpam* /
daśana-vasanam aruṇaṃ tava kṛṣṇa tanoti tanor anurūpam //
4. *vapur anuharati tava smara-saṅgara-khara-nakhara-kṣata-rekham* /
marakata-śakala-kalita-kaladhauta-liper iva rati-jaya-lekham //
5. *caraṇa-kamala-galad-alaktaka-siktam idaṃ tava hṛdayam udāram* /
darśayatīva bahir madana-druma-nava-kisalaya-parivāram //
6. *daśana-padaṃ bhavad-adhara-gataṃ mama janayati cetasi khedam* /
kathayati katham adhunāpi mayā saha tava vapur etad abhedam //

7. *bahir iva malinataraṃ tava kṛṣṇa mano'pi bhaviṣyati nūnam |*
katham atha vañcayase janam anugatam asama-śara-jvara-dūnam ||
8. *bhramati bhavān abalā-kavalāya vaneṣu kim atra vicitram |*
prathayati pūtanikaiva vadhū-vadha-nirdaya-bāla-caritram ||
9. *śrī-jayadeva-bhaṇita-rati-vañcita-khaṇḍita-yuvati-vilāpam |*
śṛṇuta sudhā-madhuraṃ vibudhā vibudhālayato'pi durāpam ||

* * *

10. *tavedaṃ paśyantyāḥ prasarad-anurāgaṃ bahir iva*
priyā-pādālakta-cchuritam aruṇa-dyoti hṛdayam |
mamādya prakhyāta-praṇaya-bhara-bhaṅgena kitava
tvad-ālokaḥ śokād api kim api lajjāṃ janayati ||
11. *antar-mohana-mauli-ghūrṇana-calan-mandāra-vibhraṃśana-*
stambhākarṣaṇa-dṛpti-harṣaṇa-mahā-mantraḥ kuraṅgī-dṛśām |
dṛpyad-dānava-dūyamāna-diviṣad-durvāra-duḥkapadāṃ
bhraṃśaḥ kaṃsa-ripor vyapohayatu vaḥ śreyāṃsi vaṃśī-ravaḥ ||

Navamaḥ Sargaḥ
MUGDHA-MUKUNDAḤ

1. *tām atha manmatha-khinnāṃ rati-ra[bha]sa-bhinnāṃ viṣāda-sampannām*
anucintita-hari-caritāṃ kalahāntaritām uvāca rahasi sakhī ||

gurjarī-rāga-yati-tālābhyāṃ gīyate | prabandhaḥ 18 ||

2. *harir abhisarati vahati madhu-pavane |*
kim aparam adhika-sukhaṃ sakhi bhavane ||
mādhave mā kuru mānini mānam aye || dhruva-padam ||
3. *tāla-phalād api gurum atisarasam |*
kiṃ viphalī-kuruṣe kuca-kalaśam ||
4. *kati na kathitam idam anupadam aciram |*
mā parihara harim atiśaya-ruciram ||
5. *kim iti viṣīdasi rodiṣi vikalā |*
vihasati yuvati-sabhā tava sakalā ||
6. *sajala-nalinī-dala-śītala-śayane |*
harim avalokaya saphalaya nayane ||
7. *janayasi manasi kim iti guru-khedam |*
śṛṇu mama vacanam anīhita-bhedam ||
8. *harir upayātu vadatu bahu-madhuram |*
kim iti karoṣi hṛdayam atividhuram ||

9. *śrī-jayadeva-bhaṇitam atilalitam |*
sukhayatu rasika-janaṃ hari-caritam ||

* * *

10. *snigdhe yat paruṣāsi yat praṇamati stabdhāsi yad rāgiṇi*
dveṣasthāsi yad unmukhe vimukhatāṃ yātāsi tasmin priye |
tad yuktaṃ viparīta-kāriṇi tava śrī-khaṇḍa-carcā viṣaṃ
śītāṃśus tapano himaṃ hutavahaḥ krīḍā-mudo yātanāḥ ||
11. *sāndrānanda-puraṃdarādi-diviṣad-vṛndair amandādarād*
ānamrair mukuṭendra-nīla-maṇibhiḥ saṃdarśitendivaram |
sva-cchandaṃ makaranda-sundara-galan-mandākinī-meduraṃ
śrī-govinda-padāravindam aśubha-skandāya vandāmahe ||

Daśamaḥ Sargaḥ
CATURA-CATUR-BHUJAḤ

1. *atrāntare'masṛṇa-roṣa-vaśām apāra-*
niḥśvāsa-niḥsaha-mukhīṃ sumukhīm upetya |
savrīḍam īkṣita-sakhī-vadanāṃ dinānte
sānanda-gadgada-padaṃ harir ity uvāca ||

deśavarāḍī-rāgāṣṭatālī-tālābhyāṃ gīyate / prabandhaḥ 19 //

2. *vadasi yadi kiṃcid api danta-ruci-kaumudī*
harati dara-timiram atighoram |
sphurad-adhara-śīdhave tava vadana-candramā
rocayatu locana-cakoram ||
priye cāru-śīle muñca mayi mānam anidānam |
sapadi madanānalo dahati mama mānasaṃ
dehi mukha-kamala-madhu-pānam || dhruva-padam ||
3. *satyam evāsi yadi sudati mayi kopinī*
dehi khara-nakhara-śara-ghātam |
ghaṭaya bhuja-bandhanaṃ janaya rada-khaṇḍanaṃ
yena vā bhavati sukha-jātam ||
4. *tvam asi mama bhūṣaṇaṃ tvam asi mama jīvanaṃ*
tvam asi mama bhava-jaladhi-ratnam |
bhavatu bhavatīha mayi satatam anurodhinī
tatra mama hṛdayam atiyatnam ||
5. *nīla-nalinābham api tanvi tava locanaṃ*
dhārayati koka-nada-rūpam |

kusuma-śara-bāṇa-bhāvena yadi rañjayasi
kṛṣṇam idam etad anurūpam //

6. *sphuratu kuca-kumbhayor upari maṇi-mañjarī*
rañjayatu tava hṛdaya-deśam /
rasatu raśanāpi tava ghana-jaghana-maṇḍale
ghoṣayatu manmatha-nideśam //

7. *sthala-kamala-gañjanaṃ mama hṛdaya-rañjanaṃ*
janita-rati-raṅga-parabhāgam /
bhaṇa masṛṇa-vāṇi karavāṇi caraṇa-dvayaṃ
sarasa-lasad-alaktaka-rāgam //

8. *smara-garala-khaṇḍanaṃ mama śirasi maṇḍanaṃ*
dehi pada-pallavam udāram /
jvalati mayi dāruṇo madana-kadanāruṇo
haratu tad-upāhita-vikāram //

9. *iti caṭula-cāṭu-paṭu-cāru mura-vairiṇo*
rādhikām adhi vacana-jātam /
jayatu jayadeva-kavi-bhāratī-bhūṣitaṃ
māninī-jana-janita-śātam //

* * *

10. *parihara kṛtātaṅke śaṅkāṃ tvayā satataṃ ghana-*
stana-jaghanayākrānte svānte parān-avakāśini /
viśati vitanor anyo dhanyo na ko'pi mamāntaraṃ
praṇayini parīrambhārambhe vidhehi vidheyatām //

11. *mugdhe vidhehi mayi nirdaya-danta-daṃśa-*
dor-valli-bandha-nibiḍa-stana-pīḍanāni /
caṇḍi tvam eva mudam udvaha pañca-bāṇa-
cāṇḍāla-kāṇḍa-dalanād asavaḥ prayānti //

12. *śaśi-mukhi tava bhāti bhaṅgura-bhrūr*
yuva-jana-moha-karāla-kāla-sarpī /
tad-udita-bhaya-bhañjanāya yūnāṃ
tvad-adhara-sīdhu-sudhaiva siddha-mantraḥ //

13. *vyathayati vṛthā maunaṃ tanvi prapañcaya pañcamaṃ*
taruṇi madhurālāpais tāpaṃ vinodaya dṛṣṭibhiḥ /
sumukhi vimukhī-bhāvaṃ tāvad vimuñca na muñca māṃ
svayam atiśaya-snigdho mugdhe priyo'yam upasthitaḥ //

14. *bandhūka-dyuti-bāndhavo'yam adharaḥ snigdho madhūka-cchavir*
gaṇḍaś caṇḍi cakāsti nīla-nalina-śrī-mocanaṃ locanam /
nāsābhyeti tila-prasūna-padavīṃ kundābha-danti priye
prāyas tvan-mukha-sevayā vijayate viśvaṃ sa puṣpāyudhaḥ //

15. *dṛśau tava madālase vadanam indu-saṃdīpanaṃ*
gatir jana-mano-ramā vijita-rambham ūru-dvayam /
ratis tava kalāvatī rucira-citra-lekhe bhruvāv
aho vibudha-yauvataṃ vahasi tanvi pṛthvī-gatā //

16. *sa prītiṃ tanutāṃ hariḥ kuvalayāpīḍena sārdhaṃ raṇe*
rādhā-pīna-payodhara-smaraṇa-kṛt-kumbhena saṃbhedavān /
yatra svidyati mīlati kṣaṇam api kṣipraṃ tad-ālokana-
vyāmohena jitaṃ jitaṃ jitam abhūt kaṃsasya kolāhalaḥ //

Ekādaśaḥ Sargaḥ
SĀNANDA-DĀMODARAḤ

1. *suciram anunayena prīṇayitvā mṛgākṣīṃ*
gatavati kṛtaveśe keśave kuñja-śayyām /
racita-rucira-bhūṣāṃ dṛṣṭi-moṣe pradoṣe
sphurati niravasādāṃ kāpi rādhāṃ jagāda //

vasanta-rāga-yati-tālābhyāṃ gīyate / prabandhaḥ 20 //

2. *viracita-cāṭu-vacana-racanaṃ caraṇe racita-praṇipātam* /
samprati mañjula-vañjula-sīmani keli-śayanam anuyātam //
mugdhe madhu-mathanam anugatam anusara rādhike // *dhruva.* //

3. *ghana-jaghana-stana-bhāra-bhare dara-manthara-caraṇa-vihāram* /
mukharita-maṇi-mañjīram upaihi vidhehi marāla-vikāram //

4. *śṛṇu ramaṇīyataraṃ taruṇī-jana-mohana-madhupa-virāvam* /
kusuma-śarāsana-śāsana-bandini pika-nikare bhaja bhāvam //

5. *anila-tarala-kisalaya-nikareṇa kareṇa latā-nikurambam* /
preraṇam iva karabhoru karoti gatiṃ prati muñca vilambam //

6. *sphuritam anaṅga-taraṅga-vaśād iva sūcita-hari-parirambham* /
pṛccha manohara-hāra-vimala-jala-dhāram amuṃ kuca-kumbham //

7. *adhigatam akhila-sakhībhir idaṃ tava vapur api rati-raṇa-sajjam* /
caṇḍi rasita-raśanā-rava-ḍiṇḍimam abhisara sarasam alajjam //

8. *smara-śara-subhaga-nakhena sakhīm avalambya kareṇa salīlam* /
cala valaya-kvaṇitair avabodhaya harim api nija-gati-śīlam //

9. *śrī-jayadeva-bhaṇitam adharī-kṛta-hāram udāsita-vāmam* /
hari-vinihita-manasām adhitiṣṭhatu kaṇṭha-taṭīm avirāmam //

* * *

10. *sā māṃ drakṣyati vakṣyati smara-kathāṃ praty-aṅgam āliṅganaiḥ*
prītiṃ yāsyati raṃsyate sakhi samāgatyeti cintākulaḥ /

sa tvāṃ paśyati vepate pulakayaty ānandati svidyati
praty-udgacchati mūrcchati sthira-tamaḥ-puñje nikuñje priyaḥ ||

11. *akṣṇor nikṣipad añjanaṃ śravaṇayos tāpiccha-gucchāvalīṃ*
mūrdhni śyāma-saroja-dāma kucayoḥ kastūrikā-patrakam |
dhūrtānām abhisāra-saṃbhrama-juṣāṃ viṣvaṅ-nikuñje sakhi
dhvāntaṃ nīla-nicola-cāru sudṛśāṃ praty-aṅgam āliṅgati ||

12. *kāśmīra-gaura-vapuṣām abhisārikāṇāṃ*
ābaddha-rekham abhito ruci mañjarībhiḥ |
etat tamāla-dala-nīlatamaṃ tamisraṃ
tat-prema-hema-nikaṣopalatāṃ tanoti ||

13. *hārāvalī-tarala-kāñcana-kāñci-dāma-*
keyūra-kaṅkaṇa-maṇi-dyuti-dīpitasya |
dvāre nikuñja-nilayasya hariṃ nirīkṣya
vrīḍāvatīm atha sakhīm iyam ity uvāca ||

varāḍī-rāga-rūpaka-tālābhyāṃ gīyate / prabandhaḥ 21 //

14. *mañjutara-kuñja-tala-keli-sadane* |
vilasa rati-rabhasa-hasita-vadane ||
praviśa rādhe mādhava-samīpam iha || *dhruva-padam* ||

15. *nava-lasad-aśoka-dala-śayana-sāre* |
vilasa kuca-kalaśa-tarala-hāre ||

16. *kusuma-caya-racita-śuci-vāsagehe* |
vilasa kusuma-sukumāra-dehe ||

17. *mṛdu-cala-malaya-pavana-surabhi-śīte* |
vilasa madana-śara-nikara-bhīte ||

18. *vitata-bahu-valli-nava-pallava-ghane* |
vilasa ciram alasa-pīna-jaghane ||

19. *madhu-mudita-madhupa-kula-kalita-rāve* |
vilasa madana-rasa-sarasa-bhāve ||

20. *madhura-tara-pika-nikara-ninada-mukhare* |
vilasa daśana-ruci-rucira-śikhare ||

21. *vihita-padmāvatī-sukha-samāje* |
kuru murāre maṅgala-śatāni |
bhaṇati jayadeva-kavi-rāja-rāje ||

* * *

22. *tvāṃ cittena ciraṃ vahann ayam atiśrānto bhṛśaṃ tāpitaḥ*
kandarpeṇa ca pātum icchati sudhā-saṃbādha-bimbādharam |
asyāṅkaṃ tad alaṃkuru kṣaṇam iha bhrū-kṣepa-lakṣmī-lava-
krīte dāsa ivopasevita-padāmbhoje kutaḥ saṃbhramaḥ ||

23. *sā sasādhvasa-sānandaṃ govinde lola-locanā /*
siñjāna-mañju-mañjīraṃ praviveśābhiveśanam //

varāḍī-rāga-yati-tālābhyāṃ gīyate / prabandhaḥ 22 //

24. *rādhā-vadana-vilokana-vikasita-vividha-vikāra-vibhaṅgam /*
jala-nidhim iva vidhu-maṇḍala-darśana-taralita-tuṅga-taraṅgam //
harim eka-rasaṃ ciram abhilaṣita-vilāsam /
sā dadarśa guru-harṣa-vaśaṃ-vada-vadanam anaṅga-nivāsam // dhr. //
25. *hāram amalatara-tāram urasi dadhataṃ parilaṃbya vidūram /*
sphuṭatara-phena-kadamba-karambitam iva yamunā-jala-pūram //
26. *śyāmala-mṛdula-kalevara-maṇḍalam adhigata-gaura-dukūlam /*
nīla-nalinam iva pīta-parāga-paṭala-bhara-valayita-mūlam //
27. *tarala-dṛg-añcala-calana-manohara-vadana-janita-rati-rāgam /*
sphuṭa-kamalodara-khelita-khañjana-yugam iva śaradi taḍāgam //
28. *vadana-kamala-pariśīlana-milita-mihira-sama-kuṇḍala-śobham /*
smita-ruci-rucira-samullasitādhara-pallava-kṛta-rati-lobham //
29. *śaśi-kiraṇa-cchuritodara-jala-dhara-sundara-sakusuma-keśam /*
timirodita-vidhu-maṇḍala-nirmala-malayaja-tilaka-niveśam //
30. *vipula-pulaka-bhara-danturitaṃ rati-keli-kalābhir adhīram /*
maṇi-gaṇa-kiraṇa-samūha-samujjvala-bhūṣaṇa-subhaga-śarīram //
31. *śrī-jayadeva-bhaṇita-vibhava-dviguṇīkṛta-bhūṣaṇa-bhāram /*
praṇamata hṛdi vinidhāya hariṃ suciraṃ sukṛtodaya-sāram //

* * *

32. *atikramyāpāṅgaṃ śravaṇa-patha-paryanta-gamana-*
prayāsenevākṣṇos taralatara-tāraṃ gamitayoḥ /
idānīṃ rādhāyāḥ priyatama-samāloka-samaye
papāta svedāmbu-prasara iva harṣāśru-nikaraḥ //
33. *bhajantyās talpāntaṃ kṛta-kapaṭa-kaṇḍūti-pihita-*
smitaṃ yāte gehād bahir avahitālī-parijane /
priyāsyaṃ paśyantyāḥ smara-paravaśākūta-subhagaṃ
salajjā lajjāpi vyagamad iva dūraṃ mṛga-dṛśaḥ //
34. *sānandaṃ nanda-sūnur diśatu mitaparaṃ sammadaṃ manda-mandaṃ*
rādhām ādhāya bāhvor vivaram anu dṛḍhaṃ pīḍayan prīti-yogāt /
tuṅgau tasyā urojāv atanu-varatanor nirgatau mā sma bhūtāṃ
pṛṣṭhaṃ nirbhidya tasmād bahir iti valita-grīvam ālokayan vaḥ //
35. *jaya-śrī-vinyastair mahita iva mandāra-kusumaiḥ*
svayaṃ sindūreṇa dvipa-raṇa-mudā mudrita iva /
bhujāpīḍa-krīḍā-hata-kuvalayāpīḍa-kariṇaḥ
prakīrṇāsṛg-bindur jayati bhuja-daṇḍo mura-jitaḥ //

36. *saundaryaika-nidher anaṅga-lalanā-lāvaṇya-līlā-juṣo*
rādhāyā hṛdi palvale manasija-krīḍaika-raṅga-sthale /
ramyoroja-saroja-khelana-rasitvād ātmanaḥ khyāpayan
dhyātur mānasa-rāja-haṃsa-nibhatāṃ deyān mukundo mudam //

Dvādaśaḥ Sargaḥ
SUPRĪTA-PĪTĀMBARAḤ

1. *gatavati sakhī-vṛnde'manda-trapā-bhara-nirbhara-*
smara-paravaśākūta-sphīta-smita-snapitādharām /
sarasa-manasaṃ dṛṣṭvā rādhām muhur nava-pallava-
prasava-śayane nikṣiptākṣīm uvāca hariḥ priyām //

vibhāsa-rāgaikatālī-tālābhyāṃ gīyate / prabandhaḥ 23 //

2. *kisalaya-śayana-tale kuru kāmini caraṇa-nalina-viniveśam* /
tava pada-pallava-vairi-parābhavam idam anubhavatu suveśam //
kṣaṇam adhunā nārāyaṇam anugatam anusara rādhike // *dhruva.* //
3. *kara-kamalena karomi caraṇam aham āgamitāsi vidūram* /
kṣaṇam upakuru śayanopari mām iva nūpuram anugati-śūram //
4. *vadana-sudhā-nidhi-galitam amṛtam iva racaya vacanam anukūlam* /
viraham ivāpanayāmi payodhara-rodhakam urasi dukūlam //
5. *priya-parirambhaṇa-rabhasa-valitam iva pulakitam atiduravāpam* /
mad-urasi kuca-kalaśaṃ viniveśaya śoṣaya manasija-tāpam //
6. *adhara-sudhā-rasam upanaya bhāmini jīvaya mṛtam iva dāsam* /
tvayi vinihita-manasaṃ virahānala-dagdha-vapuṣam avilāsam //
7. *śaśi-mukhi mukharaya maṇi-raśanā-guṇam anuguṇa-kaṇṭha-ninādam* /
śruti-yugale pika-ruta-vikale mama śamaya cirād avasādam //
8. *mām ativiphala-ruṣā vikalī-kṛtam avalokitum adhunedam* /
lajjitam iva nayanaṃ tava viramati visṛja vṛthā rati-khedam //
9. *śrī-jayadeva-bhaṇitam idam anupada-nigadita-madhu-ripu-modam* /
janayatu rasika-janeṣu mano-rama-rati-rasa-bhāva-vinodam //

* * *

10. *pratyūhaḥ pulakāṅkureṇa nibiḍāśleṣe nimeṣeṇa ca*
krīḍākūta-vilokite'dhara-sudhā-pāne kathā-kelibhiḥ /
ānandādhigamena manmatha-kalā-yuddhe'pi yasminn abhūd
udbhūtaḥ sa tayor babhūva suratārambhaḥ priyaṃ-bhāvukaḥ //
11. *dorbhyāṃ saṃyamitaḥ payodhara-bhareṇāpīḍitaḥ pāṇijair*
āviddho daśanaiḥ kṣatādhara-puṭaḥ śroṇī-taṭenāhataḥ /

hastenānamitaḥ kace'dhara-madhu-syandena saṃmohitaḥ
kāntaḥ kām api tṛptim āpa tad aho kāmasya vāmā gatiḥ ||

12. *mārāṅke rati-keli-saṃkula-raṇārambhe tayā sāhasa-*
prāyaṃ kānta-jayāya kiṃcid upari prārambhi yat-saṃbhramāt |
niṣpandā jaghana-sthalī śithilitā dor-vallir utkampitaṃ
vakṣo mīlitam akṣi pauruṣa-rasaḥ strīṇāṃ kutaḥ sidhyati ||

13. *tasyāḥ pāṭala-pāṇijāṅkitam uro nidrā-kaṣāye dṛṣau*
nirdhautādhara-śonimā vilulita-srasta-srajo mūrdha-jāḥ |
kāñcī-dāma dara-ślathāñcalam iti prātar nikhātair dṛśor
ebhiḥ kāma-śarais tad-adbhutam abhūt patyur manaḥ kīlitam ||

14. *vyālolaḥ keśa-pāśas taralitam alakaiḥ sveda-mokṣau kapolau*
kliṣṭā bimbādhara-śrīḥ kuca-kalaśa-rucā hāritā hāra-yaṣṭiḥ |
kaṅcī-kāntir hatāśā stana-jaghana-padaṃ pāṇinācchadya sadyaḥ
paśyanti satrapā sā tad api vilulitā mugdha-kāntir dhinoti ||

15. *īṣan-mīlita-dṛṣṭi mugdha-vilasat-sīt-kāra-dhārā-vaśād*
avyaktākula-keli-kāku-vikasad-dantāṃśu-dhautādharam |
śānta-stabdha-payodharaṃ bhṛśa-pariṣvaṅgāt kuraṅgī-dṛśo
harṣotkarṣa-vimukta-niḥsaha-tanor dhanyo dhayaty ānanam ||

16. *atha sahasā suprītaṃ suratānte sā nitānta-khinnāṅgī* |
rādhā jagāda sādaram idam ānandena govindam ||

rāmakarī-rāga-yati-tālābhyāṃ gīyate / prabandhaḥ 24 ||

17. *kuru yadu-nandana candana-śiśiratareṇa kareṇa payodhare* |
mṛga-mada-patrakam atra mano-bhava-maṅgala-kalaśa-sahodare ||
nijagāda sā yadu-nandane krīḍati hṛdayānandane || *dhruva.* ||

18. *ali-kula-gañjanam añjanakaṃ rati-nāyaka-sāyaka-mocane* |
tvad-adhara-cumbana-lambita-kajjala ujjvalaya priya locane ||

19. *nayana-kuraṅga-taraṅga-vikāsa-nirāsa-kare śruti-maṇḍale* |
manasija-pāśa-vilāsa-dhare śubha-veśa niveśaya kuṇḍale ||

20. *bhramara-cayaṃ racayantam upari ruciraṃ suciraṃ mama saṃmukhe* |
jita-kamale vimale parikarmaya narma-janakam alakaṃ mukhe ||

21. *mṛga-mada-rasa-valitaṃ lalitaṃ kuru tilakam alika-rajanī-kare* |
vihita-kalaṅka-kalaṃ kamalānana viśramita-śrama-śīkare ||

22. *mama rucire cikure kuru mānada manasija-dhvaja-cāmare* |
rati-galite lalite kusumāni śikhaṇḍi-śikhaṇḍaka-ḍāmare ||

23. *sarasa-ghane jaghane mama śambara-dāraṇa-vāraṇa-kandare* |
maṇi-raśanā-vasanābharaṇāni śubhāśaya vāsaya sundare ||

24. *śrī-jayadeva-vacasi rucire sadayaṃ hṛdayaṃ kuru maṇḍane* |
hari-caraṇa-smaraṇāmṛta-nirmita-kali-kaluṣa-jvara-khaṇḍane ||

* * *

25. *racaya kucayoś citram patraṃ kuruṣva kapolayor*
ghaṭaya jaghane kāñcīṃ mugdha-srajā kabarī-bharam /
kalaya valaya-śreṇīṃ pāṇau pade maṇi-nūpurāv
iti nigaditaḥ prītaḥ pītāmbaro'pi tathākarot //

26. *paryaṅkī-kṛta-nāga-nāyaka-phaṇā-śreṇī-maṇīnāṃ gaṇe*
saṅkrānta-pratibimba-saṃvalanayā bibhrad-vibhu-prakriyām /
pādāmbho-ruha-dhāri-vāridhi-sutām akṣṇāṃ didṛkṣuḥ śataiḥ
kāyavyūham ivācarann apacitau bhūyo hariḥ pātu vaḥ //

27. *tvām aprāpya mayi svayaṃvara-parāṃ kṣīroda-tīrodare*
śaṅke sundari kālakūṭam apiban mūḍho mṛḍānī-patiḥ /
itthaṃ pūrva-kathābhir anya-manasā vikṣipya vakṣo'ñcalaṃ
rādhāyāḥ stana-korakopari-milan-netro hariḥ pātu vaḥ //

28. *yad gāndharva-kalāsu kauśalam anudhyānaṃ ca yad vaiṣṇavaṃ*
yac chṛṅgāra-viveka-tattva-racanā-kāvyeṣu līlāyitam /
tat sarvaṃ jayadeva-paṇḍita-kaveḥ kṛṣṇaika-tānātmanaḥ
sānandāḥ pariśodhayantu sudhiyaḥ śrī-gīta-govindataḥ //

29. *sādhūnāṃ svata eva saṃmatir iha syād eva bhaktyārthinām*
ālocya grathana-śramaṃ ca viduṣām asmin bhaved ādaraḥ /
ye kecit para-kṛty-upaśruti-parās tān arthaye mat-kṛtiṃ
bhūyo vīkṣya vadantv avadyam iha cet sā vāsanā sthāsyati //

30. *śrī-bhojadeva-prabhavasya rāmādevī-suta-śrī-jayadevakasya* /
parāśarādi-priya-varga-kaṇṭhe śrī-gīta-govinda-kavitvam astu //

31. *sādhvī mādhvīka cintā na bhavati bhavataḥ śarkare karkaśāsi*
drākṣe drakṣyanti ke tvām amṛta mṛtam asi kṣīra nīraṃ rasaste /
mākanda kranda kāntādhara dhara na tulāṃ gaccha yacchanti bhāvaṃ
yāvac chṛṅgāra-sāraṃ śubham iva jayadevasya vaidagdhya-vācaḥ //

32. *itthaṃ keli-tatir vihṛtya yamunā-kūle samaṃ rādhayā*
tad-romāvali-mauktikāvali-yuge veṇī-bhramaṃ bibhrati /
tatrāhlādi-kuca-prayāga-phalayor lipsāvator hastayor
vyāpārāḥ puruṣottamasya dadatu sphītāṃ mudāṃ saṃpadam //

Variant readings (following manuscripts cited in the Nirṇaya-Sāgara and Kulkarni editions):

I. 27: valad *for* calad.
33: mālati-jāti-sugandhau *for* mālikayātisugandhau.
37: kalakalair *for* kalaravair.
III. 16: gīta *for* dīpti.
V. 13: nidhānam *for* nidānam.
VI. title: sotkaṇṭha-vaikuṇṭhaḥ *for* dhanya-vaikuṇṭha-kuṅkumaḥ.
VII. 5: iha *for* iti.
14: calita *for* valita.

39: nidhāya *for* vidhāya.
VIII. 11: vyapohayatu vo'śreyāṃsi *for* vipolayatu vaḥ śreyāṃsi.
X. 8: dehi *for* dhehi.
9: jayatu *for* jayati.
XI. title: sānanda *for* sāmoda.
3: vikāram *for* nikāram.
8: nijagati *for* nigadita.
15: lasad *for* bhavad.
19: madana-rasa *for* kusuma-sara.
25: parilaṃbya *for* parirabhya.
33: samāloka *for* samāyāta.
XII. 8: visṛja *for* sṛjasi.
12: mārāṅke *for* vāmāṅge.
14: vyālolaḥ *for* vyākośaḥ.
26: bhūyo *for* bhūto.

BIBLIOGRAPHY OF WORKS CITED

Alphonso-Karkala, John B. (ed.), *An Anthology of Indian Literature*, Penguin edn, Harmondsworth, 1971.

Amaruśatakam with the commentary of Vemabhūpāla, ed. Chintaman Ramchandra Devadhar, Poona, 1959.

Anaṅgaraṅga of Kalyānamalla, ed. Viṣṇuprasāda Bhandārī, Benares, 1923.

Andreas Capellanus, *The Art of Courtly Love*, trans. John J. Parry, New York, 1941.

Aristotle, '*Art*' *of Rhetoric*, trans. J. H. Freese, Loeb Library, London, 1926.

———, *The Nicomachean Ethics*, trans. H. Rackham, London, 1934.

Arnold, Edwin, 'The Indian Song of Songs' in *Indian Poetry*, pp. 1–97, London, 1881.

Atharva Veda, with the commentary of Sāyaṇa, Bombay, 1895 (see Whitney and Griffith).

Augustine, *Confessions*, trans. E. B. Pusey, Everyman edn, London, 1907.

———, *Divine Providence and the Problem of Evil* (*De Ordine*), trans. Robert P. Russell in *The Writings of Saint Augustine*, vol. I, The Fathers of the Church Series, New York, 1948.

———, *On the Trinity* (*De Trinitate*), trans. A. W. Haddan, vol. VII of *The Works of Aurelius Augustine*, Edinburgh, 1873.

Bakan, David, *The Duality of Human Existence*, Boston, 1966.

Basham, A. L., *The Wonder That Was India*, Evergreen edn, New York, 1959.

Baudelaire, Charles, *Les Fleurs du Mal* in *Œuvres Complètes*, Pléiade edn, Paris, 1951.

Bayley, John, *The Characters of Love*, 2nd edn, London, 1968.

Behari, Bankey (trans.), *The Geet Govind*, Jodhpur, 1964.

Betjeman, John and Taylor, Geoffrey (eds), *English Love Poems*, London, 1957.

Bhagavad-Gītā, see Zaehner.

Bhāgavata-Purāṇa : Le Bhāgavata Purāṇa, Sanskrit text with French trans. by Eugene Burnouf, 5 vols, Paris, 1884.

Bhaktamālā of Nābhādās (Hindi), trans. into Sanskrit by Candradatta (*sargas* 39–41) in the Nirṇaya Sāgar Press edn of the *Gītagovinda* [q.v.].

Bhaktavijaya of Mahīpati (Marathi), trans. J. E. Abbot and Pandit N. R. Godbole, Poona, 1933.

Bhakti-rasāmṛta-sindhuḥ of Rūpa Gosvāmin, vol. I, with a trans. by

Swāmī Bhakti Hṛdaya Bon Mahārāj, Vrindaban, 1965. See also Murshidabad edn, 1924.

Bhaktiratnāvalī of Viṣṇupuri in the *Sacred Books of the Hindus*, vol. VII, Pt. 3 (Sanskrit text with anonymous English trans.), Allahabad, 1912.

Bharati, Agehananda, *The Tantric Tradition*, Anchor Books edn, New York, 1970.

Bhartṛhari, *Śatakatrayādi-subhāṣita-saṃgraha: The Epigrams attributed to Bhartṛhari*, ed. D. D. Kosambi, Bombay, 1948. (See also *Śṛṅgāra-Śataka.*)

Blake, William, 'The Marriage of Heaven and Hell', in *Selected Poems of William Blake*, ed. F. W. Bateson, London, 1957.

Bolle, Kees, *The Freedom of Man in Myth*, Nashville, 1968.

Bose, M. M., *The Post-Chaitanya Sahajia Cult of Bengal*, Calcutta, 1930.

Brahma-Sūtras, Sanskrit text with English trans. By Swami Vireswarananda, 2nd edn, Almora, 1948.

Brough, John, *Poems from the Sanskrit*, Penguin edn, Harmondsworth, 1968.

Brower, R. A. (ed.), *On Translation*, New York, 1966.

Brown, Norman O., *Life Against Death*, Sphere Books edn, London, 1968.

———, *Love's Body*, New York, 1966.

Buddha-carita of Aśvaghoṣa, ed. E. B. Cowell, Oxford, 1893.

Bühler, G. (trans.) *The Laws of Manu, Sacred Books of the East* XXV, Oxford, 1886.

Campbell, Joseph, *Hero with a Thousand Faces*, Meridian edn, New York, 1956.

——— (ed.), *The Mystic Vision, Papers from the Eranos Yearbooks*, vol. 6, Princeton, 1970.

Caṇḍīdāsa, *Love Songs of Chandidas*, trans. from Bengali by D. Bhattacharya, London, 1967.

Cassirer, Ernst, *The Philosophy of Symbolic Forms*, 3 vols, trans. Ralph Manheim, New Haven and London, 1955.

Chaitanya Charitamrita of Sri Sri Krishnadasa Kaviraya Gosvamin (Bengali), trans. Nagendra Kumar Ray, 2nd edn, 6 vols, Puri, 1959.

Chakravarti, Monmohan, 'Sanskrit Literature in Bengal during the Sena Rule', *Journal and Proceedings of the Asiatic Society of Bengal*, vol. II, no. 5, 1906.

Chakravarti, S. C., *Philosophical Foundation of Bengal Vaiṣṇavism*, Calcutta, 1969.

Chandra, Moti, 'Notes' to the Lalit Kalā Akademi, Portfolio of paintings from the *Gīta Govinda*, New Delhi, 1965.

Chattarji, Suniti Kumar, 'Jayadeva Kavi', *Acarya Dhruva Commemorial*, vol. III, pp. 183–96.

Courtillier, M. Gaston (trans.), *Le Gīta-Govinda*, Paris, 1904.

Cowell, E. B. and Thomas, E. J. (trans.), The *Harṣacarita* of Bana, London, 1929.

Daniélou, Alain, *Hindu Polytheism*, New York, 1963.

Daniélou, Jean (ed.), *From Glory to Glory, Texts from Gregory of Nyssa's Mystical Writings*, London, 1962.

Dante Alighieri, *La Vita Nuova*, trans. Barbara Reynolds, Penguin edn, Harmondsworth, 1969.

Daśarūpa of Dhanaṃjaya, ed. with a trans. by G.C.O. Haas, Delhi, 1962.

Das Gupta, S. B., *Obscure Religious Cults*, 2nd edn, Calcutta, 1962.

Das Gupta, S. N., *Hindu Mysticism*, 2nd edn, New York, 1959.

———, *A History of Indian Philosophy*, 5 vols, Cambridge, 1951–5.

De, S. K., 'Ancient Indian Erotics', published with *Treatment of Love in Sanskrit Literature*, as *Ancient Indian Erotics and Erotic Literature*, Calcutta, 1959.

———, *Bengal's Contribution to Sanskrit Literature and Studies in Bengal Vaiṣṇavism*, Reprint, Calcutta, 1960.

———, 'The Doctrine of Avatāra (Incarnation) in Bengal Vaiṣṇavism', in *Bengal's Contribution to Sanskrit Literature* [q.v.], pp. 143–53.

———, *Early History of the Vaiṣṇava Faith and Movement in Bengal*, 2nd edn, Calcutta, 1942.

———, *Studies in the History of Sanskrit Poetics*, 2 vols, London, 1923.

———, *Treatment of Love in Sanskrit Literature*, 2nd edn, Calcutta, 1959. (See *Ancient Indian Erotics and Erotic Literature*.)

De Bary, W. T. (ed.), *Sources of the Indian Tradition*, vol. I, New York, 1958.

De Rougemont, Denis, *Love Declared* (trans. Richard Howard), Beacon Paperback edn, Boston, 1964.

———, *Love in the Western World* (trans. Montgomery Belgion), Fawcett Premier edn, New York, 1966.

Deussen, Paul, *The Philosophy of the Upanishads* (trans. A. S. Geiden), Dover edn, New York, 1966.

Dikshitar, V. R. R., *The Purāṇa Index*, 3 vols, Madras, 1955.

Dimock, Edward, 'Doctrine and Practice among the Vaiṣṇavas of Bengal' in Milton Singer, *Krishna: Myths, Rites, and Attitudes* (Chicago, 1968) [q.v.], pp. 41–63.

———, *The Place of the Hidden Moon*, Chicago, 1966.

Dimock, E. C. and Levertov, D., *In Praise of Krishna*, New York, 1967.

Donne, John, *Complete Poems*, Everyman edn, London, 1931.

Eastman, A. M. *et al.*, *The Norton Anthology of Poetry*, New York, 1970.

Edgerton, Franklin, *The Beginnings of Indian Philosophy*, London, 1965.

———, 'The Hindu Song of Songs' in Wilfred Schoff (ed.), *The Song of Songs: A Symposium* [q.v.], pp. 41–7.

Eliade, Mircea, *Patterns in Comparative Religion* (trans. Rosemary Sheed), Meridian edn, New York, 1963.

———, *Rites and Symbols of Initiation* (trans. Willard R. Trask), Harper Torchbook edn, New York, 1965.

———, *The Sacred and The Profane* (trans. Willard R. Trask), New York, 1959.

———, *Yoga, Immortality and Freedom* (trans. Willard R. Trask), 2nd edn, Princeton, 1969.

Empson, William, *Seven Types of Ambiguity*, New Directions, New York, n.d.

Fang, Achilles, 'Some reflections on the difficulty of translation', in R. A. Brower, ed., *On Translation* [q.v.], pp. 111–13, 1966.

Fletcher, Angus, *Allegory*, New York, 1967.

Forster, Leonard (ed. and trans.), *The Penguin Book of German Verse* (rev. edn), Harmondsworth, 1959.

Freud, Sigmund, *Beyond the Pleasure Principle* (trans. and ed. James Strachey), Bantam Classic edn, New York, 1959.

———, *Civilization and Its Discontents*, trans. J. Riviere, London, 1930.

———, *Collected Papers* (ed. and trans. Joan Riviere and James Strachey), 5 vols, London–New York, 1924–50.

———, *A General Introduction to Psychoanalysis*, trans. Joan Riviere, rev. edn, 1952.

———, *Group Psychology and the Analysis of the Ego* (1920), *The Complete Psychological Works of Sigmund Freud* (standard edn), (trans. and ed. James Strachey), London and New York.

———, *New Introductory Lectures on Psychoanalysis*, trans. and ed. James Strachey, New York, 1964.

Gardner, Helen (ed.), *The New Oxford Book of English Verse*, Oxford, 1972.

Gerow, Edwin, *A Glossary of Indian Figures of Speech*, The Hague, 1971.

Gītagovinda of Jayadeva with the commentary *Bālabodhinī* of Caitanyadāsa [also called Pujārī Gosvāmin], Calcutta, 1872.

Gītagovinda of Jayadeva, ed. Telang and Panshikar, with the commentaries of Kumbha (*Rasikapriyā*) and Śaṅkara Miśra (*Rasamañjarī*), Bombay, 1949.

Gītagovinda of Jayadeva, ed. V. M. Kulkarni with the commentary of King Mānāṅka, Ahmedabad, 1965.

Gītagovinda of Jayadeva, ed. A. Sharma, K. Deshpande and V. Sundara Sharma, with the commentaries: *Sañjīvanī* of Vanamālibhaṭṭa, *Padayotasikā* of Nārāyaṇa Paṇḍita, and *Jayantī* of Kṛṣṇa; Hyderabad, 1969.

Gītagovinda, see also Arnold, Behari, Courtillier, Greenlees, Jones, Keyt, Lassen, Varma.

Goethe and Schiller, *Correspondence Between Goethe and Schiller* (trans. L. Dora Schmitz), 3 vols, London, 1909.

Goethe, Johann Wolfgang, *Faust*, trans. Philip Wayne, Penguin edn, Harmondsworth, 1949 (Part I) and 1959 (Part II).

Gombrich, Richard, 'The Fifty Stanzas of a Thief' (a translation of the *Caura-pañcāśikā* attributed to Bilhana) in *Mahfil*, vol. VII, nos. 2 & 3, East Lansing, Michigan, 1971, pp. 175–86.

Gonda, J., *Aspects of Early Viṣṇuism*, 2nd edn, Delhi, 1969.

Gottfried von Strassburg, *Tristan*, trans. A. T. Hatto, Penguin edn, Harmondsworth, 1960.

Govinda, Lama Anagarika, *Foundations of Tibetan Mysticism*, U.S. edn, New York, 1969.

Greenlees, Duncan, *The Song of Divine Love* (a translation and commentary on the *Gītagovinda*), Madras, 1962.

Gregory of Nyssa, see Jean Daniélou.

Griffith, Ralph T. H. (trans.), *The Hymns of the Atharva-Veda* [q.v.], Benares, 1916.

Guruge, Ananda, *The Society of the Rāmāyaṇa*, Maharagama, Ceylon, 1960.

Halm, Friedrich (pseud.), *Der Sohn der Wildniss*, Vienna, 1845.

Harding, M. Esther, *Woman's Mysteries*, Rider edn, New York, 1971.

Hastings, James (ed.), *Encyclopedia of Religion and Ethics*, 12 vols, London, 1908–1921.

Hesiod, *Theogony*, trans. Norman O. Brown, New York, 1953.

Hevajra Tantra, ed. and trans. D. L. Snellgrove, 2 vols, London, 1959.

Hill, R. T. and Bergin, T. G. (ed. and trans.), *Anthology of Provençal Troubadours*, New Haven, 1957.

Honig, Edward, *Dark Conceit: The Making of Allegory*, London, 1959.

Hopkins, E. W., *Epic Mythology*, Indian edn, Varanasi and Delhi, 1968.

Hugh of Saint-Victor, *Selected Spiritual Writings*, trans. Aelfred Squire, London, 1962.

Huizinga, Johan, *Homo Ludens*, Paladin edn, London, 1970.

———, *The Waning of the Middle Ages* (trans. F. Hopman), London, 1924.

Hunt, Morton, *The Natural History of Love*, New York, 1967.

Ingalls, Daniel H. H., *Anthology of Sanskrit Court Poetry* (a translation of the *Subhāṣitaratnakoṣa* [q.v.] of Vidyākara), Cambridge, Mass., 1965 (Harvard Oriental Series 44).

———, 'Foreword' to Milton Singer (ed.), *Krishna: Myths, Rites, and Attitudes* [q.v.], pp. v–xi.

———, 'Words for Beauty in Classical Sanskrit Poetry', *Indological Studies in Honor of W. Norman Brown*, American Oriental Society, New Haven, 1962, pp. 87–107.

James, William, *The Varieties of Religious Experience*, Mentor edn, New York, 1958.

John of the Cross, St., *The Complete Works of Saint John of the Cross*, ed. and trans. E. Allison Peers, 3 vols, 2nd edn, London, 1953.

———, *Poems*, ed. with translations by Roy Campbell, Penguin edn, Baltimore, 1960.

Jones, Sir William (trans.), *The Gītagovinda*, Asiatick Researches, vol. III, pp. 185–207, Calcutta, 1792.

Jung, C. G., *Man and His Symbols*, Laurel edn, New York, 1968.

———, *Memories, Dreams, Reflections*, trans. Richard and Clara Winston, Fontana edn, London, 1967.

Kāma-Sūtra of Vātsyāyana with the commentary *Jayamaṅgala* of Yaśodhara, Benares, 1912.

Kāvyamīmāṃsā of Rājaśekhara, Baroda, 1916. (See Stchoupak.)

Kāvyaprakāśa of Mammata, ed. with an English trans. by Ganganatha Jha, Varanasi, 1967.

Kay, George (ed.), *The Penguin Book of Italian Verse*, revised edn, Harmondsworth, 1965.

Keith, Arthur Berriedale, *History of Sanskrit Literature*, Oxford, 1928.

———, *Sanskrit Drama*, Oxford, 1924.

Kennedy, Melville T., *The Chaitanya Movement*, Calcutta, 1925.

Keyt, George (trans.), *Sri Jayadeva's Gita Govinda: The Loves of Kṛṣṇa and Rādhā*, Bombay, 1940.

Konow, Sten, 'Anaṅga, the Bodiless Cupid', *Festschrift Jacob Wackernagel*, pp. 1–8, Göttingen, 1923.

Krishnamachariar, M., *History of Sanskrit Literature*, Delhi, 1970.

Kṛṣṇakarṇamṛta of Līlāśuka Bilvamaṅgala, crit. ed. with an English trans. by Frances Wilson, Philadelphia, 1975.

Lakṣmī Tantra, ed. with an English trans. by S. Gupta, Leiden, 1972.

Lal, Kanwar, *The Religion of Love*, Delhi, 1971.

Lassen, Christianus (trans. and ed.), *Gītā Govinda: Jayadevae Poetae Indici Drama Lyricum*, Bonn, 1836.

Lewis, C. S., *The Allegory of Love*, London, 1936.

Longus, *Daphnis and Chole*, trans. Paul Turner, Penguin edn, Harmondsworth, 1968.

Macauliffe, M. A., *The Sikh Religion*, 6 vols, Oxford, 1909.

Macdonell, A. A., *A History of Sanskrit Literature*, London, 1900.

———, *Vedic Mythology*, Strasbourg, 1897.

———, *A Vedic Reader for Students*, Oxford, 1917.

Mahābhārata, critical edn, Poona, 1933–69.

Mahapatra, K. N., 'New Light on Poet Jayadeva, The Author of the *Gītagovinda*', *Orissa Historical Research Journal*, vol. VII, pp. 191 ff.

Majumdar, A. K., *Caitanya: His Life and Doctrine*, Bombay, 1969.

Majumdar, Bimanbehari, *Kṛṣṇa in History and Legend*, Calcutta, 1969.

Majumdar, R. C., *History of Ancient Bengal*, Calcutta, 1971.

Manu-smṛti, ed. G. C. Haughton, London, 1825. (See also Bühler.)

Matsya Purāṇa, Ānandāśrama Sanskrit Series 54, Poona, 1907.

Meyer, Johan Jakob, *Sexual Life in Ancient India*, London, 1931.

Michaux, Guy, *Message Poétique du Symbolisme*, Paris, 1947.

Mishra, H. R., *Theory of Rasa in Sanskrit Drama*, Chhatarpur, 1964.

Mishra, K. C., *The Cult of Jagannātha*, Cuttack, 1971.

Monier-Williams, Monier, *Sanskrit-English Dictionary*, Oxford, 1899.

Morgan, Bayard Q., 'Bibliography' to R. A. Brower, ed., *On Translation* [q.v.], pp. 271–93, 1966.

Mukherji, M., 'Jayadeva, Poet and Mystic', *Journal of the Department of Letters*, University of Calcutta, 1935.

Mukherji, S. C., *A Study of Vaiṣṇavism in Ancient and Medieval India*, Calcutta, 1966.

Nāradīya-bhakti-sūtra, *Sacred Books of the Hindus*, vol. VII, Pt. 1, Allahabad, 1911. (Sanskrit text with an anonymous English translation.)

Nāṭyaśāstra of Bharatamuni with the commentary of Abhinavagupta, 3 vols, Baroda, 1934–56 (Gaekwad's Oriental Series).

Nicholson, R. A., *Studies in Islamic Mysticism*, Cambridge, 1921.

Nida, Eugene A., *Toward a Science of Translating*, Leiden, 1964.

Nietzsche, Friedrich, *Thus Spake Zarathustra*, trans. A. Tille, Everyman edn, London, 1933.

Nygren, Anders, *Agape and Eros*, trans. Philip S. Watson, Harper Torchbook edn, New York, 1969.

O'Flaherty, Wendy, *Asceticism and Eroticism in the Mythology of Śiva*, London, 1973.

———, 'Asceticism and Sexuality in the Mythology of Śiva', Pt. I, *The History of Religions* VIII, no. 4, May, 1969, pp. 300–37.

———, *Hindu Myths*, Penguin edn, Harmondsworth, 1975.

Origen, *The Song of Songs, Commentary and Homilies*, trans. R. P. Lawson, London, 1957.

Ovid, *Heroides and Amores*, ed. and trans. Grant Showerman, Loeb Classical Library, London, 1914.

Padyāvalī, compiled by Rūpa Gosvāmin, crit. ed. S. K. De, Dacca, 1934.

Pandey, S. M. and Zide, Norman, 'Surdas and his Krishna-*bhakti*', in Milton Singer, ed., *Krishna: Myths, Rites, and Attitudes* [q.v.], pp. 173–99.

Pavana-dūtam of Dhoyīka, ed. Monmohan Chakravarti with an appendix on the Sena Kings, *Journal of the Asiatic Society of Bengal*, vol. I, no. 3, 1905.

Perella, N. J., *The Kiss Sacred and Profane*, Berkeley and Los Angeles, 1969.

Pischel, *Die Hofdichter des Lakṣmaṇasena*, Göttingen, 1894.

Plato, *Euthyphro, Crito, Apology and Symposium* (the Howett trans.), Gateway edn, Chicago, 1953.

Prabodha-candrodaya of Kṛṣṇamiśra, ed. V. L. Pansikar, 2nd edn, Bombay, 1904.

Rāmāyaṇa of Valmīki, Baroda, 1960.

Randhawa, M. S., *Kangra Paintings of the Gīta Govinda*, New Delhi, 1963.

———, *Kangra Paintings on Love*, New Delhi, 1962.

Rasika-rañjana of Rāmacandra—*Rāmachandra's Ergotzen der Kenner*, ed. and trans. R. Schmidt, Stuttgart, 1896.

Reese, G., *Music in the Middle Ages*, London, 1940.

Ṛg Veda, with the commentary of Sāyaṇa, 6 vols, London, 1890–92. (See also Wilson.)

Rilke, Rainer Maria, *Duino Elegies*, trans. J. B. Leishman and Stephen Spender, 4th edn, London, 1963.

Robert, A. and Tournay, R., *Le Cantique des Cantiques*, Paris, 1963.

Robinson, H. Wheeler, 'The Soul (Christian)', in James Hastings (ed.), *Encyclopedia of Religion and Ethics* [q.v.], vol. XI, pp. 733–7.

Rūpa Gosvāmin, see *Bhakti-rasāmṛta-sindhu*, *Padyāvalī* and *Ujjvala-nīla-maṇi*.

Sadukti-Karṇāmṛta of Śrīdharadāsa, crit. ed. S. C. Banerji, Calcutta, 1965.

Sadukti-Karṇāmṛta of Śrīdharadāsa, ed. Paṇḍit Rāmāvatāra Śarmā, Lahore, 1933.

Sāhityadarpaṇa of Viśvanātha Kavirāja, Benares, 1947–8.

Saraha, see Shahidullah.

Sarkar, Ranajit, *Gītagovinda: Towards a Total Understanding*, Publikaties van het Instituut voor Indische talen en culturen No. 2, Groningen, 1974.

Sattasaī of Hāla, *Saptaśatakam des Hāla*, ed. Albrecht Weber, Leipzig, 1870.

Saundaryalaharī (*Flood of Beauty*), ed. and trans. W. Norman Brown, Cambridge, Mass., 1958 (Harvard Oriental Series 43).

Saville, Jonathan, *The Medieval Erotic Alba*, New York, 1972.

Schmidt, Richard, *Beiträge zur Indischen Erotik*, Leipzig, 1902.

Schoff, Wilfred (ed.), *The Song of Songs: A Symposium*, Philadelphia, 1924.

Sekaśubhodaya (*Holy Advent of the Sheik*) of Halāyudha, ed. with an English trans. by Sukumar Sen, Calcutta, 1963.

Sen, Sukumar, *History of Bengali Literature*, New Delhi, 1960.

Shahidullah, M., ed. and trans., *Les Chants Mystiques de Kāṇha et Saraha*, Paris, 1928.

Sharma, Suresh Candra, *Tulsidas*, Bombay, 1974(?). (An Amar Chitra Katha 'comic book'.)

Singer, Milton (ed.), *Krishna: Myths, Rites, and Attitudes*, Phoenix edn, Chicago, 1968.

———, 'The Rādhā-Krishna *Bhajanas* of Madras City', in Singer (ed.), *Krishna: Myths, Rites, and Attitudes* [q.v.], pp. 90–138.

Śiśupālavadha of Māgha, Kashi Sanskrit Series 69, Benares, 1929.

Spink, Walter M., *Krishnamandala*, Ann Arbor, 1971.

Śṛṅgāra-śataka of Bhartṛhari, Varanasi, 1967.

Śṛṅgāra-vairāgya-taraṅgiṇī of Somaprabhācārya, ed. *Kāvyamālā*, Gucchaka v, 2nd edn, 1908, pp. 124–42.

Stchoupak, Nadine and Renou, Louis (trans.), Rājaśekhara's *Kāvyamīmāṃsā* [q.v.], Paris, 1946.

Stendhal, *On Love* (trans. H.B.V.), Universal Library edn, New York, 1967.

Subhāṣitaratnakoṣa of Vidyākara, ed. D. D. Kosambi and V. V. Gokhale, Cambridge, Mass., 1957 (Harvard Oriental Series 42). (See Ingalls.)

Tantratattva, see Woodroffe (*Principles of Tantra*).

Teresa of Jesus, *The Complete Works of Saint Teresa of Jesus*, 3 vols, trans. E. Allison Peers, London, 1946.

Thompson, E. J. and Spencer, A. M. (trans.), *Bengali Religious Lyrics, Śākta*, Calcutta, 1923.

Ujjvala-nīla-maṇi of Rūpa Gosvāmin, Kāvyamālā edn, Bombay, 1913.

Underhill, Evelyn, *Mysticism*, Dutton Paperback edn, New York, 1961.

Upaniṣads: *Eighteen Principal Upaniṣads*, vol. I, ed. V. P. Limaye and R. D. Vadekar, Poona, 1958.

Urban, Wilbur, *Language and Reality*, New York, 1939.

Uttara-rāma-carita of Bhavabhūti, ed. G. K. Bhat, Surat, 1965.

Valency, Maurice, *In Praise of Love*, New York, 1958.

Varma, Monika (trans.), *The Gita Govinda of Jayadeva*, Calcutta, 1968.

Vaudeville, Charlotte, 'Evolution of Love-Symbolism in Bhāgavatism', *Journal of the American Oriental Society* LXXXII (1962), pp. 31–40.

———, *Kabīr*, Oxford, 1974.

Vedārthasaṃgraha of Rāmānuja, ed. with an English trans. by J. A. B. van Buitenen, Poona, 1956.

Viṣṇu Purāṇa, Bombay, 1902.

Waley, Arthur, *The Way and Its Power*, London, 1934.

Walker, Benjamin, *Hindu World*, 2 vols, London, 1968.

Warder, A. K., *Indian Kāvya Literature*, vol. I, Delhi, 1972.

Wellek, René and Warren, Austin, *Theory of Literature*, New York, 1949.

Wheelwright, Philip, *Metaphor and Reality*, Bloomington, 1967.

Whitman, Walt, *A Choice of Whitman's Verse*, selected by Donald Hall, London, 1968.

———, *Complete Poetry*, ed. E. Holloway, London, 1938.

Whitney, W. D. (trans.), *Atharva Veda Samhita* [q.v.], 2 vols, Cambridge, Mass., 1905.

Wilhelm, James, *The Cruelest Month: Spring, Nature, and Love in Classical and Medieval Lyrics*, New Haven and London, 1965.

Wilson, H. H., (trans.), *Ṛig-Veda Sanhita* [q.v.], 6 vols, London, 1886–88.

———(trans.), *The Vishṇu Purāṇa* [q.v.], 3rd edn, Calcutta, 1961.

Wimsatt, William, *The Verbal Icon*, British edn, London, 1970.

Wimsatt, William and Brooks, Cleanth, *Literary Criticism: A Short History*, New York, 1957.

Woḷedge, Brian (ed.), *The Penguin Book of French Verse*, vol. I (text and translations), Harmondsworth, 1961.

Woodroffe, Sir John, *Introduction to the Tantra Śāstra*, 5th edn, Madras, 1969.

———, *Principles of Tantra* (trans. and commentary on the *Tantratattva*), 2nd edn, Madras, 1955.

———, *The Serpent Power*, 5th edn, Madras, 1953.

Woods, James Haughton, *The Yoga-System of Patañjali*, Harvard University Press, Cambridge, Mass., 1927.

Yoga-sūtras of Patañjali, see Woods.

Zaehner, R. C. (trans.), *The Bhagavad-Gītā*, Oxford, 1969.

———, *Concordant Discord*, Oxford, 1970.

———, *Hinduism*, OUP Paperback edn, Oxford, 1966.

———(trans. and ed.), *Hindu Scriptures*, London, 1966.

———, *Mysticism, Sacred and Profane*, New York, 1961.

Zimmer, H., *The Art of Indian Asia*, 2 vols, New York, 1955.

———, 'The Indian World Mother', in *The Mystic Vision, Papers from the Eranos Yearbooks*, vol. 6, ed. Joseph Campbell, Princeton, 1970 [q.v.],

———, *Myths and Symbols in Indian Art and Civilization*, Harper Torchbook edn, New York, 1962.

Zohar, trans. Harry Sperling, Maurice Simon, Paul Levertov, 5 vols, London, 1931–34.

INDEX

Abhinavagupta, 52, 158
Ādi Granth, 107, 152
adultery, 52, 113–20. *See also* marriage
agape, 4, 6, 68
allegory, xii, 5, 29, 178–93
Āḻvārs, 23
amor, 1, 143
Amor, 72, 79
Arabic poetry, 31, 148
Aristotle, 8, 158
Arjuna, 89, 201
Arnold, Edwin, 180, 236
artha (material gain), 18, 64–5, 162, 222
asceticism (*tapas*), 17, 19, 54, 65, 72, 73, 77, 85, 107–8, 143, 151, 215, 222. *See also yoga*
āśramas (stages of life), 221–2
ātman. *See* self
attachment, 14, 21, 41, 66, 140, 204
Augustine, xiii, 1, 8, 69
avatāras, ('descents'), 28, 101–4, 166–7, 174, 197, 211, 214, 220, 241–3

Baudelaire, Charles, 46, 203
beauty, 19, 45, 79, 94–7, 111–13, 132, 142, 172, 192, 226
Bengal, 209, 220
Bengali Vaiṣṇavism. *See* Vaiṣṇavism; Caitanya; Sahajiyā
Bhagavad Gītā, 13, 66–7, 80, 104, 158
Bhagavat. *See* Lord
Bhāgavata Purāṇa. *See Purāṇas*
Bhaktamāla, 213, 222
bhakti/bhakta (devotion/devotee), xii, 14, 19, 21, 25, 37–9, 42, 51, 53, 61–2, 66–7, 70, 73, 78–9, 90, 96–7, 106, 108–9, 117–18, 128, 135–6, 139, 142, 145–51, 155–8, 168, 172, 175–6, 178, 184–6, 190, 198, 200, 208–9, 213–14, 216–19, 221, 223, 225–32
Bhaktivijaya, 156, 214
Bhakti Sūtras, 140, 153
Bharata, 52
Bhāravi, 163
Bhartṛhari, 19–20, 74, 131, 153
bhāva (emotion), 47–8, 50, 52–3, 58–9, 61, 63, 117, 129, 136, 163, 186, 192, 237. *See also rasa*
bhayānaka rasa. *See rasa*
Bhīma, 65
Blake, William, 12
Boehme, Jacob, 176
Brahmā, 74, 76–7, 108, 143, 197
brahman, 13–15, 24, 49, 52–3, 62, 103–4, 106, 122, 128, 152–3, 160, 168–9, 182–3, 194–5, 204
Brahmin caste, 206, 211
Bṛhadāraṇyaka Upaniṣad. *See Upaniṣads*
Buddha and Buddhism, 16, 25, 40–1, 68, 80–1, 188, 211, 220, 243*n*

Caitanya, xi, 23–5, 27, 53, 59, 68, 83–4, 95, 108, 118, 129, 135–6, 139, 143–4, 155, 176, 184–8
Caitanya-Caritāmṛta, 50, 186
chastity. *See* asceticism
Caṇḍīdāsa, 25–6, 54, 71, 120, 136, 184, 191–2
Candradatta, 229
caritas, 8, 69
caryāpadas, 188, 196
Christ, 5, 85, 128, 147, 184, 187, 203
Christianity, 100, 132, 146–8, 187
Church, 3, 5
coital postures. *See* sexuality
cupiditas, 8, 69

Dāmodaragupta, 170
Dante, Alighieri, 8–9, 142–3
Death and love, 3, 44–5, 80–3, 146–9 164, 170. *See also* love
defilements (*kleśa*), 61, 63
Degas, Edgar, 33
de Rougemont, Denis, 137
desire. *See kāma*, love
Devī. *See* Goddess
Devī-Bhāgavata Purāṇa. *See Purāṇas*
Dhanaṃjaya, 113, 161, 171
dharma (law, duty, religion), 18, 26,

64–5, 71, 114–15, 117–18, 162, 188, 226; *svadharma*, 18, 53
Dhoyī, 210–11
Donne, John, 30–1, 170

Empson, William, 199
Epics, Indian, 64–5, 74, 76, 86–7, 112, 206; *Mahābhārata*, 18, 65, 85, 89, 112, 209; *Rāmāyaṇa*, 18, 112, 119, 141, 209, *see also* Rāma
Eros/*eros*, 4, 6–10, 17, 68, 75, 80–1, 93
erotic texts (*kāma-śāstra*), 110–11, 134, 149, 196, 154–5, 161–71, 206; *Kāma-sūtra*, 18, 49, 100, 104, 161, 166; Vātsyāyana, 59, 134, 144, 149, 162, 164–7, 169, 171
European literature, 2–3, 6–7, 29, 33, 72–4, 81, 86–7, 92–3, 100, 111, 116–18, 120–21, 128, 137–8, 142, 146–8, 161, 170, 181, 187, 202–3

Fitzgerald, Edward, 234
folk tradition, Indian, 21, 36, 91, 140
Freud, Sigmund, 9, 56, 77–8, 80, 100, 120, 132, 173–4

Gaṅgāstava-prabandha, 125–7
Ganges, 218
Gītagovinda, *passim*; allegorical approaches to, 178–93; Kāmadeva in, 73–81, 83–4, 86–8; Kṛṣṇa in, 90–110 *passim*; love-in-separation in, 142–59 *passim*; Rādhā in, 110–134 *passim*; *rasa* in, 45–6, 50–2; sacred and profane love in, 27–43; *sakhī* in, 134–7; symbols, 193–206; translation of, 240–87; transliterated text of, 287–313; union in, 163–77 *passim*; vocabulary of love used, 58–9, 63–4, 67–70
God. *See* Lord
Goddess, 56, 93–4, 111, 120–31, 211; Caṇḍī, 131; Devī, 39–40, 94, 122, 129, 131; Gaṅgā, 125–7, 218; Kālī, 130–1; Lakṣmī/Śrī, 122–4, 197, 202, 204, 219–20, 222, 226; Sarasvatī (Vāc/Bhāratī), 208–9. *See also* Rādhā; *śakti*; woman
Gopīs (cowherd-women), 22, 25, 28, 37–8, 63, 66, 69, 84, 89, 91–2, 101, 118, 127, 139, 144, 146, 166, 174, 181
Gosvāmins, 145. *See* Jīva Gosvāmin; Rūpa Gosvāmin
Govardhana, 50, 210
Govindadāsa, 140
Gower, John, 3
Gregory of Nyssa, 6, 138, 187
guṇa-kīrtana (chanting the qualities), 39, 185
guru, 216

Hāla, 140
haṃsa (goose), 106
Harivaṃśa, 91, 197
Harṣacarita of Bāṇa, 207
hāsya rasa. *See rasa*
Hesiod, 75
Hiraṇyakaśipu, 166
hlādinī-śakti, 24, 109–11, 129, 186. *See also śakti*

Jagannātha, 189, 220, 222, 224, 225, 228–32
japa, 67, 73
Jayadeva, *passim*; hagiographical legends, 213–33; court poet, 206–13
jīva (soul), 109, 160–1, 183, 186. *See also* self
Jiva Gosvāmin, 62, 109
John of the Cross, 118, 147
Johnson, Samuel, 234
Jones, William, 179, 235–6.
Judaism, 11, 129
Jung, C. G., 75

Kabbalah, 11, 129
Kabīr, 228
Kālī. *See* Goddess
Kali Era, 98, 105, 156–9, 200–1, 213–15, 220, 231, 243*n*
kāma (desire), 17–18, 64–9, 72–3, 76, 78–9, 151, 162, 222, 231
Kāma-deva, 59, 71–88, 100, 134, 173, 245*n*
Kāma Śāstra/Sūtra. *See* erotic texts
Kaṃsa, 99, 100, 130, 164, 174, 253–4*n*
karuṇa. *See rasa*; love-in-separation
kavi (poet), 206–13, 218

kāvya, 19, 27, 29, 31–5, 40, 47–52, 57, 66, 69, 72, 74, 76, 87, 91, 93–4, 96, 100, 102, 110–11, 113, 116, 121–2, 134, 142–3, 145–6, 150, 158, 160–2, 168, 170–1, 178, 198–9, 201–2, 204, 206–13, 226, 234–5
Kenduli village, 209, 218, 223, 230, 255*n*
Kierkegaard, Søren, 3
Kṛṣṇa, *passim*; 89–110
Kṛṣṇadāsa Kavirāja, 84, 184
Kubjā, 60
Kumbha, 51, 69, 86, 128
kuṇḍalinī, 189–90
Kuvalayāpīḍa, 99

Lakṣmaṇasena, 101, 209, 211–12, 220, 225, 231–2
Lakṣmī. *See* Goddess
Lassen, Christian, 178
liberation (*mokṣa*), xii, 13–14, 19, 21–3, 25–6, 41–2, 60–2, 68, 70, 78, 85, 90, 108, 133, 143, 148, 156–8, 160–1, 188, 200–1, 204, 221–2
libido, 9, 56, 93
līlā. *See* play
Līlāśuka Bilvamaṅgala, 98
Lord (*Bhagavat*), 14, 62, 70, 88, 102–3, 108–9, 150, 155, 159, 163, 168–70, 174–5, 178, 182, 184, 186–7, 191–2, 203, 215–18, 220, 222–3, 225, 227–8, 231
lotus. *See* symbols
love, sacred and profane, *passim*; as aesthetic experience, 46–52; allegory of, 178–93; in *Gītagovinda*, 27–43; Indian attitudes toward, 13–27; mythology of, 71–89; physiological experience of, 43–4; religious experience of, 52–7; symbols of, 193–206; Western attitudes toward, 1–13; Sanskrit words for, 57–71. *See also agape; amor; caritas; cupiditas*; death and love; *eros*; *kāma*; Kāmadeva; love-in-enjoyment; love-in-separation; *prema(n)*; *prīti*; *rāga*; *rasa*, *śṛṅgāra*; *rati*, sexuality; *sneha*; war and love
love-in-enjoyment/union (*saṃbhoga*), xii, 35, 37, 48, 63, 70–1, 119, 129, 138, 146, 149, 151, 153, 158, 159–78 *passim*, 200
love-in-separation (*viraha/vipralambha*), xii, 35, 37, 48, 66, 70–1, 78–9, 83, 87, 93, 117, 137–59, 173, 198
Luther, 5

magic, 15
Mahābhārata. *See* Epics
Mahīpati, 213–14, 220
Maitrī Upaniṣad. *See Upaniṣads*
makara (crocodile), 74, 76, 83–4
Mallarmé, Stephane, 33
māna (pique), 71, 144–5, 150. *See* love-in-separation
Mānāṅka, 82, 123, 151
mantra, 73, 84, 151, 196
Māra, 80–1. *See also* Kāma-deva
Marie of Champagne, 113–14
Mary the Madonna, 120–1, 132, 203
marriage, 18, 113–20, 215, 220–7
Matsya Purāṇa. *See Purāṇas*
māyā (illusion), 13, 40, 54, 106, 109, 160, 201, 216
meditation, 30, 54, 73, 79, 105, 150, 151, 196, 204
metaphor, 198–99
Mīrābāī, 23
mother, 56, 93, 111, 120, 132–3. *See also* Goddess; woman

Nābhādās, 213, 219
Nabokov, Vladmir, 234
nāgaraka ('man-about-town'), 162–3
nāyaka (hero), 35, 94–6, 145, 150, 192
nāyikā (heroine), 35, 96, 110–34, 143–4, 150, 192; *abhisārikā*, 115–16, 118–19, 150; *padminī*, 112, 196; *parakīyā* and *svakīyā*, 116–20, *see also* adultery; *virahiṇī/viyoginī*, 149–53, *see also* love-in-separation. *See also* woman; Rādhā
Nanda, 119–20, 130
Neoplatonists, 8–9
nirvāṇa, 17, 41–2, 152, 163. *See also* liberation; *samādhi*

Nṛsiṃhatapinī Upaniṣad. See Upaniṣads
Nygren, Anders, 4–5

Oedipus complex, 120. *See also* Freud, Sigmund
Origen, 5
Orissa, 189, 209, 227–32
Ovid, 100

Pañcarātra literature, 127
Padmāvatī, 125, 215, 218–27, 230–2, 240*n*
Patañjali, 61. *See also yoga*, Sāṃkhya
paramātman, 103–4. *See also brahman*
Paul, 3–5, 7, 68
Paulinus of Nola, 84–5
Pavanadūta, 211–12
Petersen, Gerlac, 2
Plato, 4, 7–10, 17, 68–9, 142, 176
play, 62, 73, 77, 122, 127–8, 135–6, 161, 171–7, 185–6, 203
prakṛti (matter), 13, 122, 133, 160, 169, 183
pravāsa. *See* love-in-separation
prema(n) (love), 26, 62, 68–70, 109, 129
prīti (delight of love), 69–70
Purāṇas, xii, 21, 37, 39, 74, 76, 88, 92–3, 95–6, 105, 124, 139, 156, 206, 213–14; *Bhāgavata*, 37–9, 51, 91–2, 102–3, 108, 139, 158, 228; *Devī-Bhāgavata*, 121; *Matsya*, 83, 106; *Viṣṇu*, 37, 39, 91–2, 123–4, 127, 139, 158, 204
puruṣa, 133, 160–1, 169, 183
pūrva-rāga. *See* love-in-separation
Pūtanā, 130–1, 133, 271*n*

Rādhā, *passim*; 110–34
rāga (passion), 61–4, 70–1
rajas, 61
Rājaśekhara, 207
Rāma, 95, 119, 123, 141. *See* Epics (*Rāmāyaṇa*)
Rāmāyaṇa. *See* Epics
rasa (flavour, mood, taste), xii, 42–57, 59, 84, 92, 117, 175, 184, 190, 231, 240*n*, 246*n*, 259*n*; *bhakti rasa*, 53–4, 58; *bhayānaka rasa* (fearful mood), 47, 200; *hāsya rasa* (comic mood), 47, 52, 145; *karuṇa rasa* (tragic mood), 47, 52, 146; *śānta rasa* (peaceful mood), 47, 89*n*; *śṛṅgāra rasa* (erotic mood), 35, 47–8, 50, 52–3, 58, 89–90, 94, 98–9, 101, 129, 145–6, 167, 173–4, 198, 200; *vīra rasa* (heroic mood), 47, 89, 98–9, 101, 167, 174
rāsa dance, 37, 45, 121, 158, 176
Rasaratnākara, 70
Rasika (person of taste), 47, 50–1, 58, 62, 70, 99, 157, 182, 184, 192, 226
rati (love-pleasure), 58–61, 71. *See bhāva*
Ratnasāra, 117
Rāvaṇa, 119
remembrance (*smaraṇa*), 28–30, 42, 48, 51, 76, 97, 105, 108, 156–9
rhetorical texts of Sanskrit poetry, 47, 49, 116, 128, 134, 142, 144, 149, 154–5, 198. *See also kāvya*
Richard of Saint Victor, 6
Rūpa Gosvāmin, 59, 62

sacrifice, 11, 16, 72
Sado-masochism. *See* sexuality
Saduktikarṇāmṛta, 101, 208, 211–12
sahaja, 41, 191–2
Sahajiyā, xii, 23 25–6, 45–6, 62, 68–9, 83, 117, 129, 143, 190–3
Sāhityadarpaṇa, 146, 159
saints in Vaiṣṇavism, 213–32
Sairaṃdhrī, 60
sakhī (friend), xii, 28, 51, 134–6, 167, 201
śakti, 25, 40, 94, 104, 109, 124, 169, 175, 189, 191, 203. *See* goddess; woman
samādhi, 78–9, 151
samāja, 207–8
Sāṃkhya, 13, 61, 63, 133, 160, 183. *See yoga*
saṃsāra (empirical existence), 13, 17, 19, 41–2, 66, 108, 133, 148, 156, 161, 171, 199–201, 213
sandhyā-bhāṣā ('twilight language'), 40, 188–9, 196
Śaṅkara Miśra, 219
śānta rasa. *See rasa*
Sappho, 155

Ṣaraṇa, 210
Sedley, Sir Charles, 87
Sekāśubhodaya of Halāyudha, 231
self/soul (*ātman*), 13–14, 106, 128, 139, 148, 150, 160, 168, 175, 181–4, 192, 201, 204. *See also jīva*
sexuality, 9–12, 14–18, 25, 28, 32, 38, 40, 46, 52, 55–6, 58–9, 63, 68, 70, 72, 78, 82–3, 97, 100, 117, 128, 131, 133–4, 143, 163–78, 184, 188–9, 191, 196, 202, 221, 227; coital postures, 63, 104, 162, 167–9, 196, 252*n*; sado-masochism, 163–7. *See also* erotic texts
Sharma, Paṇḍit Sadā Śiv Rath, 189–90, 220, 227, 232
Shekhinah, 11, 129
Shelley, Percy B., 198
Shulamite, 138, 155
siddhācāryas, 40, 188
Sikhs, 107
Śiśupālavadha of *Māgha*, 166
Sītā, 119, 123, 141
Śiva, 16–17, 25, 73–4, 77–9, 108, 124, 169, 189, 191–2, 208, 211, 255*n*
sneha, 145
Song of Songs, 3, 5–6, 43, 138, 179–80, 182–4, 187, 203
Śrī. *See* Goddess
śṛṅgāra rasa. *See rasa*
Śrutidhara, 210
Stendhal, 144
sublimation, 7–9, 17–18, 39, 78, 142
suffering, 7, 28, 138, 140–1, 145–6, 148–9, 154, 165, 173, 198
suggestion (*dhvani*), 35
Suso, Blessed Henry, 138
svadharma. *See dharma*
Śvetāśvatara Upaniṣad. *See Upaniṣads*
Sylvester, Joshua, 87
symbols, 193–200; lotus, 195–9; nocturnal, 199–201; vernal, 201–5
Symposium, 10, 15

Thoreau, Henry David, 12
tamas, 199
tan-maya, 153
Tantra, 16–17, 25, 40, 45, 68, 94, 124, 127, 163, 169–70, 187–91, 196; *Hevajra*, 17; *Lakṣmī*, 124
translation, problems of, 233–8
Tristan and Isolde, 2, 128, 137
Tulsīdās, 224

Uc de Saint Circ, 3
Umāpatidhara, 210
Upaniṣads, 13, 24, 128, 143, 153, 168, 183, 194, 213; *Bṛhadāraṇyaka*, 14–16, 72; *Maitrī*, 61, 106; *Nṛsiṃhatapinī*, 228; *Śvetāśvatara*, 106
upāya and *prajñā*, 16, 25

Vaikuṇṭha, 105, 218
Vaiṣṇavism, 23–7, 46, 49, 53–4, 56–7, 60–3, 66, 69–71, 73, 79, 83–4, 94, 102, 112, 116, 127, 129, 136, 139, 142, 148–50, 159, 169, 171, 174–6, 183–4, 187, 191, 199, 202–4, 208, 211, 213. *See* Caitanya; Sahajiyā
Vallaṇa, 72
Vanamālibhaṭṭa, 103, 166–7
Vātsyāyana. *See* erotic texts
Veda, 16, 62–5, 68, 74, 85, 103–4, 117–18, 156, 192, 194, 206, 211, 228, 231; *Atharva*, 75, 157; *Ṛg*, 43, 64, 76, 131, 141*n*, 204
Vedānta, 13, 160, 168–9, 174–5, 183, 194
Vergil, 74, 84
vernacular literature of India, 23, 36, 226
Viṣṇu, 74, 102, 105–8, 122–4, 151, 156, 158, 195, 197, 202, 204, 209, 211, 213–14, 218–20, 223
Vidyāpati, 136, 169, 184
vīra rasa. *See rasa*
viraha. *See* love-in-separation
Viṣṇu Purāṇa. *See Purāṇas*
Vṛndāvana, 46, 136, 175, 186, 191, 203, 221, 247*n*
Vyāsa, 156

war and love, 52, 82–3, 100–1, 131, 133, 163–5, 169. *See also* love
Whitman, Walt, 12, 134, 147

woman, 18–20, 40, 110–34, 169, 172, 191, 202. *See* Goddess; *Gopi*; Mother; *nāyikā*; Rādhā

Yama, 83, 107, 141*n*
Yāmunācārya, 140
yoga/*yogī*, 17, 30, 37, 40–1, 51, 61, 78–9, 106–8, 116, 150–3, 170, 189–90, 195–6, 204, 221, 228, 256*n*
Yudhiṣṭhira, 65, 85

Zohar, 11–12